The Shock Absorber Handbook
Second Edition

Wiley-Professional Engineering Publishing Series

This series of books from John Wiley Ltd and Professional Engineering Publishing Ltd aims to promote scientific and technical texts of exceptional academic quality that have a particular appeal to the professional engineer.

Forthcoming titles:

Vehicle Particulate Emissions
Peter Eastwood

Suspension Analysis and Computational Geometry
John C. Dixon

Managing Reliability Growth in Engineering Design: Decisions, Data and Modelling
Lesley Walls and John Quigley

The Shock Absorber Handbook
Second Edition

John C. Dixon, Ph.D, F.I.Mech.E., F.R.Ae.S.
Senior Lecturer in Engineering Mechanics
The Open University, Great Britain

John Wiley & Sons, Ltd

This Work is a co-publication between Professional Engineering Publishing Ltd and John Wiley and Sons, Ltd.

This Work is a co-publication between Professional Engineering Publishing Ltd and John Wiley and Sons, Ltd.

Previously published as *The Shock Absorber Handbook, 1st Edition*, by The Society of Automotive Engineers, Inc, 1999, ISBN 0-7680-0050-5.

By the same author: Tires, Suspension and Handling (SAE).

Copyright © 2007   John Wiley & Sons Ltd, The Atrium, Southern Gate, Chichester,
West Sussex PO19 8SQ, England
Telephone (+44) 1243 779777

Email (for orders and customer service enquiries): cs-books@wiley.co.uk
Visit our Home Page on www.wiley.com

All Rights Reserved. No part of this publication may be reproduced, stored in a retrieval system or transmitted in any form or by any means, electronic, mechanical, photocopying, recording, scanning or otherwise, except under the terms of the Copyright, Designs and Patents Act 1988 or under the terms of a licence issued by the Copyright Licensing Agency Ltd, 90 Tottenham Court Road, London W1T 4LP, UK, without the permission in writing of the Publisher. Requests to the Publisher should be addressed to the Permissions Department, John Wiley & Sons Ltd, The Atrium, Southern Gate, Chichester, West Sussex PO19 8SQ, England, or emailed to permreq@wiley.co.uk, or faxed to (+44) 1243 770620.

Designations used by companies to distinguish their products are often claimed as trademarks. All brand names and product names used in this book are trade names, service marks, trademarks or registered trademarks of their respective owners. The Publisher is not associated with any product or vendor mentioned in this book.

This publication is designed to provide accurate and authoritative information in regard to the subject matter covered. It is sold on the understanding that the Publisher is not engaged in rendering professional services. If professional advice or other expert assistance is required, the services of a competent professional should be sought.

Anniversary Logo Design: Richard J. Pacifico

British Library Cataloguing in Publication Data

A catalogue record for this book is available from the British Library

ISBN 978-0-470-51020-9 (HB)

Typeset in 10/12 pt Times by Thomson Digital, India

Disclaimer: This book is not intended as a guide for vehicle modification, and anyone who uses it as such does so entirely at their own risk. Testing vehicle performance may be dangerous. The author and publisher are not liable for consequential damage arising from application of any information in this book.

Contents

Preface   **xiii**

Acknowledgements   **xv**

1 Introduction   **1**

    1.1 History   1
    1.2 Types of Friction   15
    1.3 Damper Configurations   17
    1.4 Ride-Levelling Dampers   33
    1.5 Position-Dependent Dampers   35
    1.6 General Form of the Telescopic Damper   37
    1.7 Mountings   42
    1.8 Operating Speeds and Strokes   47
    1.9 Manufacture   53
    1.10 Literature Review   54

2 Vibration Theory   **61**

    2.1 Introduction   61
    2.2 Free Vibration Undamped (1-dof)   61
    2.3 Free Vibration Damped (1-dof)   63
    2.4 Forced Vibration Undamped (1-dof)   68
    2.5 Forced Vibration Damped (1-dof)   71
    2.6 Coulomb Damping   74
    2.7 Quadratic Damping   77
    2.8 Series Stiffness   79
    2.9 Free Vibration Undamped (2-dof)   85
    2.10 Free Vibration Damped (2-dof)   85
    2.11 The Resonant Absorber   86
    2.12 Damper Models in Ride and Handling   87
    2.13 End Frequencies   88
    2.14 Heave and Pitch Undamped 1-dof   90
    2.15 Heave and Pitch Damped 1-dof   91
    2.16 Roll Vibration Undamped   93

2.17	Roll Vibration Damped	94
2.18	Heave-and-Pitch Undamped 2-dof	95
2.19	Heave-and-Pitch Damped 2-dof Simplified	100
2.20	Heave-and-Pitch Damped 2-dof Full Analysis	102

3 Ride and Handling — 105

3.1	Introduction	105
3.2	Modelling the Road	105
3.3	Ride	111
3.4	Time-Domain Ride Analysis	113
3.5	Frequency-Domain Ride Analysis	117
3.6	Passenger on Seat	118
3.7	Wheel Hop	119
3.8	Handling	120
3.9	Axle Vibrations	122
3.10	Steering Vibrations	124
3.11	The Ride–Handling Compromise	124
3.12	Damper Optimisation	129
3.13	Damper Asymmetry	131

4 Installation — 135

4.1	Introduction	135
4.2	Motion Ratio	135
4.3	Displacement Method	137
4.4	Velocity Diagrams	138
4.5	Computer Evaluation	138
4.6	Mechanical Displacement	138
4.7	Effect of Motion Ratio	139
4.8	Evaluation of Motion Ratio	142
4.9	The Rocker	142
4.10	The Rigid Arm	148
4.11	Double Wishbones	150
4.12	Struts	153
4.13	Pushrods and Pullrods	155
4.14	Motorcycle Front Suspensions	156
4.15	Motorcycle Rear Suspensions	160
4.16	Solid Axles	165
4.17	Dry Scissor Dampers	168

5 Fluid Mechanics — 169

5.1	Introduction	169
5.2	Properties of Fluids	170
5.3	Chemical Properties	171
5.4	Density	171
5.5	Thermal Expansion	172

5.6	Compressibility	172
5.7	Viscosity	173
5.8	Thermal Capacity	175
5.9	Thermal Conductivity	176
5.10	Vapour Pressure	176
5.11	Gas Density	176
5.12	Gas Viscosity	177
5.13	Gas Compressibility	177
5.14	Gas Absorbability	177
5.15	Emulsification	179
5.16	Continuity	188
5.17	Bernoulli's Equation	188
5.18	Fluid Momentum	189
5.19	Pipe Flow	191
5.20	Velocity Profiles	196
5.21	Other Losses	199
5.22	The Orifice	203
5.23	Combined Orifices	207
5.24	Vortices	209
5.25	Bingham Flow	212
5.26	Liquid–Solid Suspensions	212
5.27	ER and MR Fluids	214

6 Valve Design — 217

6.1	Introduction	217
6.2	Valve Types	219
6.3	Disc Valves	220
6.4	Rod Valves	221
6.5	Spool Valves	222
6.6	Shim Valves	223
6.7	Valve Characteristics	225
6.8	Basic Valve Models	227
6.9	Complete Valve Models	230
6.10	Solution of Valve Flow	235
6.11	Temperature Compensation	237
6.12	Position-Sensitive Valves	240
6.13	Acceleration-Sensitive Valves	240
6.14	Pressure-Rate Valves	243
6.15	Frequency-Sensitive Valves	245
6.16	Stroke-Sensitive Valves	245
6.17	Piezoelectric Valves	249
6.18	Double-Acting Shim Valves	249
6.19	Rotary Adjustables	250
6.20	Bellows Valves	252
6.21	Simple Tube Valves	252

6.22	Head Valves	257
6.23	Multi-Stage Valves	257

7 Damper Characteristics — 259

7.1	Introduction	259
7.2	Basic Damper Parameters	263
7.3	Mechanical Friction	265
7.4	Static Forces	268
7.5	Piston Free Body Diagram	269
7.6	Valve Flow Rates	271
7.7	Pressures and Forces	272
7.8	Linear Valve Analysis	273
7.9	Cavitation	274
7.10	Temperature	276
7.11	Compressibility	276
7.12	Cyclical Characteristics, $F(X)$	278
7.13	Extreme Cyclic Operation	282
7.14	Stresses and Strains	283
7.15	Damper Jacking	286
7.16	Noise	287

8 Adjustables — 289

8.1	Introduction	289
8.2	The Adjustable Valve	290
8.3	Parallel Hole	294
8.4	Series Hole	294
8.5	Maximum Area	294
8.6	Opening Pressure	294
8.7	Area Coefficient (Stiffness)	295
8.8	Automatic Systems	295
8.9	Fast Adaptive Systems	299
8.10	Motion Ratio	301

9 ER and MR Dampers — 303

9.1	Introduction	303
9.2	ER–MR History	303
9.3	ER Materials	309
9.4	ER Dampers	314
9.5	ER Controlled Valve	319
9.6	MR Materials	321
9.7	MR Dampers	324

10 Specifying a Damper — 333

10.1	Introduction	333

10.2	End Fittings	334
10.3	Length Range	334
10.4	$F(V)$ Curve	334
10.5	Configuration	334
10.6	Diameter	335
10.7	Oil Properties	335
10.8	Life	335
10.9	Cost	335

11 Testing   337

11.1	Introduction	337
11.2	Transient Testing	338
11.3	Electromechanical Testers	342
11.4	Hydraulic Testers	344
11.5	Instrumentation	345
11.6	Data Processing	346
11.7	Sinusoidal Test Theory	348
11.8	Test Procedure	352
11.9	Triangular Test	354
11.10	Other Laboratory Tests	356
11.11	On-Road Testing	357

Appendix A: Nomenclature	361
Appendix B: Properties of Air	375
Appendix C: Properties of Water	379
Appendix D: Test Sheets	381
Appendix E: Solution of Algebraic Equations	385
Appendix F: Units	393
Appendix G: Bingham Flow	397

References   **401**

Index   **409**

Preface to Second Edition

In view of the tremendous worldwide production of automotive dampers (shock absorbers), the former absence of a book devoted to this topic is surprising. During some years of damper design, research and commercial testing, the author has become aware of a need for a suitable book to present the fundamentals of damper design and use, for the benefit of the many designers of vehicles such as passenger cars, motorcycles, trucks, racing cars and so on, since the necessary body of knowledge is far from readily available in the research literature. Damper designers themselves will already be familiar with most of the material here, but may find some useful items, especially with regard to installation motion ratios and behaviour of the vehicle as a whole, but in any case will probably be pleased to see the basic material collected together.

As in my previous work, I have tried to present the basic core of theory and practice, so that the book will be of lasting value. I would be delighted to hear from readers who wish to suggest any improvements to presentation or coverage.

Amongst many suggestions received for additions and improvements to the first edition, there was clearly a desire that the book should be extended to cover extensively the effect of the damper on ride and handling. The extra material would, however, be vast in scope, and would greatly increase the size and expense of the book. Also, in the author's view, such analysis belongs in a separate book on ride quality and handling, where the effect of the damper can be considered fully in the context of other suspension factors.

Instead, the general character of the first edition has been retained, with its emphasis on the internal design of the damper. Considerable efforts have been made to eliminate known errors in the first edition, and substantial detailed additions and revisions have been made. In many areas the material has been reorganised for greater clarity. The variety of damper types found historically is now more fully covered, and the recent developments in magnetorheological dampers are now included. Conventional damper valve design is considered much more carefully, and more space is allocated to detailed variations in valve design, including stroke-sensitive types. Many new figures have been added. On this basis, it is hoped that the new edition will offer a worthwhile service to the vehicle design community, at least as an introduction to the complex and fascinating field of damper design.

Finally, the title *The Shock Absorber Handbook* has been controversial, as it was said that the subject was not shock absorbers and it was not a handbook. It would probably have been better to use the technically correct term *damper*, with a title such as *The Automotive Damper*. However, a change of title has been deemed impractical given that the book is well established under its original name, and it has been decided to remain with the devil that we know for this, second, edition.

John C. Dixon

Acknowledgements

Numerous figures are reproduced by permission of the Society of Automotive Engineers, The Institution of Mechanical Engineers, and others. The reference for all previously published figures is given with the figure.

1
Introduction

1.1 History

The current world-wide production of vehicle dampers, or so-called shock absorbers, is difficult to estimate with accuracy, but is probably around 50–100 million units per annum with a retail value well in excess of one billion dollars per annum. A typical European country has a demand for over 5 million units per year on new cars and over 1 million replacement units, The US market is several times that. If all is well, these suspension dampers do their work quietly and without fuss. Like punctuation or acting, dampers are at their best when they are not noticed - drivers and passengers simply want the dampers to be trouble free. In contrast, for the designer they are a constant interest and challenge. For the suspension engineer there is some satisfaction in creating a good new damper for a racing car or rally car and perhaps making some contribution to competition success. Less exciting, but economically more important, there is also satisfaction in seeing everyday vehicles travelling safely with comfortable occupants at speeds that would, even on good roads, be quite impractical without an effective suspension system.

The need for dampers arises because of the roll and pitch associated with vehicle manoeuvring, and from the roughness of roads. In the mid nineteenth century, road quality was generally very poor. The better horse-drawn carriages of the period therefore had soft suspension, achieved by using long bent leaf springs called semi-elliptics, or even by using a pair of such curved leaf springs set back-to-back on each side, forming full-elliptic suspension. No special devices were fitted to provide damping; rather this depended upon inherent friction, mainly between the leaves of the beam springs. Such a set-up was appropriate to the period, being easy to manufacture, and probably worked tolerably well at moderate speed, although running at high speed must have been at least exciting, and probably dangerous, because of the lack of damping control.

The arrival of the so-called horseless carriage, i.e. the carriage driven by an internal combustion engine, at the end of the nineteenth century, provided a new stimulus for suspension development which continues to this day. The rapidly increasing power available from the internal combustion engine made higher speeds routine; this, plus the technical aptitude of the vehicle and component designers, coupled with a general commercial mood favouring development and change, provided an environment that led to invention and innovation.

The fitting of damping devices to vehicle suspensions followed rapidly on the heels of the arrival of the motor car itself. Since those early days the damper has passed through a century of evolution, the basic stages of which may perhaps be considered as:

The Shock Absorber Handbook/Second Edition   John C. Dixon
© 2007 John Wiley & Sons, Ltd

(1) dry friction (snubbers);
(2) blow-off hydraulics;
(3) progressive hydraulics;
(4) adjustables (manual alteration);
(5) slow adaptives (automatic alteration);
(6) fast adaptives ('semi-active');
(7) electrofluidic, e.g. magnetorheological.

Historically, the zeitgeist regarding dampers has changed considerably over the years, in roughly the following periods:

(1) Up to 1910 dampers were hardly used at all. In 1913, Rolls Royce actually discontinued rear dampers on the Silver Ghost, illustrating just how different the situation was in the early years.
(2) From 1910 to 1925 mostly dry snubbers were used.
(3) From 1925 to 1980 there was a long period of dominance by simple hydraulics, initially simply constant-force blow-off, then through progressive development to a more proportional characteristic, then adjustables, leading to a mature modern product.
(4) From 1980 to 1985 there was excitement about the possibilities for active suspension, which could effectively eliminate the ordinary damper, but little has come of this commercially in practice so far because of the cost.
(5) From 1985 it became increasingly apparent that a good deal of the benefit of active suspension could be obtained much more cheaply by fast auto-adjusting dampers, and the damper suddenly became an interesting, developing, component again.
(6) From about 2000, the introduction, on high-price vehicles at least, of controllable magnetorheological dampers.

Development of the adaptive damper has occurred rapidly. Although there will continue to be differences between commercial units, such systems are now effective and can be considered to be mature products. Fully active suspension offers some performance advantages, but is not very cost effective for passenger cars. Further developments can then be expected to be restricted to rather slow detail refinement of design, control strategies and production costs. Fast acting control, requiring extra sensors and controls, will continue to be more expensive, so simple fixed dampers, adjustables and slow adaptive types will probably continue to dominate the market numerically for the foreseeable future.

The basic suspension using the simple spring and damper is not ideal, but it is good enough for most purposes. For low-cost vehicles, it is the most cost-effective system. Therefore much emphasis remains on improvement of operating life, reliability and low-cost production rather than on refinement of performance by technical development. The variable damper, in several forms, has now found quite wide application on mid-range and expensive vehicles. On the most expensive passenger and sports cars, magetorheologically controlled dampers are now a popular fitment, at significant expense.

The damper is commonly known as the shock absorber, although the implication that shocks are absorbed is misleading. Arguably, the shocks are 'absorbed' by the deflection of the tires and springs. The purpose of dampers is to dissipate any energy in the vertical motion of body or wheels, such motion having arisen from control inputs, or from disturbance by rough roads or wind. Here 'vertical' motion includes body heave, pitch and roll, and wheel hop. As an agglomeration of masses and springs, the car with its wheels constitutes a vibrating system that needs dampers to optimise control behaviour, by preventing response overshoots, and to minimise the influence of some unavoidable resonances. The mathematical theory of vibrating systems largely uses the concept of a linear damper, with force proportional to extension speed, mainly because it gives equations for which the solutions are well understood and documented, and usually tolerably realistic. There is no obligation on a damper to exhibit such a characteristic; nevertheless the typical modern hydraulic damper does so approximately. This is because the vehicle and damper manufacturers consider this to be desirable for good physical

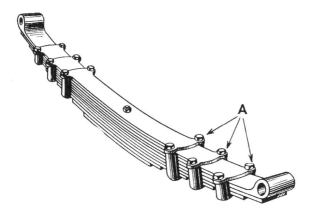

Figure 1.1.1 Dry friction damping by controlled clamping (adjustable normal force) of the leaf spring (Woodhead).

behaviour, not for the convenience of the theorist. The desired characteristics are achieved only by some effort from the manufacturer in the detail design of the valves.

Damper types, which are explained fully later, can be initially classified as

(a) dry friction with solid elements;
   (i) scissor;
   (ii) snubber;
(b) hydraulic with fluid elements;
   (i) lever-arm;
   (ii) telescopic.

Only the hydraulic type is in use in modern times. The friction type came originally as sliding discs operated by two arms, with a scissor action, and later as a belt wrapped around blocks, the 'snubber'. The basic hydraulic varieties are lever-arm and telescopic. The lever-arm type uses a lever to operate a vane, now extinct, or a pair of pistons. Telescopics, now most common, are either double-tube or gas-pressurised single-tube.

The early days of car suspension gave real opportunities for technical improvement, and financial reward. The earliest suspensions used leaf springs with inherent interleaf friction. Efforts had been made to control this to desirable levels by the free curvature of the leaves. Further developments of the leaf spring intrinsic damping included controlled adjustment of the interleaf normal forces, Figure 1.1.1, and the use of inserts of various materials to control the friction coefficients, Figure 1.1.2.

Truffault invented the scissor-action friction disc system before 1900, using bronze discs alternating with oiled leather, pressed together by conical disc springs and operated by two arms, with a floating body. The amount of friction could be adjusted by a compression hand-screw, pressing the discs together more or less firmly, varying the normal force at approximately constant friction coefficient. Between 1900 and 1903, Truffault went on to develop a version for cars, at the instigation

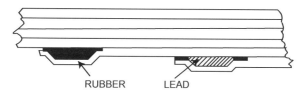

Figure 1.1.2 Leaf spring inserts to control the friction coefficient and consequent damping effect.

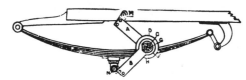

Figure 1.1.3 An advertisement from 1904 for the early Truffault designed dry friction scissor damper manufactured by Hartford.

of Hartford in the US, who began quantity production in 1904, as in Figures 1.1.3–1.1.5. Truffault, well aware of the commercial potential, also licensed several other manufacturers in Europe, including Mors and Peugeot in France, who also had them in production and use by 1904. A similar type of damper was also pressed into service on the steering, Figure 1.1.6, to reduce steering fight on rough roads and to reduce steering vibrations then emerging at higher speeds and not yet adequately understood.

Figure 1.1.7 shows an exploded diagram of a more recent (1950s) implementation from a motorcycle. This is also adjustable by the hand-screw. Subsequent to the Truffault–Hartford type, The Hartford Telecontrol (the prefix *tele* means remote) developed the theme, Figure 1.1.8, with a convenient Bowden cable adjustment usable by the driver *in situ*. A later alternative version, the Andre Telecontrol, had dry friction scissor dampers, but used hydraulic control of the compression force and hence of the damper friction moment.

In 1915, Claud Foster invented the dry friction block-and-belt snubber, Figure 1.1.9, manufactured in very large quantities by his Gabriel company, and hence usually known as the Gabriel Snubber. In view of the modern preference for hydraulics, the great success of the belt snubber was presumably based on low cost, ease of retrofitment and reliability rather than exceptional performance.

Introduction

Figure 1.1.4 The Andre–Hartford scissor-action dry friction damper.

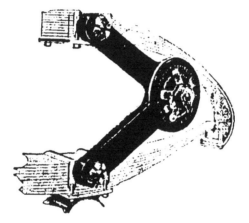

Figure 1.1.5 Installation of a dry-friction scissor damper on three-quarter-elliptic leaf springs (from Simanaitis, 1976).

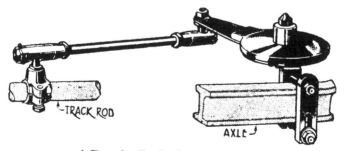

A Damping Device for Preventing Wheel Wobble on Rigid-Type Front Axles.

Figure 1.1.6 Use of the Truffault–Hartford rotary dry friction damper on steering.

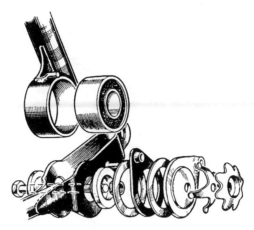

Figure 1.1.7 The Greeves motorcycle front suspension from around 1950 had a rubber-in-torsion spring, using an integral rotary dry friction damper easily adjustable by hand.

The spring-loaded blocks are mounted on the body, in particular on the chassis rails in those days, with the leather belt being fixed to the wheel upright or axle. In upward motion of the suspension, the snubber has no effect, but the spring-loaded blocks take up any slack. Any attempt by the suspension to extend will be opposed by the belt which has considerable friction where it wraps over itself and around the blocks. Hence the action is fully asymmetrical. The actual performance parameters do not seem to have been published. Some theoretical analysis may be possible, derived from the standard theory of wrapped circular members, with friction force growing exponentially with wrapping angle, for prediction of the force in relation to block shape, spring force and stiffness and belt-on-belt and belt-on-block coefficients of friction. The overall characteristic, however, seems to be an essentially velocity-independent force in extension, i.e. fully asymmetrical Coulomb damping. The characteristics could have been affected in service conditions by the friction-breaking effect of engine vibrations.

An early form of hydraulic contribution to damping was the Andrex oil-bath damper, Figure 1.1.10. This had metal and leather discs as in the dry damper, but was immersed in a sealed oil bath. There may also have been a version with separated metal discs relying on oil in shear. Another version, Figure 1.1.11, was adjustable from the dashboard, with oil pressure transmitted to the dampers to control the normal force on the discs, or perhaps in some cases to adjust the level of oil in the case. The pressure gauge in Figure 1.1.11 suggests that this type was controlling the normal force.

Figure 1.1.8 The Hartford Telecontrol damper was adjustable via a Bowden cable, and hence could be controlled easily from the driving seat, even with the vehicle in motion.

Introduction

Figure 1.1.9 The Gabriel Snubber (1915) used a leather strap around sprung metal or wooden blocks to give restraint in rebound only (from Simanaitis, 1976).

The early development timetable of dampers thus ran roughly as follows:

1901: Horock patents a telescopic hydraulic unit, laying the foundations of the modern type.
1902: Mors actually builds a vehicle with simple hydraulic pot dampers.
1905: Renault patents an opposed piston hydraulic type, and also patents improvements to Horock's telescopic, establishing substantially the design used today.
1906: Renault uses the piston type on his Grand Prix racing cars, but not on his production cars. Houdaille starts to develop his vane-type.
1907: Caille proposes the single-lever parallel-piston variety.

Figure 1.1.10 The Andrex multiple discs-in-oil-bath damper.

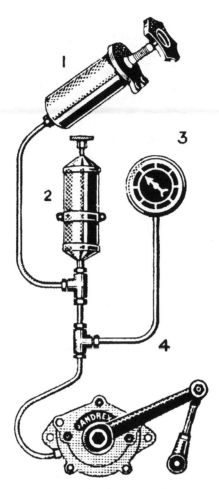

Figure 1.1.11 The adjustable version of the Andrex oil-bath damper included pump, reservoir and pressure gauge.

1909: A single-acting Houdaille vane type is fitted as original equipment, but this is an isolated success for the hydraulic type, the friction disc type remaining dominant.
1910: Oil damped undercarriages come into use on aircraft.
1915: Foster invents the belt 'snubber' which had great commercial success in the USA.
1919: Lovejoy lever-arm hydraulic produced in the USA.
1924: Lancia introduces the double-acting hydraulic unit, incorporated in the front independent pillar suspension of the Lambda. The Grand Prix Bugatti uses preloaded nonadjustable drum-brake type.
1928: Hydraulic dampers are first supplied as standard equipment in the USA.
1930: Armstrong patents the telescopic type.
1933: Cadillac 'Ride Regulator' driver-adjustable five-position on dashboard.
1934: Monroe begins manufacture of telescopics.
1947: Koning introduces the adjustable telescopic.
1950: Gas-pressurised single-tube telescopic is invented and manufactured by de Carbon.
2001: Magnetorheological high-speed adjustables introduced (Bentley, Cadillac).

Introduction

The modern success of hydraulics over dry friction is due to a combination of factors, including:

(1) Superior performance of hydraulics, due to the detrimental effect of dry Coulomb friction which is especially noticeable on modern smooth roads.
(2) Damper life has been improved by better seals and higher quality finish on wearing surfaces.
(3) Performance is now generally more consistent because of better quality control.
(4) Cost is less critical than of old, and is in any case controlled by mass production on modern machine tools.

During the 1950s, telescopic dampers gradually became more and more widely used on passenger cars, the transition being essentially complete by the late 1950s. In racing, at Indianapolis the hydraulic vane type arrived in the late 1920s, and was considered a great step forward; the adjustable piston hydraulic appeared in the early 1930s, but the telescopic was not used there until 1950. Racing cars in Europe were quite slow to change, although the very successful Mercedes Benz racers of 1954–55 used telescopics. Although other types are occasionally used, the telescopic hydraulic type of damper is now the widely accepted norm for cars and motorcycles.

It was far from obvious in early days that the hydraulic type of damper would ultimately triumph, especially in competition with the very cost-effective Gabriel snubber of 1915. The first large commercial successes for the hydraulic types came with the vane-type, developed from 1906 onwards by Maurice Houdaille. The early type used two arms with a floating body, a little like the dry friction scissor damper. The later type still used vanes, but had a body mounted on the vehicle sprung mass, operated by an arm with a drop link to the leaf spring suspension, Figures 1.1.12–1.1.14.

The 1919 Motor Manual (UK, 21st edition) devoted less than one of its three hundred pages to dampers, suggesting that the damper was not really considered to be of great importance in those days, stating that:

> These devices, of which there are a great number on the market, are made for the purpose of improving the comfortable running of the car, more especially on roughly-surfaced roads. The present system of springing is

Figure 1.1.12 The Houdaille rotary vane damper, the first large quantity production hydraulic damper. This originated in 1909 and was double-acting from 1921.

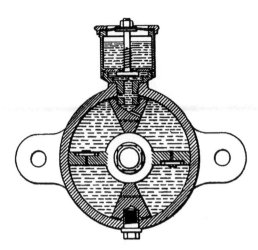

Figure 1.1.13 Cross-section of slightly different version of Houdaille rotary vane damper (from Simanaitis, 1976).

admittedly not perfect, and when travelling on rough roads there is the objectionable rebound of the body after it passes over a depression in the road, which it is desirable should be reduced as much as possible. The shock from this rebound is not only uncomfortable for the passengers, but it has a bad effect on the whole car. Hence these shock absorbers are applied as the best means available so far to check the rebound. They are made on various principles, generally employing a frictional effect such as is obtainable from two hardened steel surfaces in close contact. Another principle is that of using the fluid friction of oil, practically on the lines of any of the well-known dash-pot devices, viz., a piston moving in a cylinder against the resistance offered by the oil contained within it, the oil passing slowly through a small aperture into another chamber. This type of device is probably the best solution of the problem.

Up to 1920 hydraulic dampers were single acting, in droop only, but from 1921 a more complex valve system allowed some damping in bump too. At this point the operating characteristics of the

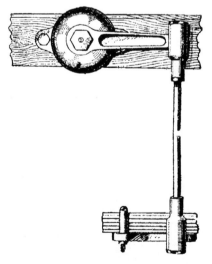

Figure 1.1.14 An early configuration of hydraulic damper, a rotary vane device with a drop arm to the axle. Note the wooden chassis rail (artist's impression, The Motor Manual, 1919).

Introduction

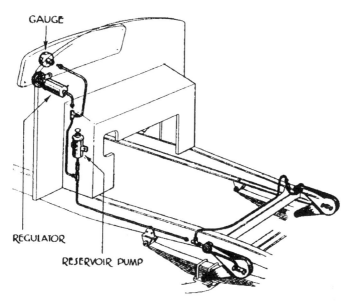

Figure 1.1.15 Layout of the hydraulically remote Andre Telecontrol damper, shown here on a front axle (The Motor Manual, 1939).

hydraulic damper had largely reached their modern form. More recent developments have had more to do with the general configuration, so that the lever-operated type has given way to the telescopic piston type which is cost-effective in manufacture, being less critical with regard to seal leakage, and has better air cooling, although lacking the conduction cooling of a body-mounted lever-arm damper. Most importantly perhaps, the telescopic type lends itself well to the modern form of suspension in terms of its mounting and ease of installation.

The 1939 Motor Manual (UK, 30th edition), devoted three pages to dampers, perhaps indicating the increased recognition of their importance for normal vehicles. An illustration was included of the Andre–Hartford dry friction scissor, and also one of the Luvax vane damper, shown later. There was also a diagram of the hydraulically adjusted, but dry action, version of the Andre Telecontrol system, as seen in Figure 1.1.15. That writer was moved to offer some additional explanation of damping and 'shock absorbing' in general, stating that:

> Whatever form of springing is employed, it is always considered necessary to damp the suspension by auxiliaries, which have become known as shock absorbers. This term is unfortunate, because it is the function of the springs to absorb shocks, whereas the 'shock absorbers' serve the purpose of providing friction in a controlled form which prevents prolonged bouncing or pitching motions, by absorbing energy. A leaf spring is inherently damped by the friction between the leaves, and it may, therefore, seem strange that after lubricating these leaves friction should be put back into the system by the use of shock absorbers. The explanation is that leaf friction is not readily controllable, whereas the shock absorber imparts a definite and adjustable degree of damping to the system.
>
> The most popular type of shock absorber is an hydraulic device which is bolted to the frame and is operated by an arm coupled to the axle. Four such devices are ordinarily fitted. When relative movement occurs between the axle and the frame, the arm on the shock absorber spindle is oscillated, and this motion is conveyed to a rotor, which fits within a circular casing. Oil in the casing in made to flow through valves from one side of the rotor to the other and so creates hydraulic resistance which damps the oscillations. In some cases the valves are arranged to give 'double action', the damping then being effective on both deflection and rebound. In other cases single-acting devices are used which can check rebound only. As a rule the action of the shock absorbers can be adjusted by means of a screw, which alters the tension of a spring and so varies the load on a ball valve.

The hydraulic shock absorber has the important merit of increasing its damping effect when subject to sudden movements, but suffers from the defect of providing very little resistance against slower motions, such as rolling. Consequently, for sports cars many users prefer frictional shock absorbers, of the scissor (constant resistance) type, of which the most famous is the Andre–Hartford.

The final comment above is significant in a modern context, regarding the preferred velocity–force relationship, which is a regressive shape with a 'knee', rather than simply linear.

The Lancia Lambda of 1925 had sliding pillar suspension, Figure 1.1.16, now almost extinct (except, e.g. Morgan) and regarded as primitive, but highly successful at the time. It was noted for the fact that its oil-filled cylinders required no maintenance, and was very reliable. This is an attractive option for a light vehicle, because it is such a compact and light system, although lacking the ability of modern suspensions to be adjusted to desired handling characteristics by detailed changes to the geometry.

Although dry friction snubbers remained in wide use through to the 1930s, hydraulic fluid-based dampers were in limited use from very early days and continued to grow in popularity. An early successful version in the USA was made by Lovejoy, Figure 1.1.17.

Difficulties with sealing and wear of vane lever arm types led to the lever arm parallel piston system as in the Lovejoy and in the Armstrong, Figure 1.1.18, in which the valve may also easily be made

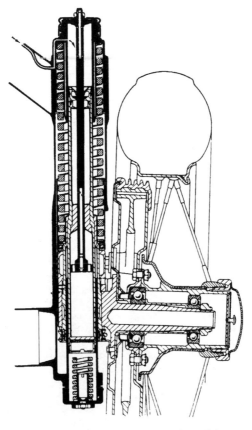

Figure 1.1.16 The Lancia Lambda sliding-pillar system had the spring and damper sealed into one unit (Lancia, 1925).

Figure 1.1.17 The Lovejoy lever-arm hydraulic damper, first produced in 1919.

interchangeable. This would still be a usable design today. Some economy of parts may be achieved by lengthening the bearing and using the lever as the load-carrying suspension arm, Figure 1.1.19. This can be taken further by putting the axle in double shear, so that the lever becomes an A-arm (wishbone), Figure 1.1.20.

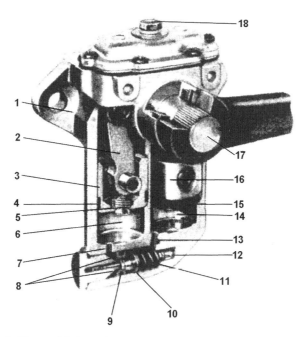

Figure 1.1.18 The double parallel-piston damper was the ultimate lever-arm configuration, overcoming the problems of the vane lever-arm type (Lucas) (see also Figure 1.3.7).

Figure 1.1.19 The simple lever-arm damper can be reinforced to carry suspension loads by lengthening the bearing rod.

However, despite the many creative innovations in lever arms, it seems that the telescopic is now almost universally preferred. At the front this has become the ubiquitous telescopic strut, partly because of the convenience of final assembly.

Figure 1.1.20 The A-arm (wishbone) suspension arm is lighter than a single arm when large loads are to be resisted, and adapts well to a double-shear connection to a lever-arm damper.

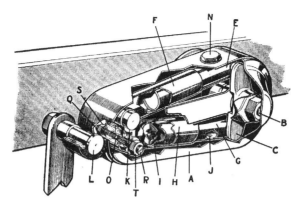

Figure 1.1.21 The Armstrong 'double telescopic lever arm' damper.

An interesting development was the Armstrong 'double telescopic lever arm', Figure 1.1.21, in which two telescopic dampers operate horizontally, fully immersed in an oil bath, with an external structure like a conventional lever arm type. Possibly this was done to combine the Armstrong-type telescopic into a unit that could be used interchangeably with its lever-arm competitors. An advantage of this layout over a plain telescopic is that any amount of damping is easily arranged in compression and rebound independently, with each damper of the pair acting in one direction only, without concern for oil cavitation.

As a final remark on the very early historical development, it may be noted that the dry friction scissor damper and the snubber were remarkably persistent. They were light in weight and low in cost, and perhaps more reliable than the early vane hydraulics which probably suffered from quality control problems and oil leakage. The parallel-piston lever-arm damper was functionally very good, and the fact that it has been superseded by the hydraulic telescopic, and the strut in particular at the front, is mainly due to the final assembly advantages of these, rather than any functional gain in the areas of ride and handling. In steering, the rack system has a better reputation than the old steering boxes, but it is hard, if not impossible, to tell the difference in practice. Similarly, the triumph of the telescopic damper system is not simply due to technical deficiencies of the older systems. The popular new direct acting telescopics that were ultimately to dominate were typified by the Woodhead–Monroe as in Figure 1.1.22.

1.2 Types of Friction

The purpose of a damper, or so-called 'shock absorber', is to introduce controlled friction into the suspension system. In this context, it is possible to identify three distinct types of friction:

(1) dry solid friction;
(2) fluid viscous friction;
(3) fluid dynamic friction.

Any of these types may be used to give suspension damping, but their characteristics are totally different.

Dry solid friction between ordinary hard materials has a maximum shear friction force which is closely proportional to the normal force at the surface:

$$F_\text{F} \leq \mu_\text{F} F_\text{N}$$

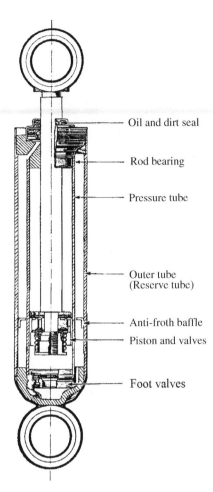

Figure 1.1.22 Cross-section of a typical telescopic damper showing the general features, shown without the dust shroud (Woodhead–Monroe).

where μ_F is the coefficient of limiting friction. For hard materials this is approximately constant over a good range of F_N, and relatively independent of the contact area. This is called Coulomb friction. However it is generally sensitive to temperature, reducing as this increases. Also it is sensitive to the sliding velocity in an undesirable way. For analysis it is common practice to consider there to be a static coefficient of friction μ_S available before any sliding occurs, and a dynamic value μ_D once there is relative motion. The dynamic value is lower, perhaps 70% of the static value.

Coulomb friction is undesirable in a suspension, provided that there is sufficient friction of desirable type, because it locks the suspension at small forces, and gives a poor ride on smooth surfaces, once known in the USA by the colourful term 'Boulevard Jerk'. Hence, nowadays, in order to optimise ride quality every effort is made to minimise the Coulomb friction, including the use of rubber bushes rather than sliding bushes at suspension pivot points.

Fluid friction is considered in detail in a later chapter, but basically viscous friction is proportional to the flow rate, and in this sense is an attractive option. Unfortunately, fluid viscosity is very sensitive to temperature. Fluid dynamic friction, arising with energy dissipation from turbulence, is proportional

Introduction

to the flow rate squared, which is undesirable because it gives forces too high at high speed or too low at low speed. However it depends on the fluid density rather than the viscosity, so the temperature sensitivity, although not zero, is much less than for viscous damping.

Much of the subtlety of damper design therefore hinges around obtaining a desirable friction characteristic which is also consistent, i.e. not unduly sensitive to temperature. This is achieved by using the fluid-dynamic type of friction, with pressure-sensitive variable-area valves to give the desired variation with speed.

1.3 Damper Configurations

There have been numerous detailed variations of the hydraulic damper. The principal types may be classified as:

(1) lever vane (e.g. Houdaille);
(2) lever cam in-line pistons (e.g. Delco Lovejoy);
(3) lever cam parallel pistons, (e.g. Delco);
(4) lever rod piston (e.g. Armstrong);
(5) telescopic.

These and some other types are further illustrated by the variety of diagrams in Figures 1.3.1–1.3.29.

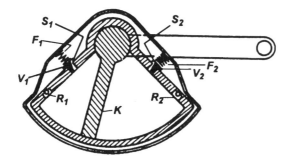

Figure 1.3.1 Double-acting vane type damper (Fuchs, 1933).

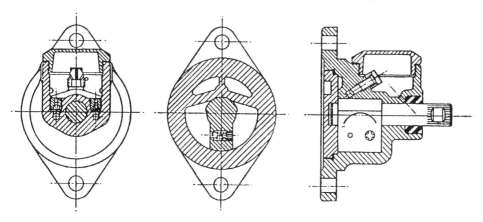

Figure 1.3.2 Early vane-type damper (Kinchin and Stock, 1951/1952).

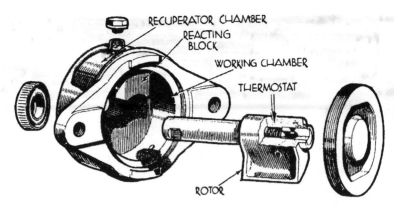

Figure 1.3.3 The Luvax rotary vane hydraulic damper, which featured thermostatic compensation of variation of oil properties. This was a genuine improvement on earlier vane types. The vane shape results in a radial force that takes up any freedom in the bearing in a way that minimises vane leakage (The Motor Manual, 1939).

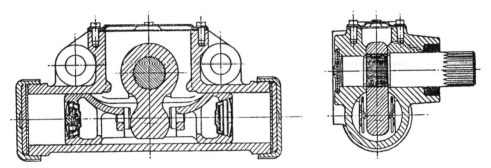

Figure 1.3.4 Lever-operated piston-type damper with discharge to recuperation space (Reproduced from Kinchin and Stock (1951) pp. 67–86 with permission).

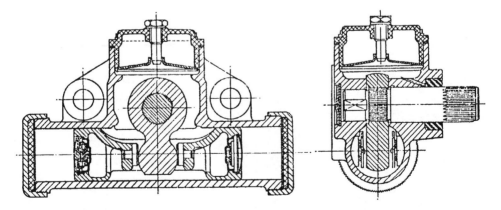

Figure 1.3.5 Lever-operated piston-type damper with pressure recuperation (Reproduced from Kinchin and Stock (1951) pp. 67–86 with permission).

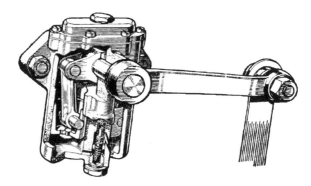

Figure 1.3.6 Double-piston lever-arm damper with removable valve (Armstrong).

Most passenger cars now have struts at the front. These combine the damping and structural functions, with an external spring. The main advantage, compared with double wishbones, is fast assembly line integration of pre-prepared assemblies. There are some disadvantages. The main rod must be of large diameter to give sufficient rigidity and bearing surface to accept running and cornering loads. The piston is subject to side loads, and must have a large rubbing area. These tend to add Coulomb friction. The top strut mounting must transmit the full vertical suspension force, so it is less easy to put a good compliance in series with the damper. The large dimensions mean larger oil flow rates and less critical valves, although wear may still be a problem in some cases.

Gas springing has been used for many years, two of the main exponents in passenger cars being Citroen and British Leyland/BMC/Austin/Morris. The gas is lighter than a metal spring, but requires containment. The damping function is then integrated with the spring units, as in Figure 1.3.22 *et seq*.

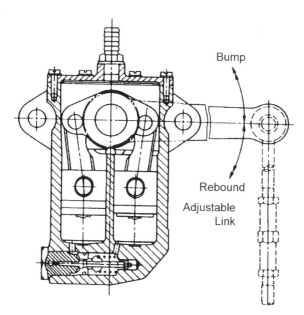

Figure 1.3.7 The classic lever-arm parallel piston type shown in engineering section, with different valve position. Reproduced from Komamura and Mizumukai (1987) History of Shock Absorbers, *JSAE*, 41(1), pp.126–131.

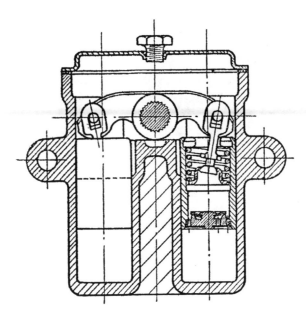

Figure 1.3.8 Lever-operated parallel-piston type damper with valves in the pistons (Reproduced from Kinchin and Stock (1951) pp. 67–86 with permission).

Front-to-rear interconnection allows reduction of the pitch frequency, which is particularly useful on small cars. BMC used simple rubber suspensions with separate dampers, and Hydrolastic and Hydragas with integrated damping.

The most common form of adjustable damper has a rotary valve with several positions each having a different orifice size. Some form of rotational position control, e.g. a stepper motor, is fixed to the top, controlling the piston valve through a shaft in the hollow rod, as seen in Figure 1.3.25. The more recent type uses magneto-rheological liquid, and is discussed separately.

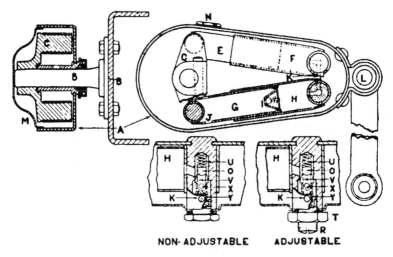

Figure 1.3.9 Double-telescopic lever-arm configuration showing details for standard fixed valve and for the manually adjustable *in situ* version (Armstrong).

Introduction

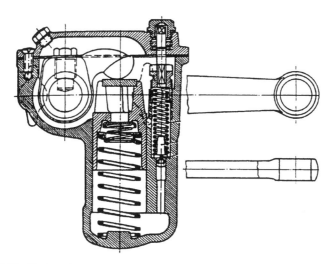

Figure 1.3.10 Single-acting lever-arm piston damper with easily changed valving (Fuchs, 1933).

Steering dampers are much smaller and lighter duty units, and usually operate in the horizontal position. Double tube dampers are not practical in this role. Figures 1.3.26 and 1.3.27 show two versions. In the first, the rod volume and oil thermal expansion are catered for by a spring-loaded free piston. In the second, there is an equalisation chamber having an elastic tube. This separates the oil and the gas, instead of a piston, reducing leakage problems.

In summary of vehicle damper types, then, the vane type is rarely used nowadays because the long seal length is prone to leakage and wear, and it therefore requires very viscous oil which increases the temperature sensitivity. The various lever and piston types are occasionally still used, but the construction implies use of a short piston stroke (in effect an extreme value of motion ratio),

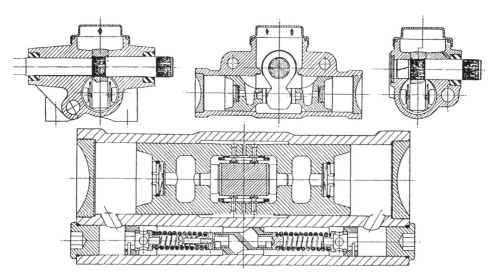

Figure 1.3.11 Lever-operated piston-type damper with end-to-end discharge (Reproduced from Kinchin and Stock (1951) pp. 67–86 with permission).

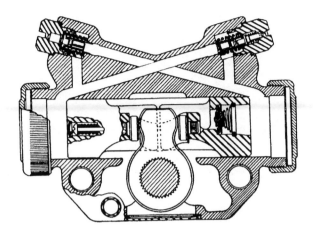

Figure 1.3.12 Double-acting lever-actuated damper with convenient alteration of characteristics by change of valve plugs, *ca* 1935 (Delco–Lovejoy).

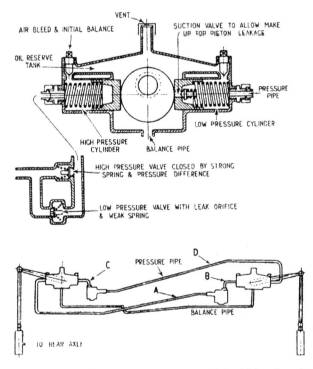

Figure 1.3.13 The German Stabilus damping system for commercial vehicles. Actuation was by the central eccentric circular cam, driven by a drop arm to each wheel. The two plugs at the top of each unit allow independent adjustment of bump and rebound forces. This forms a conventional independent system of unusual actuation. In addition, the two sides of an axle are interconnected through a balance pipe and by relief valves effective in roll only (asymmetrical action).

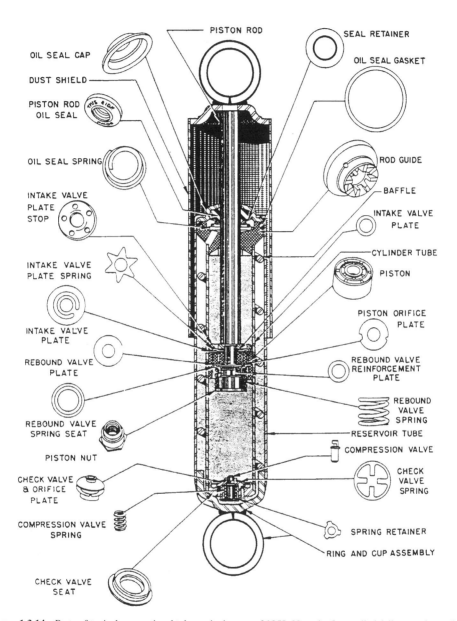

Figure 1.3.14 Parts of typical conventional telescopic damper of 1950. Note the four-coil air/oil separation rod in the reservoir to discourage the effects of agitation. Reproduced from Peterson (1953) *Proc. National Conference on Industrial Hydraulics*, 7, 23–43.

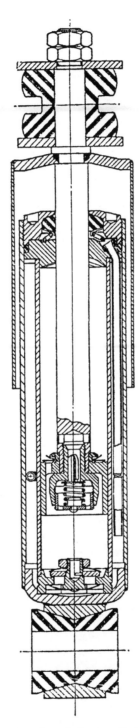

Figure 1.3.15 Standard form of direct-acting telescopic damper with double tubes (Reproduced from Kinchin and Stock (1951) pp. 67–86 with permission).

Introduction

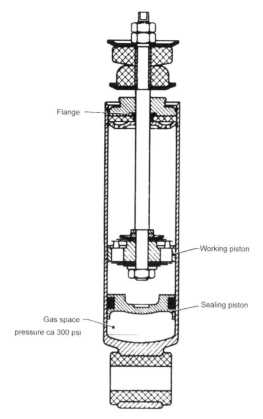

Figure 1.3.16 DeCarbon type of telescopic damper with floating secondary ('sealing') piston and high pressure in the gas chamber. The secondary piston must have sufficient free movement to accommodate the rod displacement volume and oil thermal expansion. A disadvantage is that the main rod seal is continuously subject to high pressure so good manufacturing quality is required to prevent long-term leakage. Also, in some applications (off road), the single tube is prone to damage.

so the forces and pressures need to be very high. Again this can create sensitivity to leakage. The lever types have the advantage that the damper body can be bolted firmly to the vehicle body, assisting with cooling. Another advantage is that there is no internal volume change due to the motion.

However, the lever type has now been almost entirely superseded by the telescopic type, which has numerous detail variations, and may be classified in several ways. The main classification concerns the method by which the insertion volume of the rod is accommodated. This is a major design problem because the oil itself is nowhere near compressible enough to accept the internal volume reduction of 10% or more associated with the full stroke insertion. Although this displacement volume seems to be a major disadvantage of the telescopic damper compared with the lever type, even the lever arm damper must allow for thermal expansion of the oil, which is significant, so the disadvantage in this respect is not great.

There are three basic telescopic types, as in Figure 1.3.28:

(1) the through-rod telescopic;
(2) the double-tube telescopic;
(3) the single-tube telescopic.

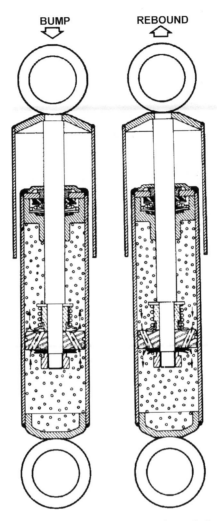

Figure 1.3.17 To eliminate the free piston, an emulsified oil may be used, distributing the expansion and rod-accommodation volume throughout the main oil volume. Overall length is reduced. On standing, the gas separates, but quickly re-emulsifies on action. The valves must be rated to allow for the passage of emulsion rather than liquid oil (Woodhead).

The through-rod telescopic avoids the displacement volume difficulty by passing the rod right through the cylinder. However this has several disadvantages; there are external seals at both ends subject to high pressures, the protruding free end may be inconvenient or dangerous, and there is still no provision for thermal expansion of the oil. However it is a simple solution which has the merit that it can be used in any orientation. This type has proved impractical for suspension damping, but is sometimes used for damping of the steering.

In the double-tube type of telescopic, a pair of concentric tubes are used, the exterior annulus containing some gas to accommodate the rod displacement volume. Hence it must be used the correct way up. In the single-tube type, some gas may be included, which normally forms an emulsion with the oil; alternatively the gas is separated by an independent floating piston (de Carbon type) as shown in

Introduction

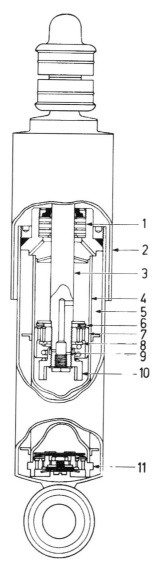

Figure 1.3.18 A double-tube damper showing various features. 1 seal; 2 shroud; 3 rod; 4 inner cylinder; 5 annular foot and gas chambers; 6 piston compression valve; 7 piston; 8 extension valve; 9 parallel hole feed; 10 adjuster; 11 foot valve. The adjuster screws along the rod compressing the valve preload spring. This is achieved by fully compressing the damper to engage the adjuster in the bottom of the cylinder, and rotating the two body parts (Koni).

Figure 1.3.28 (c). The rod is usually fitted with a shroud, of metal or plastic, or possibly a rubber boot, to reduce the amount of abrasive dirt depositing on the rod, which otherwise may cause premature seal wear.

Any internal pressure acts on the rod area to give a suspension force, normally lifting the vehicle. Such pressurisation is avoided as far as possible in the double-tube damper, which minimises leakage.

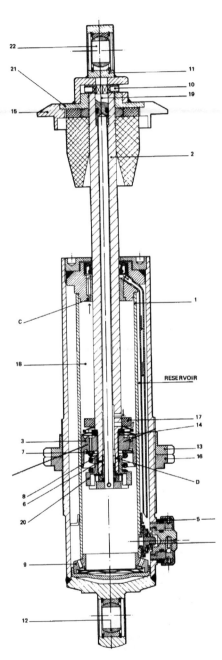

Figure 1.3.19 Detailed section of an adjustable double-tube racing damper. 1 inner cylinder; 2 hollow rod; 3 piston; 5 foot valve adjuster; 7 spring seat; 8 valve preload spring; 9 foot sintering with extension feed hole; 10 extension adjustment point; 11 mounting; 12 lower ball joint; 13 adjustable spring perch; 14 compression shim pack; 15 upper spring seat; 16 lower spring perch lock; 17 compression shim pack backing; 18 extension chamber; 19 extension adjustment member; 20 extension valve spring seat; 21 upper fixture; 22 upper ball joint (Koni).

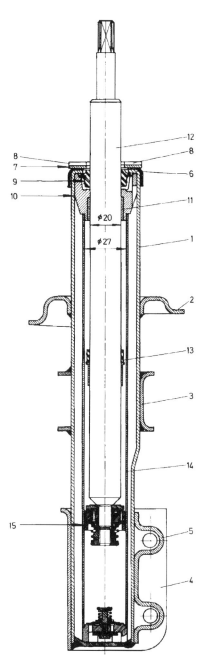

Figure 1.3.20 Sectional view of a front strut for a small car. Piston diameter 27 mm, rod diameter 20 mm. 1 outer cylinder; 2 spring seat; 3 guard; 4, 5 wheel hub fixture; 6 rolled closure; 7, 8 bump stop seat; 9 seal; 10 upper moulding; 11 bearing; 12 rod; 13 stroke limiter (?); 14 inner (working) cylinder; 15 piston (Fiat/Monroe).

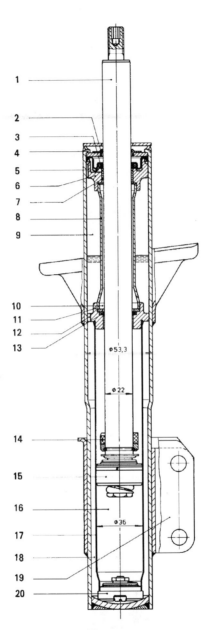

Figure 1.3.21 Sectional view of a front strut for a larger car. Piston diameter 36 mm, rod diameter 22 mm. 1 rod; 2 seal; 3 bush; 4 rolled closure; 5 bush; 6 top moulding; 7 bearing bush; 8 sleeve; 9 gas chamber; 11 centre moulding; 12 hole; 13 seal; 14 impact guard; 15 piston; 16 compression chamber; 17 inner (working) cylinder; 18 annular foot chamber; 19 wheel hub fixture; 20 foot valve (Boge).

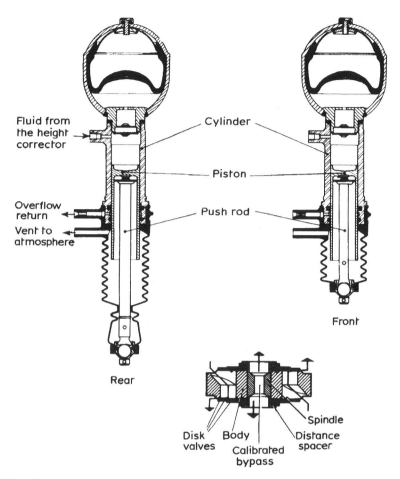

Figure 1.3.22 Citroen air suspension. The valve, with two shim pack valves, is fixed in position. It is not in the piston, which ideally would pass no oil. The gas is held in the elastomeric rolling seal bag. Nitrogen is used, reducing oxidation ageing of the rubber.

The pressurised single-tube type may suffer from loss of pressure with failure of correct function in compression due to cavitation behind the piston.

Suda *et al.* (2004) have proposed a nonhydraulic EM (electromagnetic) damper, of general configuration as in Figure 1.3.29. Actuation of the EM damper rotates the ball screw nut which drives an electrical generator through a planetary gearbox. An alternative arrangement uses a rack and pinion for the mechanical drive. The obvious advantage of an EM damper is controllability—the damper force depends on the generator and its electrical load. An external power supply is not needed, because the damper can generate its own electrical supply. A suggestion that energy from suspension motions can usefully be recovered to save fuel seems optimistic, as the average damper power dissipation is only a few watts for the whole vehicle. The Suda prototype successfully demonstrated appropriate characteristics with a damping coefficient around 1.6 kN s/m, and was tested with encouraging performance on the rear of a passenger car. The concept is an interesting alternative to ER and MR dampers, but it remains to demonstrate its life and manufacturing economics.

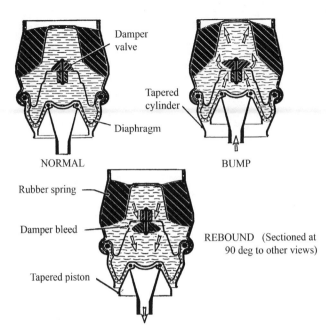

Figure 1.3.23 The BMC Hydrolastic system had a somewhat similar operation to the Citroen gas system, but replaced the gas spring by a rubber-in-shear spring already well proven. Based on Campbell, C., *Automobile Suspensions*, Chapman and Hall, 1981.

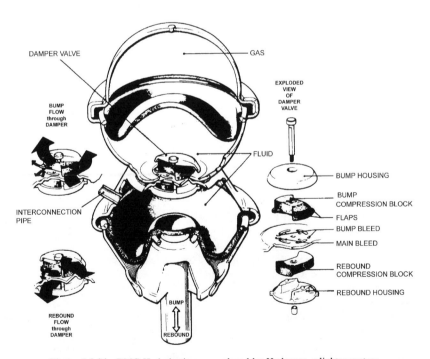

Figure 1.3.24 BMC Hydrolastic was replaced by Hydragas, a lighter system.

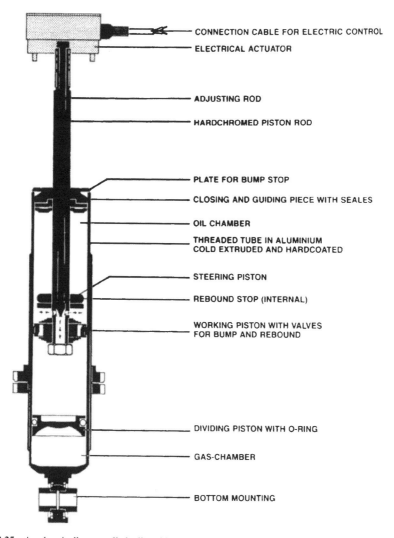

Figure 1.3.25 An electrically controlled adjustable damper. In this example, the basic construction is a single tube de Carbon configuration (Bilstein).

1.4 Ride-Levelling Dampers

One common problem with vehicles is that the load variation is a significant fraction of the kerb weight, perhaps 40%, particularly for small cars. This causes variation of the suspension performance with load condition. Many efforts have been made to overcome this. The most basic factor is the ride height, which varies, in particular at the rear. The telescopic damper offers the obvious possibility of making compensating adjustments to restore the ride height by simple pressurisation, Figures 1.4.1–1.4.5 illustrate some efforts along these lines. The operation of a self-levelling system can be very slow acting without detriment, so the pump may be very low power. It

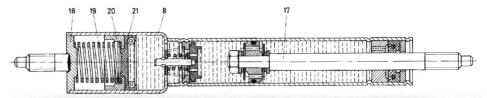

Figure 1.3.26 Steering damper of basic de Carbon layout, having a free piston separating the oil from the gas chamber, but with spring assistance. The piston has two shim packs. The foot valve has a coil spring blow-off valve (Stabilus).

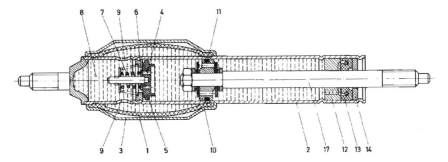

Figure 1.3.27 With similar internals to the last example, this steering damper uses a rubber oil/gas separator, achieving somewhat shorter overall length (Stabilus).

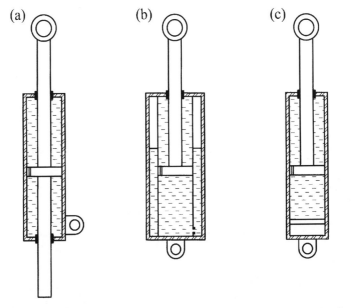

Figure 1.3.28 Basic types of telescopic damper: (a) through-rod; (b) double-tube; (c) single-tube (with floating piston).

Introduction

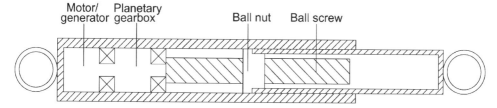

Figure 1.3.29 Electromagnetic damper configuration (after Suda *et al.*, 2004).

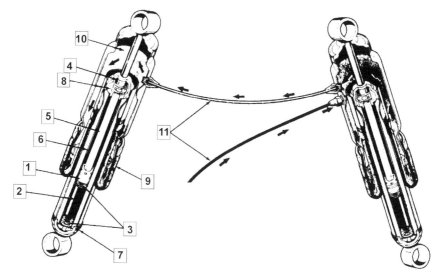

Figure 1.4.1 With an upper gas chamber, as in most dampers, the provision of a pump, a control system and a few pipes allows ride height compensation at modest cost (Delco).

is even possible to use the damper action when in motion to pump the damper up to a standard mean position.

1.5 Position-Dependent Dampers

Ordinary passenger cars have so far rarely used dampers with designed position dependence (other than indirectly, through the effect of the rubber mounting bushes), although they have been widely used on motorcycles and aircraft undercarriages. Figure 1.5.1 shows an example motorcycle front fork in which the sometimes problematic dive under braking is controlled by an internal bump stop which closes an orifice. This greatly softens the impact and allows weaker springing with improved ride quality. This basic method of position dependence by the sliding of a tapered needle in a hole to vary an orifice area has been used for aircraft undercarriages. In the Telefork the further feature is added that the rather short and blunt rubber 'needle' C entering orifice A can distort under pressure.

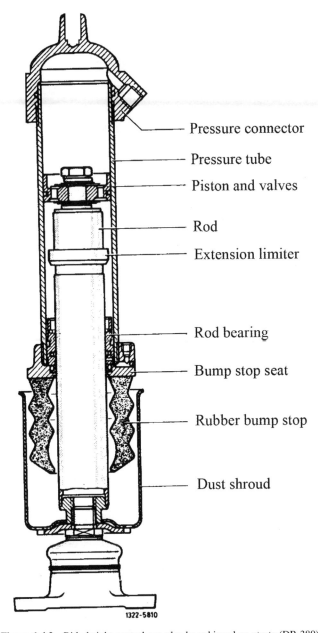

Figure 1.4.2 Ride height control can also be achieved on struts (DB 380).

Figure 1.5.2 shows a bus and truck damper in which extension is limited by the entry of the supplementary piston into the top cap, with hydraulic restraint.

Position or stroke dependence of damper force has recently become a development topic, and is discussed in detail later.

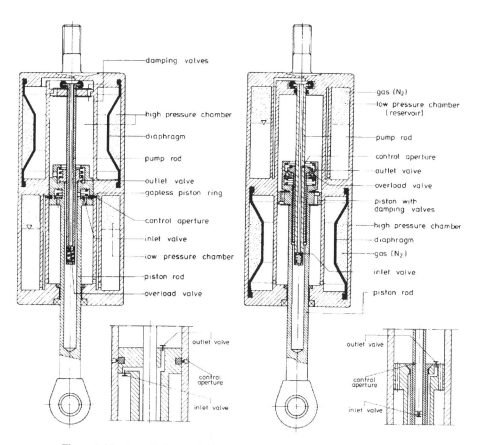

Figure 1.4.3 Boge Hydromat (left) and Nivomat (right) height-adjusting dampers.

1.6 General Form of the Telescopic Damper

A general form of the telescopic damper is shown in Figure 1.6.1, where there is a separate reservoir (chambers 0 and 1). Chamber 0 contains air, possibly pressurised, separated by a floating piston from chamber 1. Chamber 2 is called the compression chamber, at high pressure during compression, and chamber 3 is called the extension, expansion or rebound chamber, at high pressure during extension. During compression, fluid is displaced from the main cylinder (chamber 2 and 3) into the reservoir, through a restriction of given characteristics, the compression foot valve. By the Principle of Fluid Continuity, in normal noncavitating operation, neglecting compressibility, the quantity of fluid displaced through the foot valve is equal to the volume of the piston rod entering the main cylinder. During compression, fluid also passes through the piston from chamber 2 to chamber 3, through the piston compression valve. During damper extension, removal of the piston rod from the main cylinder requires a flow from the reservoir into the main cylinder, through the foot extension valve. Also, fluid passes through the piston extension valve, valve 4, from chamber 3 to chamber 2.

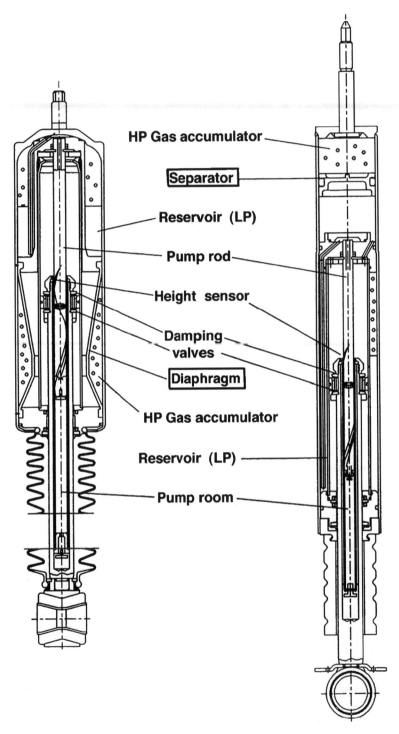

Figure 1.4.4 Mannesman Sachs height adjusting (type 1).

Introduction

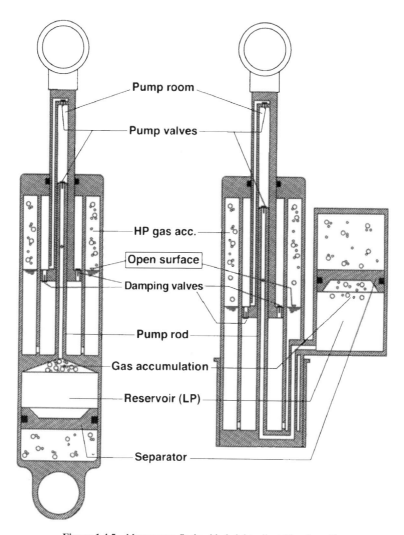

Figure 1.4.5 Mannesman Sachs ride-height adjustables (type 2).

The four valve flow rates are therefore:

(1) Q_{PE} piston valve in extension,
(2) Q_{PC} piston valve in compression,
(3) Q_{FE} foot valve in extension,
(4) Q_{FC} foot valve in compression.

These are denoted by the normal directions of flow. In extreme operation (with significant compressibility or cavitation), the actual flow direction through the valves may be momentarily abnormal, so, for example, there could be a positive flow Q_{PE} in the piston extension valve at a moment when the damper is actually compressing, immediately after reversing direction.

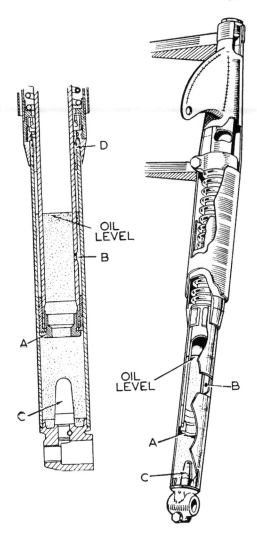

Figure 1.5.1 The BSA Telefork motorcycle spring-damper unit.

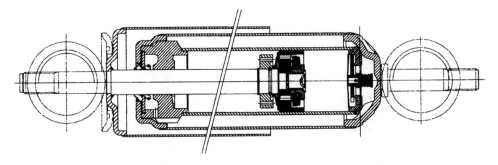

Figure 1.5.2 Truck damper with hydraulic extension limiter (Duym and Lauwerys, 2001).

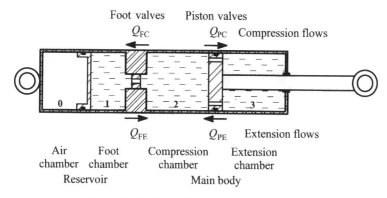

Figure 1.6.1 General form of telescopic damper.

The basic damper characteristics therefore depend on the piston and rod diameters, and on the characteristics of the four valves. Additional factors are the reservoir pressure (affecting cavitation) and the fluid properties: density, viscosity, temperature, vapour pressure, gas absorption, emulsification and compressibility, and so on. Further factors are friction at the rod bearing and seal, at the main piston and at the reservoir piston, and also leakage paths, e.g. from chamber 3 to the reservoir in many practical implementations (e.g. double-tube type).

In a real damper, the layout may actually be as shown in Figure 1.6.1, usually with the foot valve flow passing through a flexible pipe joining chambers 1 and 2; such remote reservoir dampers are used in racing, but are rare on passenger vehicles. With the layout of Figure 1.6.1, a pressurisation of 1.5 MPa might be used, with the foot compression valve providing extra pressure in the main cylinder to prevent cavitation during compression. (1 MPa, one megapascal, is about 10 atmospheres or 150 psi.) Alternatively, the reservoir may be incorporated directly into the main cylinder (in effect, the foot valves 1 and 2 then having zero resistance), it then being essential to have a high pressure, e.g. 3 MPa, to produce the required flow through valve 3 during compression without cavitation in chamber 3. In such cases a floating piston may still be used. Alternatively, this may be omitted, the gas being allowed to mix freely giving an emulsion, requiring appropriate valve calibration.

Another common configuration is to place the reservoir concentrically around the main cylinder giving the double-tube damper, usually unpressurised. This is less good for cooling, but the working inner cylinder carrying the piston is better protected against impact damage from flying stones. This type depends on gravity to separate the contained air and liquid, so it must be used with the cylinder underneath and the piston rod above, unless the gas is separated physically from the liquid, e.g. by an air bag in the reservoir. A spiral insert in the reservoir, or spiral rolling of the outer tube, can help to minimise mixing of air and oil. Extremely violent conditions can cause emulsification and loss of damping force, but this is not normally a problem for passenger cars. An advantage of this type is that it is possible to include a small annular chamber between the rod bearing (inside) and seal (outside), connected to the reservoir. The rod seal is then subject to the reservoir pressure only. The valving is arranged so that in normal operation, even in compression, the extension chamber pressure is always greater than the reservoir pressure, so the fluid leakage through the bearing (clearance 0.02–0.05 mm) is always outwards, circulating the fluid and hence improving cooling. This also prevents air being drawn back into the extension chamber (3), and ensures that any aeration gas brought through the piston valve collects at the top and is passed back to the outer reservoir.

Releasing the leaked fluid into the gas region greatly encourages emulsification, so a drain tube may be used to carry it down into the liquid at the bottom of the reservoir. This also ensures that, should there be a low extension chamber pressure causing reverse flow from the reservoir air chamber (0) into

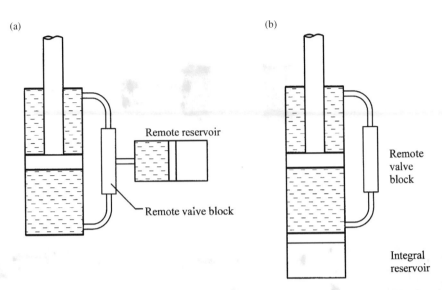

Figure 1.6.2 Two possible systems with remote valves: (a) with remote reservoir; (b) with integral floating piston.

the extension chamber (3), liquid rather than air is drawn back in. Otherwise, gas can pass very rapidly through the bearing bush, much more rapidly than liquid, with subsequent loss of damping function.

Numerous other configurations are possible, especially where one or two remote reservoirs are used. It is possible to use the piston for displacement only, with no piston valve, passing the liquid through an external circuit, giving excellent access for adjustability of the valves. This can be arranged, for example, with a remote reservoir, as in Figure 1.6.2(a), or with an integral secondary piston, as in Figure 1.6.2(b), or using emulsified gas. This concept can usefully be extended by adding oil take-off points near to the middle of the stroke, allowing position sensitivity with an extra flow and lower force when the piston is near to the centre position. This is known to offer a better combination of ride and handling. With external valves this can be controlled properly, instead of just having a bypass channel.

Adaptive dampers typically have rotary barrels with several holes of different sizes to adjust the resistance. These are of similar general configuration to the ones described above. The electrorheological variable damper uses a distinct configuration described later. The magnetorheological types are typically similar to the de Carbon floating secondary piston type, and, again, are considered in more detail later.

1.7 Mountings

To discourage the transmission of small-amplitude higher-frequency vibrations (noise, vibration and harshness) the ends of a damper are each mounted through a rubber bush (with the exception of a few very specialised cases such as racing cars). These bushes also permit some rotational motion and misalignment of the mounting points. The two principal configurations of end mounting are:

(1) axially threaded rod, as in Figure 1.7.1 (and 1.3.10);
(2) transverse eye, as in Figure 1.7.2 (and 1.3.10).

The former type is ideal for mounting directly into the top of a wheel arch. It is therefore frequently used on conventional passenger cars at the rear, and also for the top of front strut suspensions. In the

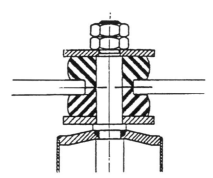

Figure 1.7.1 Axially threaded rod mounting (Reproduced from Kinchin and Stock (1951) pp. 67–86 with permission).

latter case, the bottom mounting of the strut is usually a ball joint, forming the lower defining point of the steering axis. The threaded rod is easily formed by machining the end of the damper rod or by inserting a stud.

The transverse eye of Figure 1.7.2. uses a concentric rubber bush with a bolt through. Preferably it is mounted in double shear, but frequently it simply uses a single bolt or stud into the side of the wheel upright or into the suspension arm. Figures 1.7.3–1.7.5. show some other common types of mounting.

The effect of the rubber mounting bushes is to put a nonlinear compliance in series with the damper, giving the complete unit a characteristic which depends upon the displacement amplitude, for a test at a given velocity amplitude. Small-amplitude motions with high frequencies are more readily met by bush compliance, hence reducing the transmission of such motions. More substantial movements relating to deflections of the suspension in handling movements (roll and pitch), or gross suspension movements in ride at the sprung mass natural frequency, are little affected by the bushes because their small compliance is effective for only a small deflection. The bushes are therefore important in introducing some stroke sensitivity to the transmitted forces, keeping the damping high for large amplitudes, as found in handling motions such as roll in corners, and high for large body motions on rough roads, but desirably reducing the damping at small amplitudes to improve the ride on smooth roads.

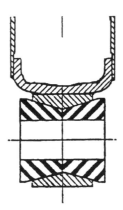

Figure 1.7.2 Transverse eye or integral sleeve mounting (Reproduced from Kinchin and Stock (1951) pp. 67–86 with permission).

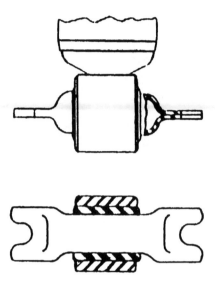

Figure 1.7.3 Integral bar mount (Jackson, 1959).

The axial-rod mounting lends itself to an axially asymmetrical form of compliance bushing, as in Figure 1.7.6 and 1.7.7. The asymmetry may be achieved by differing thickness, area or material properties, and differing axial preload distance. With a deflection such that one of the two bushes has expanded to reach axial freedom, that bush then contributes zero further stiffness. If the preload is small, the essential result is that the stiffness is different for damper compression and extension, a feature that can be turned to advantage, particularly on strut suspensions.

The basic damper characteristics are normally considered to be those when the mountings are rigid, not soft bushed, and that is how they are normally tested.

Struts require a more robust mounting than dampers alone, as shown by the examples in Figure 1.7.8 *et seq*. For front struts there must also be provision for steering action at this point.

It is advantageous to separate the seat force exerted by the spring and the damper. This is natural in many suspension designs, but not always automatic. Figure 1.7.9 above shows how the separation may be achieved. This is known as a 'dual-path mounting'. Figure 1.7.12 gives a further example.

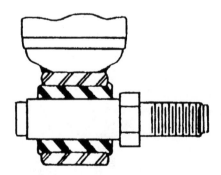

Figure 1.7.4 Integral stud mount (Jackson, 1959).

Figure 1.7.5 Integral bayonet stud mount (Jackson, 1959).

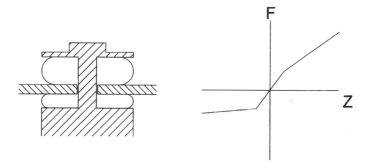

Figure 1.7.6 Asymmetrical type of damper or strut bushing.

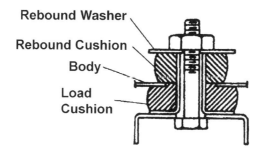

Figure 1.7.7 Asymmetrical damper bushing (Puydak and Auda, 1966).

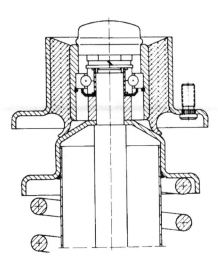

Figure 1.7.8 Strut top mounting (INA).

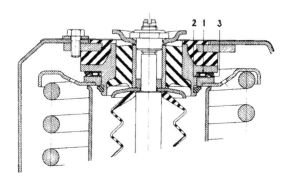

Figure 1.7.9 Strut top mounting (Peugeot).

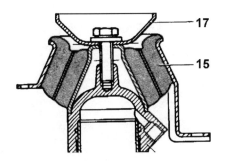

Figure 1.7.10 Strut top mounting (Daimler-Benz 380).

Introduction

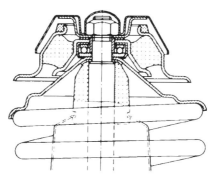

Figure 1.7.11 Strut top mounting (Audi).

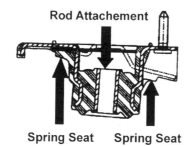

Figure 1.7.12 Dual-path damper/strut mounting (Lewitske and Lee, 2001).

1.8 Operating Speeds and Strokes

The suspension is brought into operation by:

(1) dropping, e.g. falling off a jack, driving off a kerb, or in a drop test;
(2) ride motions, in response to road roughness;
(3) transient longitudinal motions giving pitch change in acceleration or braking;
(4) transient handling response, mainly roll velocity during corner entry and exit.

The above are, in general, likely to be combined in real cases. The vehicle motion effectively defines the suspension wheel bump speed V_{SB}, but evaluation of the actual damper speed requires consideration of the particular installation geometry (Chapter 5) to determine the velocity ratio (motion ratio) V_D/V_{SB}.

The suspension bump velocities may be estimated under the following headings:

(1) drop test;
(2) ride motions;
(3) longitudinal acceleration transients;
(4) lateral acceleration transients;

(5) combined effects;
(6) damper failure speeds.

These will be dealt with in more detail as follows.

Drop Test

The vehicle is released to fall freely from height h_D above the position at which the wheels touch the ground. The wheels may initially be in the full droop position, simulating the situation where the vehicle leaves the ground, e.g. a rally car passing over a crest. Where it is intended to simulate a simple road step-up impact it is better to restrain the wheels to the normal ride position. When the vehicle is below that point at which contact of wheel and ground just occurs, the springs and dampers will act, and there will be relatively little further speed increase, unless the drop is from a very low initial position. This can be studied more accurately, either analytically or by time stepping, but for a simple high-drop analysis the impact speed V_I is given by energy analysis of the fall as

$$\tfrac{1}{2}mV_I^2 = mgh_D$$

where m is the mass and g is the gravitational field strength, leading to

$$V_I = \sqrt{2gh_D}$$

This is independent of the vehicle mass. Example values are given in Table 1.8.1, where it may be seen that even a small drop of 50 mm gives an impact speed of 1 m/s, which is quite high in damper terms.

A passenger car may experience a drop of 100 mm or so occasionally, for example by driving over a kerb, with a corresponding impact speed of 1.4 m/s. Rally cars may be expected to have drops of as much as one metre as a result of driving at high speed over a crest, with vertical-component impact speeds of 5 m/s or more in extreme cases. The tyre deflection will absorb some of this impact, but it is evident that very high damper speeds may occur in this way.

Free Droop Release

Considering the vehicle body to be temporarily fixed, a wheel may be suddenly released so that the suspension is forced into droop motion (jounce) under the action of the main suspension spring. This

Table 1.8.1 Impact speeds resulting from suspension drop test

h_D (m)	V_I (m/s)
0.05	0.990
0.10	1.401
0.20	1.981
0.40	2.801
1.00	4.429
2.00	6.264

could occur in practical use if one wheel suddenly passes over a wide hole. If the suspension was previously in a normal position, the free force on the wheel equals the normal suspension force, about $mg/4$. This will give a free extension velocity of 2 m/s or more, large in damper terms. This is the speed at which the car would settle on its dampers if the springs could be instantly removed:

$$V_R = \frac{mg}{\sum C_D} = \frac{g}{2\zeta}\sqrt{\frac{m}{\sum K}} = \frac{g}{2\zeta\omega_{NH}}$$

where the spring stiffnesses K and the damping coefficients C_D are the effective values at the wheels. The natural heave frequency in rad/s is ω_{NH}.

In principle this is a feasible experiment, by removing the springs and holding the body, followed by a sudden release. If the wheel is in a compressed position at the moment of release then the potential speed is even greater. In practical conditions, of course, the wheel inertia plays a part.

Ride Motions

The vehicle is stimulated by a wide spectrum of frequencies from road roughness. The main suspension response occurs at the natural heave frequency f_{NH} of the vehicle, around 1.4 Hz for a passenger car. For a sinusoidal motion in displacement, the velocity and acceleration are also sinusoidal. The amplitudes of these are called the displacement amplitude, the velocity amplitude and the acceleration amplitude. The abbreviated term 'amplitude' means the displacement amplitude. Twice the displacement amplitude is the distance between extreme positions, and is called the stroke. The suspension bump velocity amplitude V_{SB} depends on the heave displacement amplitude Z_H according to

$$V_{SB} = \omega_{NH} Z_H = 2\pi f_{NH} Z_H$$

where ω_{NH} is the radian natural frequency (rad/s) of the body in heave.

Table 1.8.2 gives some example values. The total range of body motion, the stroke, is twice the amplitude, so at 1 Hz a total bump stroke of 0.16 m, essentially the full range of suspension movement corresponding to an amplitude of 80 mm, is needed to give a bump speed of 0.5 m/s. This is a very severe ride motion. Ride amplitudes of under 20 mm are normal, with corresponding suspensions bump speeds up to 0.15 m/s. Naturally, this depends upon the quality of road and the vehicle velocity, although for passenger cars these are compensating, as speed will be reduced by the driver if the ride motions become severe.

In the case of rally cars, which are often required to travel at high speed on rough roads, the suspension velocities are correspondingly higher. For racing cars, the natural heave frequency is higher, especially for ground effect vehicles where it may be around 5 Hz or more, even for very stiff suspensions, because of the tyre compliance. Logger data shows that the ride response is strong at this

Table 1.8.2 Bump speed amplitudes in suspension ride (at $f_{NH} = 1$ Hz)

Z_H (m)	V_{SB} (m/s)
0.005	0.031
0.010	0.063
0.020	0.126
0.040	0.251

natural frequency, but the amplitude is necessarily very small, so the actual suspension ride velocities are still quite modest, generally 0.1–0.2 m/s, although worse over particular bumps.

Longitudinal Acceleration Transients

Variations of longitudinal acceleration cause (angular) pitch motions, and occur by sudden application or removal of engine power or brake action, with a step change of A_X. This is most easily demonstrated when the brakes are kept firmly on as the vehicle actually comes to a halt, giving a distinct jerk with a subsequent damped pitching motion. The avoidance of this discomfort factor by ramping off the braking force at the end is called 'feathering' the brakes. This case is a fairly easy one to analyse. Consider a longitudinal deceleration A_X on a simple vehicle having wheelbase L with centre of mass at the mid point, centre of mass height H_G, and suspension stiffness wheel rate K_W at each wheel. An angular pitch angle θ radians gives a suspension bump displacement

$$z_{SB} = \tfrac{1}{2}L\theta$$

Hence, the restoring pitch moment M_P is

$$M_P = 2K_W z_{SB} L = K_W L^2 \theta$$

and the pitch angular stiffness K_P is

$$K_P = K_W L^2$$

A longitudinal acceleration A_X gives a longitudinal load transfer moment

$$M_{AX} = mA_X H_G$$

with an associated pitch angle

$$\theta = \frac{M_{AX}}{K_P} = \frac{mH_G}{K_W L^2} A_X$$

and a suspension deflection

$$z_S = \frac{1}{2}L\theta = \frac{mH_G}{2K_W L} A_X$$

The natural frequency f_{NP} of pitch motion depends upon the pitch stiffness and also on the sprung mass (body) second moment of mass in pitch, I_{PB}:

$$f_{NP} = \frac{1}{2\pi}\sqrt{\frac{K_P}{I_{PB}}}$$

with a value for most vehicles approximately equal to the heave frequency, at around 1 Hz for a passenger car. Hence the estimated suspension bump velocity for this amplitude and frequency $V_{SB,Ax}$ is

$$V_{SB,Ax} = \omega_{NP} z_{SB}$$
$$= 2\pi f_{NP}\left(\frac{mH_G}{2K_W L}\right) A_X$$

where ω_{NP} is the natural frequency of the body in pitch and z_{SB} is the displacement of the suspension in bump.

Realistic physical values for a passenger car give a pitch angular deflection of about $0.4°/\text{m s}^{-2}$ ($4°/g$) with a suspension deflection/acceleration rate of 10 mm/m s^{-2} ($100 \text{ mm}/g$). At a pitch natural frequency of 1 Hz the consequent suspension velocity is about $50 \text{ mm s}^{-1}/\text{m s}^{-2}$ ($0.5 \text{ m s}^{-1}/g$). Hence the associated suspension velocity for pitch motions is typically up to 0.2 m/s for normal braking transients and up to 0.5 m/s for extremes. The most severe case is a sudden switch from acceleration to braking. For a 5 m/s² (0.5g) change of A_X, the suspension stroke is about 50 mm with a velocity of 0.25 m/s.

A similar analysis may be performed for other kinds of vehicle. Racing cars, despite their high natural pitch and heave frequencies, tend to have smaller suspension travel velocities because of the limited suspension movements, plus the use of anti-dive suspension geometry, which reduces the pitch angles and velocities in proportion.

Lateral Transients (roll)

Sudden changes of lateral acceleration cause roll motions that can be analysed in a similar way to longitudinal motions. A lateral acceleration causes a suspension roll angle

$$\phi_S = k_{\varphi S} A_y$$

where $k_{\varphi S}$ is the suspension roll angle gradient, of about 0.014 rad/m s^{-2} ($0.80°/\text{m s}^{-2}$, $8°/g$, 0.14 rad/g). The corresponding suspension deflection z_S is

$$z_S = \tfrac{1}{2}\phi_S T$$

with ϕ_S in radians and a track (tread) T of about 1.5 m. For a roll natural frequency f_{NR} the suspension velocity amplitude estimate becomes

$$V_{SB,Ay} = 2\pi f_{NR} z_{SB} = \pi f_{NR} \phi_S T$$

Realistic values for a passenger car are a natural roll frequency of 1.5 Hz and a suspension deflection of 10 mm/m s^{-2} ($100 \text{ mm}/g$). Entering a normal corner of lateral acceleration 3 m/s² (0.3g), the suspension deflection is about 30 mm and velocity 0.30 m/s. An extreme corner entry gives 80 mm stroke and 0.60 m/s, depending on the time for steering wheel movement.

It is essential to have good roll damping in this speed range. This is the basic reason why nonlinear $F(V)$ characteristics are used—to give adequate roll damping without excess forces at higher speed.

Combinations

The above suspension velocities are frequently combined. In general a handling motion is mixed with some ride motion, so the handling motion does not always occur in a progressive manner, there may even be velocity reversals.

Overall, for a passenger car the suspension bump velocities are up to 0.3 m/s in normal driving, up to 1 m/s in harder driving or poorer roads, with values in the range 1 to 2 m/s occurring relatively rarely.

Table 1.8.3 indicates a possible classification of bump velocity ranges for passenger cars. A racing car classification would generally be lower, especially for a ground effect vehicle such as Formula 1, or Indy. However, even ground effect racing cars are subject to driving over angled kerbs, to dropping from jacks and so on.

Table 1.8.3 Possible classification of suspension bump velocities (m/s)

	Passenger car	Racing car (F1, Indy)
Very low	Below 0.1	Below 0.025
Low	0.1 – 0.2	0.025 – 0.050
Medium	0.2 – 0.4	0.050 – 0.080
High	0.4 – 1.0	0.080 – 0.200
Very high	1.0 – 2.0	0.200 – 0.400
Extreme	Above 2.0	Above 0.400

Damper Failure Speeds

The damper will suffer permanent damage if subject to excessive velocities, although these limits are not widely advertised. The physical survival limit speed is that which will not quite cause any metal yielding or, more likely first, damage to the valves or seals. The limit speed is typically around 5 m/s.

Velocity and Stroke

Fukushima *et al.* (1983) suggested that dampers should have a stroke-dependent characteristic, such that for a given velocity a longer stroke would give a greater force. More information on this is given in Chapter 3. Table 1.8.4 shows the velocities and strokes found by Fukushima, with some extra analysis, where F/V is the desirable C_D, and F/S is the desirable 'stiffness' relationship if the force were produced by a spring. Slow steering is a 'figure of 8' course with large lateral acceleration, but gentle entry.

Perhaps the clearest form of analysis for this problem is to plot a diagram of damper velocity against amplitude or stroke, showing the active points. This is done in Figure 1.8.1. Points on one line radiating from the origin have the same frequency, since the velocity and displacement amplitudes of a sinusoidal motion are related by $V = 2\pi f X$. With suitable scales for a particular vehicle, the main diagonal is the natural frequency of the vehicle body in heave type motions, around 1.4 Hz. The basic handling manoeuvres, i.e. changes of A_X and A_Y, operate along this diagonal according to severity, or somewhat below it, according to the rate of application of the controls. Driving along a straight road gives heave motions along the main diagonal, with good-quality roads at the bottom left, and rough roads, or fast driving, further up. Bumps act on the bottom of the wheel and are transmitted through to the body at essentially the wheel hop frequency. Isolated bumps therefore lie on a steep line at about 12 Hz, on the left.

To avoid transmission of bumps, but to give good damping for handling evidently could be achieved by having frequency-dependent damping, within the meaning of this diagram, with low damping

Table 1.8.4 Suspension stroke, velocity and damper force

Manoeuvre	Stroke (mm)	V (m/s)	F (N)	F/V (kN s/m)	F/S (kN/m)
Slow steering	50	0.02	48	2.4	1.0
Rapid steering	45	0.12	288	2.4	6.4
Lane change	25	0.06	144	2.4	5.8
Single bumps	4	0.40	120	0.3	30.0
Brake/accelerate	50	0.20	400	2.0	8.0
Smooth road	3	0.08	8	0.1	2.7
Rough road	12	0.60	600	1.0	50.0

(Based on Fukushima *et al.*, 1983)

Introduction

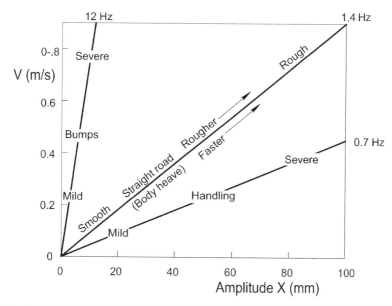

Figure 1.8.1 Damper speed versus damper amplitude for various ride and handling activities. Handling 'frequency' depends on the rate of control application.

above, say, 3 Hz. On the other hand, to also have low damping as desired on smooth roads, stroke-dependent damping with low damping for strokes below about 10 mm is desirable.

1.9 Manufacture

The complete manufacturing process involves:

(1) material production;
(2) individual component production;
(3) assembly;
(4) oil-filling;
(5) pressurisation (in some cases);
(6) insertion of bushes;
(7) external finishing;
(8) testing at various stages as appropriate.

Initial material production is of course not the province of the damper manufacturer, who will simply buy in materials, but may heat-treat them.

The production method of the tubes depends upon the price range of the damper. Cheaper ones may have tubes rolled from low-carbon steel with a resistance-welded seam. More expensive ones may avoid the seam by using extruded tubing, and in any case racing dampers, and possibly more passenger cars in the future, will have aluminium cylinders to reduce the weight. Steel tubes are annealed and then cold-drawn to final diameter, giving accurate sizing and cold-worked strength. The internal finish is important to give a good seal with the piston and also to give a low wear rate of the piston seal; the internal finishing process, e.g. honing, varies with manufacturer and budget.

The piston rod is normally made from high-carbon steel with cold-forged ends, often with some drilled passages, possibly threaded at the top, and case-hardened or chromed on rubbing areas. The life of the rod-bush bearing and of the main rod seal is sensitive to the rod surface finish, which should be 0.1 µm r.m.s. or better, achieved by grinding and lapping. It is now normal to use a hard-chrome-plated rod which enhances both wear resistance and corrosion resistance.

Pistons themselves may be forged or sintered. Piston seals vary considerably. Some have iron piston rings not unlike those in an engine, with a small end gap, necessary for thermal expansion. Plastic piston-type rings are also used, typically with stepped-overlapped ends, from Teflon (PTFE) or nylon, etc. Similar designs are used in Tufnol and other paper and cloth-reinforced phenolic materials. Square-section rubber rings and O-ring seals are common. Finally, some plastic seals are moulded solid to the piston with no apparent attempt to allow for self-adjustment to fit, possibly expected to run-in to a minimal correct working clearance, with a wide contact area. Racing pistons are usually machined aluminium, turned and drilled, with a plastic piston ring

Valve manufacture of course varies with design. At the low-cost high-volume end of the market, where possible, stamping and pressing are used, followed by automated or manual assembly of these parts plus coil springs, washers, *et cetera*.

For common double-tube dampers, the top and bottom cylinder ends are pressed from steel sheet, typically with a mounting bush ring spot-welded onto the bottom one. The lower cup is welded to the cylinder and foot valve inserted. The piston, piston valves, rod, rod-bush housing, top cup and its seal and bearing are assembled. The piston and oil are inserted and the top cup welded, spun or crimped into place. The rubber bushes are then pressed in, and, finally, paint is applied.

With a worldwide production of several hundred million dampers per year, there is obviously a considerable difference between the methods used for price-sensitive mass-production units and those for higher-price small-volume precision units produced for specialised applications such as racing and rallying.

1.10 Literature Review

Books on vehicle suspension or vehicle dynamics nearly always have some basic description of damper (shock absorber) design and operating characteristics, although this is invariably limited in scope e.g.:

(1) Norbye (1980), 10 pages descriptive;
(2) Campbell (1981), 8 pages descriptive, one equation;
(3) Bastow (1987), 10 pages on the effect of damping, 19 pages of description of dampers;
(4) Dixon (1991, 1996), 10 pages on history plus comments on effect on handling;
(5) Gillespie (1992), 7 pages on dampers and their effect;
(6) Milliken and Milliken (1995), 50 pages, almost all on vehicle behaviour rather than damper design itself;
(7) Haney and Braun (1995), 32 pages, with the emphasis on racing car dampers;
(8) Reimpell and Stoll (1996), 36 pages, on damper design, construction and characteristics.

Within the context of a complete book on ride, or on handling, or on vehicle dynamics as a whole, such limited treatment is understandable.

Although a literature search reveals several thousand papers referring to 'damper' or 'shock absorber' in the context of vehicles, rather few of these are about the design and characteristics of the damper itself. A full list of these is given in the references.

The earliest paper appears to be that by Hadley (1928), which deals with mechanical friction dampers, seeking to obtain a suitable characteristic by position dependence. The ability of a damper to control large amplitudes adequately whilst not overdamping small motions is called the 'range' of the damper, and was a leading problem of that era. It seems surprising now that mechanical types were considered more promising than hydraulics.

Weaver (1929) investigated damper characteristics by applying them to a vibrating system, actually a 272 kg mass on a 31 kN/m spring giving 1.71 Hz undamped natural frequency, with initial deflections of typically 76 mm, and producing free vibration $F(X)$ plots, which are roughly spiral for less than critical damping. This is a simple and cheap method of investigation, of some value for comparative purposes, but not as useful as an $F(X)$ curve for a controlled displacement pattern, preferably sinusoidal, as obtained on a modern testing machine.

Nickelsen (1930) states that by 1930 nearly all cars had some kind of damper as standard equipment, and therefore, of course, by implication some still did not. He concentrates on the double-acting hydraulic type, recommending that no more than 20% of the resistance should be in bump. $F(X)$ curves are shown, and temperature fade considered, although with the apparent belief that resistance is simply proportional to viscosity.

James and Ullery (1932) discuss the problems of Coulomb friction against the advantages of hydraulic damping with a suitable force–speed relationship. Various force–displacement graphs ('card diagrams') are shown, illustrating various effects and faults, and the loss of damping effect with temperature rise is eliminated by a variable orifice comprising a plug valve regulated by a bimetallic strip. Oil viscosities of 20–200 centistokes were in use. They emphasise the need for a smooth characteristic, building up with speed, and the need for temperature compensation.

Fuchs (1933) gives $F(X)$ diagrams for single-acting and double-acting hydraulics, and for the old friction disc type, and also shows a characteristic in the phase plane (V–X) for a position–dependent and force–dependent damper.

Kindl (1933) proposed to regulate the damper by a valve controlled by body inertia, so that wheel-only movements are free, but the body motion is damped. Unfortunately his acceleration sensor and controller was a 680 g mass for each damper. However, with electrically controlled dampers, and modern sensors and control systems, this is an idea whose time may now be ripe.

Schilling and Fuchs (1941) show force–speed diagrams for year 1940 passenger car dampers, some of which are similar to modern dampers, showing a predominance of rebound force over bump.

Connor (1946) distinguishes various types of damping effect (Coulomb, real dry friction, viscous, degenerate viscous, 'hydraulic' damping proportional to V^2, and solid material hysteresis). Typical damping ratios are stated to be 0.15–0.20, and the temperature rise in a telescopic damper is stated to be about 50°C on extremely rough roads, although only about 20°C for rotary and cam types because of conductivity into the vehicle body.

The related area of buffer design is considered by Brown (1948), Tatarinov (1948, 1949) and Brown (1950 a, b). For buffering, the resisting force is commonly made a function of position. This is not normally done for car dampers, although it has been used on motorcycle front forks, where it gives a bump stop effect, but is regarded as superior to an elastic bump stop because of reduced spring back.

Kinchin and Stock (1951/1952), in a comprehensive paper, discuss various details, such as the physical layout of various hydraulic types, fluids, testing, operating problems such as aeration, fade, noise problems, flow resistances, and force–displacement $F(X)$ diagram (work diagram) shapes.

Peterson (1953) describes the introduction of a damper designed to have a linear characteristic (in contrast to many of the previous ones which were almost Coulomb in character, as may also be seen in some of the characteristics shown by Schilling and Fuchs (1941). Prototypes had tubes intended to give a simple viscous pressure drop, but actually giving force proportional to speed to the power 1.25. More conventional drilled and coined holes were used for production. The higher forces obtained at high speeds were considered a substantial improvement for rough roads.

Cline (1958) discusses various aspects of dampers in general terms, giving some example $F(X)$ curves, with brief comments on the effect of valve changes. He also shows a photograph of the then new Gabriel dampers with electrical remote adjustment.

Hoffman (1958) was amongst the first to do analytical work on the damper. He shows various experimental $F(V)$ and $F(X)$ curves for different types of damper, and used an analogue computer to study vehicle ride for various dampers.

Jackson (1959) discusses the basic principles of telescopic dampers and their manufacture, and the effect of valving, stating that control of unsprung mass vibrations had become the greater problem once independent suspension was introduced. The standard fluid is stated to be straight mineral oil with viscosity somewhat less than SAE 5, but this was considered unsatisfactory because of viscosity change with temperature causing difficulties at very low temperatures. High-viscosity-index oils were just coming into use. Curves of damping force against temperature are given. A freon–filled nylon cell for preventing aeration problems is described.

Eberan-Eberhorst and Willich (1962) show experimental $F(X)$ curves for various strokes, and $F(X)$ and $F(V)$ curves with lag and hysteresis. The experimentally observed temperature effect on force was about -0.1%/K for compression and -0.3%/K for extension. A Plexiglas damper was built, allowing cavitation to be observed directly.

Ellis and Karbowniczek (1962) deal mainly with buffers rather than vehicle dampers. However, they draw attention to possibilities for separation of the air from the fluid to prevent aeration, including a spring-loaded free piston, a diaphragm, a gas-filled bladder, or a compressible closed-cell solid foam.

Speckhart and Harrison (1968) return to the idea by Kindl (1933) of using an inertia-regulated valve, this time much lighter and built into the damper piston, the intention being to reduce vehicle ride jerk (rate of change of acceleration).

Polak and Burton (1971) discuss damper construction in general terms, giving attention to possible designs of amplitude sensitive dampers, to a multi-frequency 'seismically' actuated damper (mounted on the wheel only, not to the body), and propose a configuration with a conventional damper coupled between suspension and engine. They also mention the possibility of achieving adjustable damping by the use of certain electrostatically sensitive solid/oil mixtures which have variable viscosity, i.e. use of electrorheological liquids.

Wössner and Dantele (1971) compare the pressurised single-tube type (with and without a free separator piston) with the unpressurised dual-tube type, and give experimental results on the cooling effect of airstreams.

Mitschke and Riesenberg (1972) discuss damper temperature rise on various types of road, and the consequences for fluid viscosity and damper force.

Mayne (1973) gives an analysis of the effect of liquid and mechanical compliance on buffer performance.

Jennings (1974) gives 21 $F(X)$ loops for commercially available motorcycle front forks and rear dampers, which prove to be characterised by extreme compression/extension asymmetry.

Cline (1974) gives a simple review of the application of hydraulic dampers to recreational vehicles, considering briefly how to achieve some of the particular characteristics required.

Schubert and Racca (1974) describe an unconventional 'elastomeric-pneumatic isolator with orifice-type relaxation damping', proposing application to motor vehicles.

Simanaitis (1976) gives a brief history and a discussion of some of the operating principles and problems ('The Dutch call them schokdempers; the French, amortisseurs; and the English, dampers. Indeed, many early automotive designers called them unnecessary ...'). (As a matter of interest it is *Stossdämpfer* in German and *ammortizzatori* in Italian.) Oil viscosities are given as SAE 5 to 10. Aeration is stated to be the cause of compression lag (although actually this can also be cavitation or desorption of air). Manufacturing methods are outlined.

Dalibert (1977) considers the effect of some oil properties on damper performance. The sensitivity of peak force to temperature was found to be about -0.3 %/°C, giving a reduction of 35% at 130°C, which is stated to be the approximate maximum temperature encountered with hard driving on bad roads. Maximum safe viscosity is stated to be 4000 cs at the lowest temperature, and a minimum of 4 to 7 cs at 100 °C, depending on design. Noise problems are considered in some detail.

Segel and Lang (1981) report a detailed investigation (additional information being given in Lang's PhD thesis (1977), using an 82 parameter analogue computer model that gave quite good agreement with the experimental data obtained. Compressibility of the liquid, plus a slight additional effect from cylinder compliance, was found to give hysteresis in the $F(V)$ curve at higher frequencies (above 1 Hz). Presumably series rubber bushes would markedly increase this effect, but were not mentioned. Gas compressibility and absorption were considered; this could not readily be handled by the analogue computer, but the use of an effective vapour pressure of up to 70 kPa (instead of the actual value of less than 2 kPa) was found to give realistic results. Orifice discharge coefficients were investigated, but for simplicity it was considered that a constant value of 0.7 was acceptable. Testing was actually done with a square acceleration wave, giving a triangular speed waveform and a piecewise parabolic displacement waveform.

Van Vliet and Sankar (1981) studied motorcycle forks and rear dampers, using analogue and digital simulations respectively to obtain good agreement with experimental $F(X)$ diagrams.

Arndt et al. (1981) consider seal design with tests of friction especially in the context of smaller, lighter vehicles where the problem seems more critical, advocating a lip design with a lubrication groove, allowing improved ride comfort and reduced pressurisation.

Ohtake et al. (1981) also consider seal design, with tests of friction and durability, and analysis of relevant parameters, considering variation of optimum design with details of the application.

Fukushima et al. (1983) advocate stroke-sensitive damping, and present the arguments in its favour, including analysis of ride and handling motions.

Fukushima et al. (1984) reiterate the points on stroke-sensitive damping, and consider a vortex valve type which has a resistance that depends on stroke as well as speed, being greater for larger stroke. They made successful tests of a demonstration unit.

Steeples et al. (1984) describe a damper testing facility, with the emphasis on durability testing.

Morman (1984) considers the mathematical modelling of dampers, expressed in terms of governing differential equations.

Holman (1984) considers a rotary-type damper configured to give better immunity to stone impact damage, intended for military and off-road vehicles.

Yukimasa et al. (1985) consider the design of oil and gas seals, sealing quality and frictional characteristics of damper seals.

Vannucci (1985) considers damper noise problems with special reference to the McPherson strut with integral damper.

Sugasawa et al. (1985) studied theoretically the optimum damping for ride and handling independently for a two-degrees-of-freedom system, and tested an automatically adjusting damper system using sensors for control inputs (accelerator, brakes, steering) and an ultrasonic ride height measurement to appraise the road quality.

Browne and Hamburg (1986) measured damper temperatures, and also the simultaneous forces and velocities of dampers on the vehicle, to obtain the energy dissipation rates. For passenger cars on normal roadways this was found to be 3–60 W, and about 12 W on average.

Karadayi and Masada (1986) consider factors such as directional asymmetry, dry friction, hysteresis, compressibility and backlash in a nonlinear model, aiming to create a simple damper model (i.e. fast computing) suitable for use in vehicle simulations. The total nonlinear compressibility is treated as piecewise linear, giving in effect a backlash plus a series stiffness, with Coulomb friction and asymmetric linear damping. The model gives quite good agreement with the general character of the real $F(V)$ curves, although the quantitative agreement is not particularly good, which presumably is the result of, rather oddly, omitting the most important nonlinearity of all, the resistance characteristics of the fluid valves. The usual end-fitting rubber bushes are not mentioned, but presumably could readily be incorporated in this model.

Young (1986) gives details of the internal configuration of various types of aircraft undercarriage dampers, which typically have free-piston gas separation and positional dependence of the damping force.

Hall and Gill (1986) describe a CSMP digital simulation of a dual-tube damper, including the effect of valve mass and valve damping. The foot compression valve (compression control) is a side exit spool valve, giving a fluid momentum force that gives Coulomb friction between the spool and its guide—another possible source of $F(V)$ curve hysteresis. Gas compressibility is dealt with by adopting a high effective vapour pressure of 41–85 kPa. In the simulation, a time step not exceeding 10 µs was necessary to avoid instability, corresponding to 0.02° of phase angle. Leakages and temperature effects were neglected, and a constant discharge coefficient of 0.7 was used. With an empirically selected effective vapour pressure, the trend of results was correct, but the $F(V)$ behaviour was too oscillatory, either because of inadequate theoretical valve damping or inadequate experimental sensor frequency response.

Soltis (1987) describes the Ford PRC (Programmed Ride Control) automatic damper adjustment system, particularly considering the handling conditions when high damping is desirable, and the advantages of the use of steering wheel angle sensing.

Su et al. (1989) theoretically analysed an adaptive damper, concluding that significant improvements in ride performance could be achieved.

Gvineriya et al. (1989) investigated the extension of gas spring units to include gas suspension damping, considering it to be a viable option.

Lemme and Furrer (1990) describe a self-powered system for remotely adjusting dampers by hydraulic means.

Hennecke et al. (1990a, b) describe an in-production adaptive damping system, with three states independently for each axle according to road conditions and driver. This system uses body sensors for vertical acceleration above the axles, steering angle, and longitudinal speed and longitudinal acceleration (rather than brake line pressure and engine output), with electronic control of solenoid valves.

Hagele et al. (1990) tested variable dampers with fast-acting solenoid valves and electrorheological liquids, but found the latter as yet unsatisfactory.

Fan and Anderson (1990) tested a bus damper complete with its rubber mounting bushes, naturally obtaining a large hysteresis in the $F(V)$ curve. Modelling equations are presented, with the bushes represented as effective compressibility, and good correlation with experiments is shown, with frequency-dependent hysteresis.

Rakheja et al. (1990) studied a 'sequential' damper similar to a conventional positive damper with multi-stage asymmetrical valving proposed to be mounted externally to facilitate adjustment, concluding that it offered significant possible improvements in ride over a conventional damper.

Lemme (1990) considers the advantages of hydraulic control for variable dampers over electrical control, considering in detail the design of such a damper, with the idea that action of any one damper could produce a pressure to control all four damper settings, without other sensors.

Kumagai et al. (1991) studied the internal Coulomb friction of strut bearings relating to the transmission of NVH (noise, vibration and harshness).

Patten et al. (1991) described how fast semi-active dampers can take advantage of the phase relationship between front and rear axle disturbances to give a significant improvement in ride quality.

Shiozaki et al. (1991) propose a variable damper with piezoelectric sensor and actuator built into the damper rod, with rapid response (3 ms) and high actuation force for an axial valve, favouring a 'normally hard' setting with switching to soft when appropriate, e.g. hitting a bump. The short piezoelectric actuator motion is amplified by stacking the elements, and then amplifying the total motion with an inverted hydraulic jack method to give 2 mm of motion at the valve.

Tamura et al. (1992) consider the development of materials for the rod-guide bush of strut type suspensions, which carries a large side load with unreliable lubrication, but needs to have low friction and long life, and is subject to other problems such as cavitation erosion. Various PTFE/Sn/Pb/Cu/PbF$_2$ sintered materials have been used.

Petek (1992a, b, c) describes the design principles of and tests on an experimental design of variable damper using ER (electrorheological) liquid, with durability tests and a retrofit vehicle road test. The ER effect is claimed to be fast in action (time constant about 5 ms), with good control power input efficiency (about 4 W), and giving lower cost and better reliability than mechanical valves, although the desired operating temperature range of -40 to $+120$ deg C may be a problem.

Nall and Steyn (1994) performed experimental evaluation of various control strategies for two-state dampers under practical conditions. It was concluded that such dampers offered significant ride improvements, and that the usual theoretical assessment of adaptive dampers did not adequately represent real conditions.

Pinkos et al. (1993) espouse ERM (electrorheological-magnetic, i.e. MR magnetorheological) liquid, which operates at low voltage. This was used in a lever-arm configuration test damper for road tests.

Baracat (1993) gives a theoretical analysis of damper forces and compares the predictions with experimental results.

Reybrouck (1994) considers modelling of conventional dampers to allow for force dependencies not just on velocity, but also on position, acceleration, temperature etc. for application in ride quality simulations, obtaining good agreement with experimental results.

Fash (1994) studied the application of a neural net computation model to data from dampers installed on vehicles. Correlation with the forces was superior to that of a simple linear model.

Besinger et al. (1995) describe a seven parameter damper model, particularly developed for use with studies of heavy vehicles. The model is compared with test damper data, and the effect of variation of the parameters on vehicle ride is investigated by simulation.

Sturk et al. (1995) developed and tested a high-voltage control unit for an electrorheological variable damper, and experimentally investigated the effect of control strategies on ride with a quarter-car rig.

Whelan (1995) investigated the use of triangular position waveform testing.

Cafferty et al. (1995) performed damper testing with excitation by a random waveform.

Audenino and Belingardi (1995) considered models of varying complexity for motorcycle dampers in comparison with experimental data.

Angermann (1995) investigated the use of aluminium bodies for dampers, and other means of weight reduction for passenger cars, already common on racing cars.

Petek et al. (1995) tested a complete electrorheological system on a vehicle, giving information on the support systems and control logic, concluding that such systems would be good if the working temperature range required for general use could be achieved.

La Joie (1996) discusses detailed damper modelling and comparison with experimental results, concentrating on racing dampers.

Haney (1996) made a bench comparison of several adjustable racing car dampers and presented their experimentally measured characteristics.

Ryan (1996) considers the merits of remote reservoir dampers with high shaft-displacement forces.

Warner and Rakheja (1996) investigated friction and gas spring characteristics, varying with temperature, with a particular emphasis on the significance of the latter for critical racing car ride heights.

Feigel and Romano (1996) describe a directly controlled electromagnetic valve for a damper.

Duym (1997) considers representation of damper characteristics to allow for the presence of hysteresis effects on the $F(V)$ curve, concluding that velocity and acceleration make a more satisfactory combination of variables than velocity and position.

Cafferty and Tomlinson (1997) discuss representation of damper characteristics by frequency domain techniques.

Tavner et al. (1997) consider test procedures for switchable dampers on the vehicle.

Kutsche *et al.* (1997) discuss pneumatically controlled variable truck dampers.

Lee (1997) analyses the double-acting shim valve monotube damper, using dimensional analysis and finite element analysis to deal with the nonlinear shim deflections.

Meller (1999) describes self-energising self-levelling systems and their incorporation into hydraulic dampers.

Herr *et al.* (1999) used a CFD (computational fluid dynamic) model to study flow in damper valves and to predict complete damper forces, obtaining good agreement with experimental results.

Els and Holman (1999) drew attention to the advantages of the lever-arm rotary damper in heavy duty applications.

Lion and Loose (2002) performed a thermomechanical analysis of dampers with experimental tests.

Choi (2003) tested a design of ER damper in which the ER effect was used to control the main valve rather than providing all the extra resistance itself.

Yamauchi *et al.* (2003) investigated a noisy strut suspension vibration arising from rod bending stimulated by damper piston friction against the tube, successfully obtaining a simple design criterion and testing a double-piston solution giving improved alignment.

Guglielmino and Edge (2004) investigated a dry friction telescopic damper using hydraulic control of the normal force and hence of the friction, claiming various advantages.

Smith and Wang (2004) investigate the possible application of 'inerters' in vehicle suspensions.

Suda *et al.* (2004) studied the possible use of an electromagnetic damper in telescopic form, with a high pitch shaft driving a ball nut which drives an electric motor/generator through a gearbox.

Lee and Moon (2005) reported on tests of a position-sensitive damper with a longitudinally grooved pressure cylinder to relax the damping around the central position.

Ramos *et al.* (2005) reported on a thermal model of double-tube dampers.

Yung and Cole (2005) described wavelet analysis of the high frequency (30–500 Hz) NVH characteristics of dampers.

Kasteel *et al.* (2005) described detailed modelling of the damper and its valves.

Alonso and Comas (2006) studied cavitation problems in dampers.

2

Vibration Theory

2.1 Introduction

The various masses, springs and dampers of the complete vehicle combine to make a complex vibrating system, stimulated by road roughness and control inputs. Full coverage of the theory of the ride behaviour of a vehicle would require a book in itself; introductions may be found in, for example, Gillespie (1992). A full understanding of the rôle of the damper in vehicle dynamics really requires a thorough understanding of both ride and handling. However a basic appreciation requires only an understanding of simple one and two degree of freedom vibration theory, of which a brief review is therefore included here.

The heave and pitch motions of a car body constitute a 2-dof (two degree-of-freedom) system, i.e. the position of the body requires two parameters for its specification. One combination is ride height z at the centre of mass plus pitch angle θ. Another possible combination is the front and rear ride heights. In general, such a system should be analysed as a complete 2-dof system. However, the modes of vibration found by the 2-dof analysis can be analysed separately as one degree-of-freedom (1-dof) systems. Also, 1-dof theory is an essential basis for understanding the 2-dof system, so this chapter reviews free vibration of undamped and damped systems in 1-dof and 2-dof, with analysis of forced vibration showing why damping is necessary, and its disadvantages. This analysis uses classical theory of linear systems with linear damping, proportional to speed. Further sections consider other forms of damping (Coulomb, quadratic), the so-called resonant absorber, and look at the idealised damper models commonly used in vehicle dynamics analysis. Finally, vehicle heave and pitch vibrations are examined in detail.

2.2 Free Vibration Undamped (1-dof)

Figure 2.2.1 shows the most basic possible system, a mass m, considered capable of vertical translation only, position defined by z, connected by a spring of linear stiffness K to the ground. Damping is neglected at this stage. When the mass is displaced by a distance z from its equilibrium position, then, by definition of K, there is a spring force, positive in the z direction, given by

$$F_z = -Kz$$

The negative sign indicates that for positive spring stiffness K the force is opposite in direction to the displacement; it is therefore a restoring force. This force gives the body an acceleration, expressed by Newton's second law as

$$m\ddot{z} = F_z$$

The Shock Absorber Handbook/Second Edition   John C. Dixon
© 2007 John Wiley & Sons, Ltd

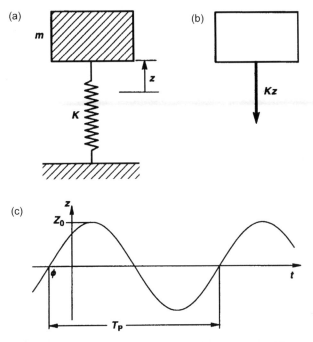

Figure 2.2.1 Basic undamped 1-dof (degree-of-freedom) vibrating system, position parameter z from the equilibrium position: (a) definition sketch; (b) free body diagram of mass; (c) free vibration motion $z(t)$.

Hence, the equation of motion for the system of Figure 2.2.1, by substituting for F_z, becomes

$$m\ddot{z} + Kz = 0$$

This is a well-known differential equation. Dividing by the mass m gives the standard form

$$\ddot{z} + \frac{K}{m}z = 0$$

For a constant stiffness K this is a linear equation and easily solved. It may be expressed as

$$\ddot{z} + \omega_N^2 z = 0$$

where ω_N is the natural frequency in radians /second (rad/s). Hence

$$\omega_N^2 = \frac{K}{m}$$

which must be a positive quantity. The mass m is necessarily positive; K is positive provided that it exerts a restoring force. The solution to the differential equation can be expressed in various ways. One way is

$$z = Z_0 \sin(\omega_N t + \phi)$$

Physically this means that, when displaced and then released, the mass oscillates freely at its natural frequency, sinusoidally at the constant amplitude Z_0, Figure 2.2.1(c), with a constant frequency, and an initial phase angle ϕ, which depends on the displacement and velocity conditions at release. The natural frequency can be expressed, in radians per second, by

$$\omega_N = \sqrt{\frac{K}{m}}$$

or in hertz (cycles per second) by

$$f_N = \frac{1}{2\pi}\omega_N = \frac{1}{2\pi}\sqrt{\frac{K}{m}}$$

The natural frequency therefore depends on the stiffness and the mass. If the mass increases, at constant stiffness, then the natural frequency reduces. This is a significant problem because of the mass variation with passengers and luggage, particularly for small cars where the relative size of the variation tends to be greater. Therefore 'rising rate' springs are often used. If the stiffness has a suitable positional dependence so that it is proportional to the load mass then the frequency of small amplitude vibrations will be independent of ride height.

For the linear system, the period, i.e. the time of one oscillation, T_P, is

$$T_P = \frac{1}{f_N}$$

The amplitude Z_0 and phase angle ϕ depend upon the conditions at $t = 0$, the so-called initial conditions, namely the displacement and velocity at this time. For example, if the initial displacement is 50 mm and the initial velocity is zero, then Z_0 will be 50 mm and the phase angle will be zero. In Figure 2.2.1(c) there is also some velocity at $t = 0$. The displacement is zero at $t = -\varphi/\omega_N + 2\pi N$.

Rather remarkably, this analysis shows that for this basic system, the free vibration may have any amplitude, and the frequency is independent of the amplitude. This conclusion is correct for the system as idealised. For a real vehicle this is a good approximation for normal motions, but, of course, is limited by the practical range of the suspension. Even if the basic spring action is effectively linear, if the amplitude is so large as to cause the bump or droop stops to come into play then the natural frequency would rise because of the effective increase of stiffness. A further extreme, with a wheel lifting off the ground, will reduce the system stiffness and reduce the frequency.

As a final remark on Figure 2.2.1, the specific stiffness k_{SS} of the spring, i.e. of the suspension, is

$$k_{SS} = \frac{K}{m} = \omega_N^2$$

which has units of N/m per kg, i.e. N/m·kg, or rad^2/s^2. The radian undamped natural frequency, then, is the square root of the specific stiffness.

2.3 Free Vibration Damped (1-dof)

The suspension is present in order to reduce the discomfort arising from road roughness. However, as will be shown later, the vehicle is very prone to vibrate at its natural frequency shown above, around 1–2 Hz in practice; to control this, some damping must be added. Damping is dissipation of energy when there is movement. In a mechanism such as a car suspension, damping is deliberately introduced by the incorporation of the dampers. There is extra damping from material hysteresis in the rubber bushes and in some cases from Coulomb friction at sliding joints. In a structure of metal or concrete there is some, usually small, damping from material hysteresis and from microscopic sliding at

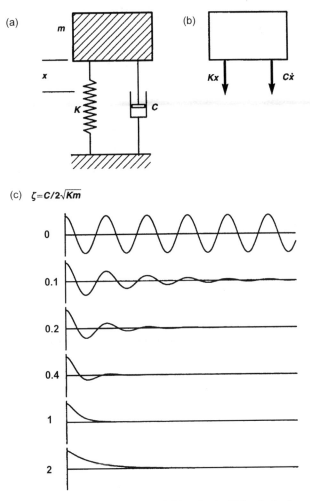

Figure 2.3.1 Basic linearly-damped 1–dof system: (a) definition sketch; (b) free body diagram; (c) free vibration of system with linear damping for various damping ratios ζ.

material joints. This causes vibrations to diminish with time unless they are continually stimulated from outside. Useful vibration analysis may often be performed without consideration of the damping, e.g. to find undamped natural frequencies and mode shapes, but in some cases it is desirable to include it. Often it is modelled as simply linear, proportional to speed, because this can be analysed more easily than other forms. Linear damping is also called viscous damping.

Consider therefore the system shown in Figure 2.3.1, which now has a linear damper included, exerting a damping force

$$F_D = -C\dot{z}$$

where the minus sign indicates positive damping, i.e. the force is opposite to the direction of motion. Parameter C is called the damping coefficient, with basic units $N/m\,s^{-1} \equiv Ns/m$, and for practical vehicle values expressed in kN s/m.

Consider the body to have been disturbed, and now in free motion at position z and velocity $\dot{z}$. The total force on the body is

$$F_Z = -C\dot{z} - Kz$$

By Newton's second law, the consequent acceleration $\ddot{z}$ is given by

$$m\ddot{z} = F_Z = -C\dot{z} - Kz$$

so the equation of motion becomes

$$m\ddot{z} + C\dot{z} + Kz = 0$$

In standard form this becomes

$$\ddot{z} + \frac{C}{m}\dot{z} + \frac{K}{m}z = 0$$

Using the Heaviside operational notation D for d/dt, or assuming a solution of the form e^{Dt} and substituting, this becomes

$$D^2 z + \frac{C}{m} Dz + \frac{K}{m} z = 0$$

Dividing by z then gives the algebraic characteristic equation of the differential equation. This algebraic characteristic equation is

$$D^2 + \frac{C}{m} D + \frac{K}{m} = 0$$

The physical nature of the solution depends upon whether D is real (nonoscillatory) or complex (damped oscillatory). D is found by the usual standard form of quadratic equation solution:

$$D = \frac{b \pm \sqrt{b^2 - 4ac}}{2a}$$

giving

$$D = -\frac{C}{2m} \pm \sqrt{\left(\frac{C}{2m}\right)^2 - \frac{K}{m}}$$

$$= \alpha \pm \sqrt{\alpha^2 - \omega_N^2}$$

where α (units of s^{-1}) is the real part of the root, called the damping factor, negative in value, and ω_N is the undamped natural frequency (rad/s).

The mathematical solution will be complex, which physically means damped oscillatory, if $\omega_N > -\alpha$ (alpha itself is negative), in which case there is an undamped natural frequency ω_N, a damping ratio ζ and a damped natural frequency ω_D. The basic dynamical equation in these terms is

$$\ddot{z} + 2\zeta\omega_N \dot{z} + \omega_N^2 z = 0$$

Note the distinction between damping coefficient C (Ns/m), damping factor $\alpha\,(\text{s}^{-1})$ and damping ratio ζ (nondimensional). The system free behaviour may be expressed by the damped natural frequency ω_D and the damping ratio. Comparing the various forms of the characteristic equation gives:

$$\omega_N = \sqrt{\frac{K}{m}}$$

$$\zeta = -\frac{\alpha}{\omega_N} = \frac{C}{2m\omega_N} = \frac{C}{2\sqrt{mK}}$$

$$\omega_D = \sqrt{\omega_N^2 - \alpha^2} = \omega_N\sqrt{1-\zeta^2}$$

where it may be seen that damping reduces the damped natural frequency compared with the undamped case.

The actual displacement at time t (the solution of the differential equation) may be expressed as

$$z = Z_0 e^{\alpha t} \sin(\omega_D t + \phi)$$

The physical vibration occurs at the damped frequency. Parameters α and ω_D depend upon the system properties. The amplitude Z_0 and phase angle ϕ depend upon the initial conditions of position and velocity.

For $\omega_N < -\alpha$, there will instead be two real solutions to D, and a nonoscillatory response with two time constants, τ_1 and τ_2:

$$-\frac{1}{\tau_1} = D_1 = \alpha + \sqrt{\alpha^2 - \omega_N^2}$$

$$-\frac{1}{\tau_2} = D_2 = \alpha - \sqrt{\alpha^2 - \omega_N^2}$$

The actual displacement is then

$$z = Z_1 e^{-t/\tau_1} + Z_2 e^{-t/\tau_2}$$

where the time constants τ_1 and τ_2 depend on the system properties, whereas the amplitudes Z_1 and Z_2 depend upon the initial conditions of z and $\dot{z}$.

The basic ride motion of most vehicles is under-damped, i.e. less than critically damped, because subcritical damping provides the best balance of ride and handling, so the oscillatory solution is the main one of interest here:

$$z = Z_0 e^{\alpha t} \sin(\omega_D t + \phi) \quad \text{(damped)}$$

This may be compared with the one found for undamped motion

$$z = Z_0 \sin(\omega_N t + \phi) \quad \text{(undamped)}$$

There are two important differences to note:

(1) The damped natural frequency ω_D is reduced from the undamped natural frequency ω_N, the former falling to zero when the damping reaches critical.

(2) There is an additional exponential term $e^{\alpha t}$ ($\exp(\alpha t)$) which causes the free oscillation to die away, which is, of course, the intended purpose of damping. Parameter α in the exponent is the damping factor, ($\alpha = -\zeta\omega_N$), and is negative in value. Here α is used for the real part of the complex root. Some books use this parameter as the negative of the real part of the root, hence referring to the damping factor for practical cases as positive. There is little danger of confusion in practice because a positive real part of the root implies instability with divergence, so normal suspension case roots always have negative real part.

The previously observed result for undamped motion remains true: the initial amplitude Z_0 and the phase angle ϕ depend upon the initial position and initial velocity (at time $t = 0$), rather than on the inherent system parameters (m, K, C).

It is particularly important to distinguish clearly between the three damping parameters:

(1) damping coefficient C (N s/m),
(2) damping factor α (s^{-1})
(3) damping ratio ζ (dimensionless)

These parameters are related by:

$$-\alpha = \zeta\omega_N = \frac{C}{2m}$$

The effect of damping ratio on the free vibration from an initial displacement is shown in Figure 2.3.1(c). From a practical point of view, it may be seen that the most rapid and controlled return is obtained with damping close to critical (i.e. $\zeta = 1$). Passenger cars may have an effective mean damping coefficient of around 0.3 in heave, because although the control is less good the softer dampers give less discomfort, whereas a racing car would be better with a higher effective damping ratio, ideally approaching 1.0. In other words, depending on road conditions, the optimum overall ride may occur for a damping ratio around 0.2 whereas the optimum handling occurs for a damping ratio of perhaps 0.8; the actual value used within this spectrum depends upon the ride/handling compromise adopted for the particular vehicle. For a damping ratio of 0.2, the damped natural frequency is only about 2% less than the undamped value. At $\zeta = 0.4$ the reduction is almost 8%, and at 0.8 it is 40% lower than when undamped.

In the event of excessive damper wear, which can cause severe piston and seal leakage or loss of damping fluid, the damping ratio may drop below 0.1. From Figure 2.3.1(c) it is apparent that this will give poor control of oscillations. This causes unpleasant 'wallowing' of the vehicle, and is hazardous on rough roads or at high speed on normal roads through loss of control.

The above analysis is readily applied to a vehicle to obtain the approximate effective damping coefficients required. For a vehicle of mass 1400 kg and a basic undamped natural frequency of 1.4 Hz, 8.8 rad/s, the total suspension stiffness is $K = m\omega^2 = 108$ kN/m, an average of 27 kN/m (154 lbf/in) at each of four wheels. To obtain a damping ratio ζ of 0.4, the damping factor required is $\alpha = -\zeta\omega = -3.52$ s^{-1}. The total damping coefficient is therefore $C = -2\alpha m = 9.86$ kNs/m (56 lbf.s/in). At each of four wheels this is 2.5 kNs/m (14.3 lbf.s/in). This is as seen at the wheels – the dampers themselves must allow for the installation motion ratio.

The condition of critical damping ($\zeta = 1.0$) is the one which just prevents overshoot. Damping beyond this value causes a slower return to the equilibrium position. However Figure 2.3.1(c) deals only with free vibration and is not the whole story. Over-critical damping is actually used for control of some vibrations, although it hardly arises in the case of motor vehicles.

The suspension specific stiffness is

$$k_{SS} = \frac{K}{m} = \omega_N^2$$

The specific damping coefficient is defined as

$$c_{SD} = \frac{C_D}{m} \text{ (Ns/m.kg)}$$
$$= -2\alpha \text{ (s}^{-1}\text{)}$$

Hence, the damping factor α is negative one half of the specific damping coefficient.

The system considered so far has had no additional external forces. This free-motion behaviour is called the 'natural' or 'free' response, in contrast to a 'forced' response which is continuously stimulated. A free response follows an initial disturbance from equilibrium, usually a displacement but possibly also or alternatively a velocity, perhaps due to an impact at $t = 0$. The subsequent motion is along the general lines of Figure 2.3.1(c).

It may be required to deduce the damping ratio from such a motion trace obtained experimentally. Using the ratio R of two amplitudes one full cycle apart, for linear damping, by solution of the equation of motion,

$$\zeta = \frac{1}{\sqrt{1 + \left(\frac{2\pi}{\log_e R}\right)^2}}$$

For low damping ratio, ζ preferably less than 0.1, not really applicable to vehicle ride motions, it may be expressed approximately as

$$\zeta \approx \frac{-\log_e R}{2\pi}$$

Alternatively, in terms of the ratio of amplitudes over one half cycle, r

$$\zeta = \frac{1}{\sqrt{1 + \left(\frac{\pi}{\log_e r}\right)^2}}$$

Both r and R are the ratio of the later amplitude over the earlier one, i.e. their values will be less than 1.0, so $\log_e(R)$ and $\log_e(r)$ are always negative.

The above analysis may be extended to N complete cycles with total amplitude ratio R_N, giving

$$\zeta = \frac{1}{\sqrt{1 + \left(\frac{2\pi N}{\log_e R_N}\right)^2}}$$

2.4 Forced Vibration Undamped (1-dof)

Vehicle ride motions are stimulated by the profile of the road as the vehicle passes over it. Ride testing may also be performed in the laboratory using one hydraulic actuator under each wheel to simulate a road surface. The position-time function of the actuator may be derived from a physical road profile and a notional speed of travel. In ride analysis the road can be subject to spectral analysis, the vehicle response across the frequency spectrum then being the product of the stimulus spectrum and the vehicle ride response function. At any particular frequency, or in a narrow frequency band, the road has an effective amplitude at that frequency, and the vehicle has a response ratio applicable to that frequency. The vehicle may usefully be analysed in terms of its behaviour at individual frequencies when stimulated by a notionally sinusoidal road; this is so-called frequency domain analysis, in contrast to a computer time–stepping study of response to a suitable complex road profile, which is a time domain analysis.

Figure 2.4.1(a) shows the simple undamped vehicle 1-dof heave model, now with a road surface at a variable position z_R above a horizontal reference plane, with z_R varying in time as far as the vehicle is concerned, or imagined on an hydraulic test rig. Consider a particular frequency f_R and sinusoidal stimulus of amplitude Z_R, so that the effective frequency of the road profile, as seen by the moving vehicle, is

$$\omega_R = 2\pi f_R$$

with a sinusoidally varying road surface height

$$z_R = Z_R \sin \omega_R t$$

The vehicle body, i.e. mass m, is at a general position (above equilibrium) z. The total force on the mass is

$$\sum F_Z = -K(z_B - z_R) = m\ddot{z}$$

The equation of motion becomes

$$m\ddot{z} + Kz_B = Kz_R = K(Z_R \sin \omega_R t)$$

This is similar to the undamped free vibration case, but now there is also a forcing term Kz_R on the right hand side. Again this is a standard equation with known solution. It is linear, so the free and forced motions are independent, and the complete solution is the sum of the separate solutions. Of interest here is primarily the new forced vibration solution. This is a known standard result: the forced vibration is also sinusoidal, and at the same frequency as the forcing function, so the forced response of the body is

$$z_B = Z_B \sin \omega_R t$$

where Z_B is the amplitude of the body motion. Differentiating twice,

$$\ddot{z} = -\omega_R^2 Z_B \sin \omega_R t$$

Substituting in the equation of motion then gives the amplitude of the forced response as

$$Z_B = \frac{Z_R}{\left(1 - \dfrac{m\omega_R^2}{K}\right)}$$

The ratio of the response amplitude of the body to the forcing amplitude of the road surface is called the transmissibility factor of the suspension, T_S:

$$Z_B = T_S Z_R$$

$$T_S = \frac{1}{\mathrm{abs}\left(1 - \dfrac{m\omega_R^2}{K}\right)}$$

(a)

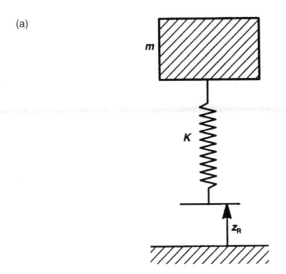

(b)

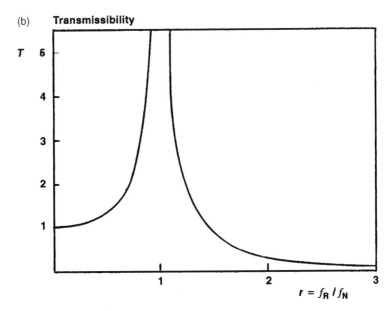

Figure 2.4.1 (a) Undamped 1-dof model vehicle stimulated by road displacement, (b) Transmissibility of an undamped 1-dof suspension.

The term 'magnification factor', sometimes incorrectly used in this context, is properly reserved for the response ratio of an oscillating force stimulus on the body itself, e.g. engine vibration. Where the forced vibration is transmitted across the suspension, as in the case of road stimulus, then the term 'transmissibility factor' should be used. The distinction becomes important when damping is included, as the transmissibility and magnification factors then have different values.

Vibration Theory

The undamped natural radian frequency is $\omega_N = \sqrt{K/m}$. The undamped suspension transmissibility factor T_S may be expressed conveniently by using the forcing frequency ratio

$$r = \frac{\omega_R}{\omega_N} = \frac{f_R}{f_N}$$

The transmissibility factor then becomes simply

$$T_S = \text{abs}\left(\frac{1}{1 - r^2}\right)$$

This is shown in Figure 2.4.1(b). The remarkable feature of this analysis is that it predicts an infinite amplitude of response if the vehicle is stimulated at its natural frequency. This is the phenomenon called resonance. A more encouraging feature of the suspension transmissibility graph is that for frequencies above $\sqrt{2}$ times the natural frequency the response amplitude is less than the stimulus amplitude, and their ratio improves as the frequency increases. Hence this simple model predicts successful functioning of the suspension in isolating the passengers from high-frequency road height fluctuations, and indicates that a low natural frequency is desirable (low specific stiffness K/m) for this purpose.

The predicted infinite resonant response would evidently be a major problem, because the road provides a wide-spectrum stimulus which will certainly include some stimulus at the natural frequency. This is a major reason for providing damping, because, as will be seen in the next section, damping can reduce the resonant response to an acceptable level, albeit at some loss of isolation performance at high frequencies.

From a practical point of view, the real response could never be infinite, being limited by the suspension range of motion, and would take some time to develop, but the essential conclusion remains; without some form of damping the response at resonant frequency will be excessive. Test driving a car without dampers or with badly worn dampers at speed on an undulating road confirms dramatically, and dangerously, that this is so. In practice, the method of control used is to provide appropriate hydraulic dampers.

2.5 Forced Vibration Damped (1-dof)

Figure 2.5.1(a) shows the single-degree-of-freedom forced vibration model, now including a linear damper, otherwise as before.

$$\sum F_Z = -K(z_B - z_R) - C(\dot{z}_B - \dot{z}_R) = m\ddot{z}_B$$

The equation of motion is therefore

$$m\ddot{z}_B + C\dot{z}_B + Kz_B = Kz_R + C\dot{z}_R$$

This is a standard linear differential equation. For a sinusoidal forcing stimulus z_R the forced response is at the forcing frequency, with an amplitude given by the transmissibility times the stimulus amplitude. There is also a phase difference, but this is not of importance in this context. The presence of the damping makes for a more complex transmissibility factor expression; to make this less unwieldy, it can be expressed in terms of the forcing frequency ratio r as defined earlier:

$$r = \frac{\omega_R}{\omega_N} = \frac{f_R}{f_N}$$

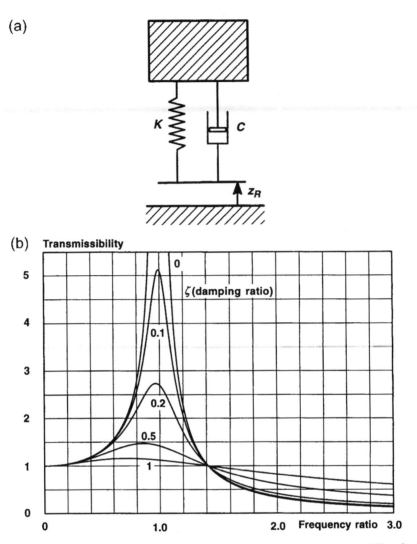

Figure 2.5.1 (a) Damped 1-dof model vehicle stimulated by road displacements; (b) transmissibility of a damped 1-dof. suspension.

giving the suspension transmissibility as

$$T_S = \sqrt{\frac{(1+4\zeta^2 r^2)}{(1-r^2)^2 + 4\zeta^2 r^2}}$$

Considering zero damping, $\zeta = 0$, then T_S can be seen to reduce to the expression found in the previous section. Figure 2.5.1(b) shows the variation of the transmissibility factor with frequency ratio for several values of damping ratio ζ. Here it may be seen how the damping reduces the resonant response as desired. However, for frequency ratios in excess of $\sqrt{2}$ the damping increases the response compared with the undamped case, making the suspension worse in this respect. In practice this problem is ameliorated by incorporating rubber bushes in the damper mountings, softening the effect of the dampers for small amplitudes, which correlate with high frequencies.

It may also be seen in Figure 2.5.1(b) that the peak transmissibility actually occurs at a frequency ratio dependent upon ζ, slightly less than 1.0 and reducing as the damping ratio increases. This is the true resonant frequency for maximum displacement amplitude. With damping, the maximum velocity amplitude and maximum acceleration amplitude do not occur at the same frequency as the maximum displacement amplitude. At a frequency ratio of $r = 1$ (nominal resonance only), an adequate approximation here, the transmissibility is

$$T_{S,r=1} = \sqrt{1 + \frac{1}{4\zeta^2}}$$

A common approximation for $T_{S,r=1}$ is $1/(2\zeta)$, but this is only good for low damping values ($\zeta < 0.1$), and not accurate for vehicle suspensions. Figure 2.5.2 and Table 2.5.1 show the actual peak value of transmissibility versus damping ratio. For comparison, the simplified equation gives 1.944 at $\zeta = 0.3$ and 1.302 at $\zeta = 0.6$.

From Figure 2.5.2 it is apparent that as the damping ratio increases from 0.1 to 0.4 the peak transmissibility reduces considerably, but beyond this value the rate of improvement is relatively modest. This provides some logic to support the use of practical vehicle damping ratios found by experience.

The above analysis is sufficiently accurate for most purposes. The exact maximum transmissibility is not conveniently expressed explicitly. By differentiation it is easily shown that the frequency ratio of amplitude resonance is

$$r = \frac{\sqrt{\sqrt{(1 + 8\zeta^2)} - 1}}{2\zeta}$$

This value can then be substituted in the expression for the transmissibility to obtain the peak value.

Table 2.5.1 Peak transmissibility

Damping ratio (ζ)	Peak transmissibility (T_{max})
0.10	5.123
0.20	2.733
0.30	1.995
0.40	1.655
0.50	1.468
0.60	1.353
0.70	1.276
0.80	1.223
0.90	1.184
1.00	1.155
1.20	1.114
1.50	1.078
2.00	1.047

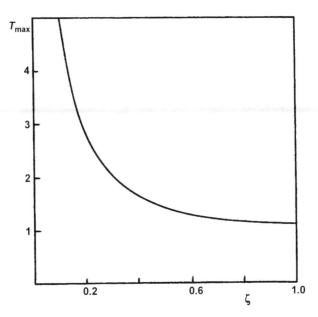

Figure 2.5.2 Peak transmissibility versus damping ratio.

2.6 Coulomb Damping

The Coulomb friction model (C. A. de Coulomb, 1736–1806) is of some practical importance because it is quite a good model of some common frictional behaviour, in particular dry friction in which there is negligible lubrication. The basic model for Coulomb friction is that the maximum friction force F_F depends only on the normal force F_N at the sliding contact, and is not dependent on speed:

$$F_{Fmax} = \mu_F F_N$$

where μ_F is the coefficient of limiting friction. Implicit in this model is that μ_F does not depend upon the normal force or on the sliding speed. By extrapolation, the term Coulomb friction is applied to any case where the friction force takes some value in a limited range regardless of speed, even if an actual F_N cannot be identified:

$$F_F < F_{Fmax}$$

A common extension of the above model allows a special form of variation. When the sliding speed is zero, there is a maximum static friction force F_{FS}, which is greater than the constant force, independent of speed, that occurs when the sliding speed is nonzero, the latter case being the dynamic friction force

$$F_{FD} < F_{FS}$$

In a case where a normal force F_N can be identified, then there are corresponding coefficients of limiting static and dynamic friction μ_{FS} and μ_{FD}. The latter may be around 0.7 times the former.

Figure 2.6.1 shows the sprung mass model with some Coulomb damping, the symbol for this being the arrowhead onto a line. The maximum Coulomb frictional force is F_{CF}. A representative value for a modern passenger car might be 200 N in total, 50 N at each suspension. This friction force can act in either direction.

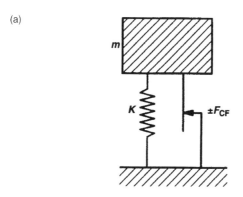

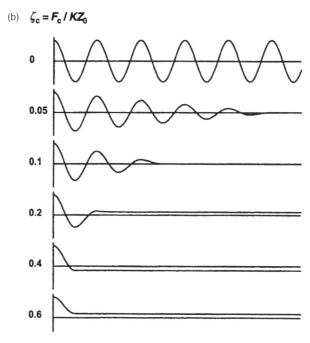

Figure 2.6.1 (a) Definition sketch of 1-dof vibration model with Coulomb friction; (b) free vibration of system with Coulomb friction damping.

The first consequence is that the static ride height of the vehicle is now indeterminate within a band, because the Coulomb friction can oppose the spring force. The body ride position z_B may be anywhere in the range

$$z_{CF} = \pm \frac{F_{CF}}{K}$$

with a value of a few millimetres.

The second consequence is that, considering the suspension to be in its central position, forces less than F_{CF} will not deflect the suspension at all. Hence road stimuli in forced vibration with small amplitude will be fully transmitted to the body, causing poor ride. Hence Coulomb type friction is to be minimised.

The actual friction force in Figure 2.6.1 is statically indeterminate if the suspension is not moving. A complete expression of the friction force is

$$-F_{CF} < F_F < +F_{CF} \quad (\dot{z} = 0)$$
$$F_F = +F_{CF} \quad (\dot{z} < 0)$$
$$F_F = -F_{CF} \quad (\dot{z} > 0)$$

The friction force will oppose the velocity and therefore provide a damping effect, but the analysis is nonlinear and discontinuous, and therefore not very convenient.

Figure 2.6.2 shows details of the subsequent motion when the body is released from an initial deflection Z_0 at zero velocity. In the first half-wave of motion the velocity is always negative, so the friction force is constant and equal to $+F_{CF}$. In the second half-wave again F_F is constant, but equal to $-F_{CF}$. Hence, the frequency is not modified by Coulomb damping, the damped frequency equals the undamped frequency, unlike the case of linear damping. The first half-wave is symmetrical about a displacement equal to

$$+z_{CF} = \frac{F_{CF}}{K}$$

The second half-wave has positive velocity with friction force $-F_{CF}$, and is symmetrical about the position $-z_{CF}$. Hence each half-cycle causes an amplitude reduction $2z_{CF}$, and this will continue until the body comes to rest somewhere within the friction band. If the initial deflection is many times the friction band, then the result is an oscillation appearing to have an amplitude reducing linearly with time, with a definite finish to motion, unlike the linear damping which has an exponential decay without a well-defined finish.

Motion traces of the type of Figure 2.6.2 can be expressed nondimensionally, as in Figure 2.6.1(b). Dimensional analysis gives the expected result that the shape of the curve depends on t/T_N or $\omega_N t$, with a nondimensional displacement expressed by

$$\frac{z}{Z_0} = f\left(\frac{Z_0}{z_{CF}}, \omega_N t\right)$$

or

$$\frac{z}{Z_0} = f\left(\frac{Z_0 K}{F_{CF}}, t\sqrt{\frac{K}{m}}\right)$$

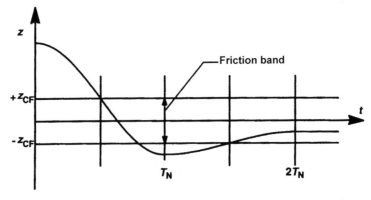

Figure 2.6.2 Free response with Coulomb friction.

The nearest equivalent of the linear damping ratio used in standard analysis is the nondimensionalised damping force, a Coulomb damping ratio

$$\zeta_C = \frac{F_{CF}}{KZ_0}$$

which depends upon the initial amplitude. In terms of this variable,

$$\left(\frac{z}{Z_0}\right) = f(\zeta_C, \omega_N t)$$

It is difficult to generalise about the forced vibration of such a system, because the motion is nonlinear and superposition does not apply. For large-amplitude stimulation, where large stimulation means substantially greater than the static friction band, the general motion is similar to that of a linearly damped system. Superimposed high frequencies may actually be better isolated than for a linear system because velocity has no effect, provided that the motion is continuous in its direction, so excellent isolation of high frequencies may be achieved in practice. However when the stimulus is small, such as on a good road, the suspension is liable to remain locked and provide no isolation (or, in practice, the entire car with its wheels vibrates as a unit at small amplitude and high frequency on the tyre stiffness—so-called Boulevard Jerk).

The Coulomb damping force F_{CF} is easily deduced from a motion trace, by

$$F_{CF} = \tfrac{1}{2} K(Z_1 - Z_2)$$

where Z_1 and Z_2 are the amplitudes one half-cycle apart, but of course this is applicable only to the case of effectively pure Coulomb damping.

2.7 Quadratic Damping

Quadratic damping has a force magnitude which is proportional to the square of the speed V:

$$F_{QD} = C_Q V^2$$

where C_Q is the quadratic damping coefficient with units Ns^2/m^2 (or kg/m). However, this equation gives a positive force for positive or negative velocity which does not normally correspond with physical reality. Usually the force opposes the motion, in which case, correctly, with positive C_Q,

$$F_{QD} = -\text{sgn}(V) C_Q V^2$$

or

$$F_{QD} = -C_Q V \, \text{abs}(V)$$

where sgn(V) is the sign of V, or $+1$ for positive V and -1 for negative V, and abs(V) is the absolute value (magnitude) of V. Occasionally physical problems arise with negative quadratic damping but this is not applicable to normal cars.

Quadratic damping is generally the result of dynamic energy dissipation in fluid flow, and would arise, for example, in a damper with a simple orifice allowing fluid flow, and is therefore of real practical interest. Figure 2.7.1(a) shows a 1-dof system with quadratic damping. There does not seem to be a standard symbol for a quadratic damper, so the standard linear damper symbol is used, a simple dashpot which could well be quadratic in reality, marked with C_Q.

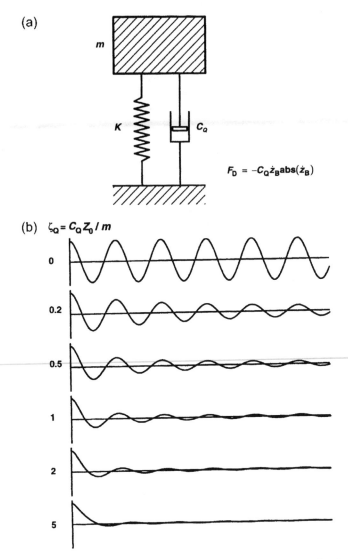

Figure 2.7.1 (a) Definition sketch of 1-dof system with quadratic damping; (b) free vibration of system with quadratic damping.

Because of the nonlinear, therefore nonsuperpositional, nature of quadratic damping, it does not lend itself to concise general solutions. Qualitatively, at a given amplitude the velocity is proportional to frequency, so high frequencies are better damped, but transmitted more, than low ones. At a given frequency, the velocity is proportional to the amplitude, so the effective damping is proportional to amplitude squared. Hence, compared with linear damping, a position–time trace is more damped initially but very little damped at low amplitude, so the remnant vibration is very slow to disappear, as seen in Figure 2.7.1(b).

Dimensional analysis shows that a nondimensional quadratic damping ratio ζ_Q can be defined as

$$\zeta_Q = \frac{C_Q Z_0}{m}$$

where the quadratic damping ratio is seen to be proportional to the initial amplitude. Hence

$$\frac{z}{Z_0} = f(\zeta_Q, \omega_N t)$$

Figure 2.7.1(b) shows $z_B(t)$ for various quadratic damping ratios. From the above analysis, the quadratic damping ratio applicable depends not only on the parameters of the system, but also on the initial conditions, namely the initial amplitude or displacement. Approximately quadratic damping is easy to achieve, but generally undesirable in practice because of harshness at high suspension speeds.

2.8 Series Stiffness

Figure 2.8.1 shows a system of a parallel spring and damper, K_S and C_S, with an additional series stiffness K_T. Evidently, in the context of vehicles K_S and C_S are the suspension stiffness and damping coefficient as seen for suspension motion. The series stiffness K_T is predominantly the tyre vertical stiffness. For a passenger vehicle, K_T is substantially more than K_S, typically by a factor of eight, although the effect to be considered here is still significant. For a ground-effect racing car the tyre vertical stiffness may be similar to the suspension spring effective stiffness, and the compliance of the links and body may add further to the effective series compliance, which may ultimately the be greater than that of the spring action. The tyre deflection has relatively small damping, as does any associated suspension or body deflection, so the case of zero damping of the series compliance is investigated here. The displacements from equilibrium are z_B for the body and z_T for the tyre. The difference is the suspension droop deflection

$$z_{SD} = z_B - z_T$$

Also note that the tyre deflection is given by

$$z_T = z_B - z_{SD}$$

For simplicity, there are no masses involved. This system is of further interest in that it has been proposed as a mechanical analogue of viscoelastic materials such as tyre rubber. In materials science, a material model with a spring and linear damper in parallel is called a Voigt element or Maxwell element. When such an element is in series with an additional spring, the combination is called a Zener element. This is a common representation of a viscoelastic material, or a 'standard linear solid'.

For a basic body-motion damping analysis, the frequency is well below the wheel hop frequency, so the masses may be neglected. Consider a sinusoidal body motion

$$z_B = Z_B \sin \omega t$$

Here ω is any forcing frequency (in rad/s), in general not the natural frequency. Assuming a harmonic solution for the suspension droop deflection z_{SD} of the form

$$z_{SD} = A \sin \omega t + B \cos \omega t$$

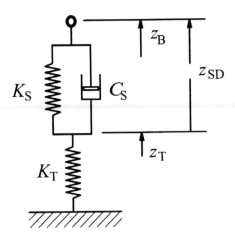

Figure 2.8.1 Suspension system $K_S C_S$ with series spring K_T.

then the tyre deflection is

$$z_T = (z_B - A)\sin \omega t + (-B)\cos \omega t$$

The suspension droop velocity, by simple differentiation, is

$$\dot{z}_{SD} = \omega A \cos \omega t - \omega B \sin \omega t$$

In the absence (neglect) of wheel mass, the net vertical wheel force is zero, so

$$\Sigma F_{Z,\text{wheel}} = K_S z_{SD} + C_S \dot{z}_{SD} - K_T z_T = m_W A_{Zw} \approx 0$$

Substituting, and separately equating the sine and cosine coefficients, gives

$$K_{ST} A - \omega C_S B = K_T Z_B$$
$$\omega C_S A + K_{ST} B = 0$$

where K_{ST} is simply $K_{ST} = K_S + K_T$. Solving the above two equations simultaneously for the unknown values A and B gives

$$A = \frac{K_T K_{ST} Z_B}{K_{ST}^2 + (\omega C_S)^2}$$

$$B = \frac{-\omega C_S K_T Z_B}{K_{ST}^2 + (\omega C_S)^2}$$

The complete suspension response amplitude Z_S is given by

$$Z_S^2 = A^2 + B^2$$

Vibration Theory

By substitution, the suspension/body amplitude response ratio R is given by

$$R^2 = \left(\frac{Z_S}{Z_B}\right)^2 = \frac{K_T^2}{(K_{ST}^2 + \omega^2 C_S^2)}$$

The mean power dissipation for sinusoidal motion of the linear suspension damper is

$$P = \tfrac{1}{2}\omega^2 Z_S^2 C_S$$

The effective damping coefficient as would be applicable at the body amplitude is C_E. For a given body amplitude Z_B, the mean power dissipation is therefore

$$P = \tfrac{1}{2}\omega^2 Z_B^2 C_E = \tfrac{1}{2}\omega^2 R^2 Z_B^2 C_S$$

The effective damping coefficient as seen at the body is then

$$C_E = R^2 C_S$$

or

$$C_E = \frac{K_T^2 C_S}{K_{ST}^2 + \omega^2 C_S^2}$$

The damping 'efficiency' (effectiveness) is then

$$\eta_D = \frac{C_E}{C_S} = \frac{K_T^2}{K_{ST}^2 + \omega^2 C_S^2}$$

The mean power dissipation over a whole cycle is

$$P = \tfrac{1}{2} Z_B^2 K_T^2 \left(\frac{C_S \omega^2}{K_{ST}^2 + C_S^2 \omega^2}\right)$$

Writing

$$\Omega = \frac{\omega C_S}{K_{ST}}$$

and

$$P_0 = \frac{Z_B^2 K_T^2}{2 C_S}$$

then

$$\frac{P}{P_0} = \frac{\Omega^2}{1 + \Omega^2}$$

At small frequencies, the dissipation rises with Ω^2 and ω^2. At higher frequencies it tends to the asymptotic value P_0.

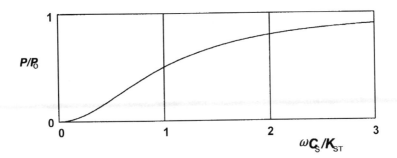

Figure 2.8.2 Variation of system power dissipation with frequency, with a series spring.

Of more significance is the variation of damping effect with change of damping coefficient. This has a maximum value. Taking the derivative of P with respect to C_S, and setting this to zero, the optimum suspension damping coefficient is

$$C_{S,opt} = \frac{K_{ST}}{\omega}$$

which gives the maximum power dissipation

$$P_{max} = \frac{\omega Z_B^2 K_T^2}{4 K_{ST}}$$

Writing

$$C_S = r_D C_{S,opt}$$

the power dissipation relative to the maximum may be written conveniently as

$$\frac{P}{P_{max}} = \frac{2 r_D}{1 + r_D^2}$$

This is shown in Figure 2.8.3. Writing the power ratio as

$$r_P = \frac{P}{P_{max}} = \frac{2 r_D}{1 + r_D^2}$$

then the variation of effective damping as the damper coefficient is increased depends on

$$\frac{dr_P}{dr_D} = \frac{2(1 - r_D^2)}{(1 + r_D^2)^2}$$

Evidently, this becomes negative for r_D greater than 1.0. The power ratio r_P and sensitivity are shown in Table 2.8.1.

Considering some practical values, for one suspension of a passenger car:

$$m = 350 \, \text{kg}$$
$$K_S = 25 \, \text{kN/m}$$
$$K_T = 175 \, \text{kN/m}$$

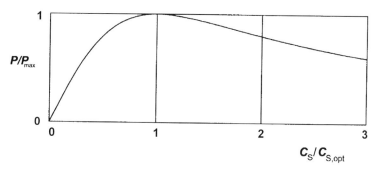

Figure 2.8.3 Variation of system power dissipation with damping coefficient, with a series spring.

$$C_S = 2\,\text{kN s/m}$$
$$K_R = 21.9\,\text{kN/m}$$
$$\omega_N = 7.906\,\text{rad/s}$$
$$K_{ST} = 200\,\text{kN/m}$$

Therefore for body heave vibrations at the natural frequency

$$C_{S,opt} = \frac{K_{ST}}{\omega_N} = 25.3\,\text{kNs/m}$$
$$\frac{C_S}{C_{S,opt}} = 0.079$$

Table 2.8.1 Damping effectiveness with series stiffness

C/C$_{opt}$	r$_P$	dr$_P$/dr$_D$
0.000	0.000	2.000
0.100	0.198	1.941
0.200	0.385	1.775
0.300	0.550	1.532
0.400	0.690	1.249
0.500	0.800	0.960
0.600	0.882	0.692
0.700	0.940	0.459
0.800	0.976	0.268
0.900	0.994	0.116
1.000	1.000	0.000
1.250	0.975	−0.171
1.500	0.923	−0.237
1.750	0.861	−0.250
2.000	0.800	−0.240
2.250	0.742	−0.221
2.500	0.689	−0.200
2.750	0.642	−0.179
3.000	0.600	−0.160

and damper effectiveness is fairly normal:

$$\frac{C_E}{C_S} = 0.761$$

$$C_E = 1.52\,\text{kN s/m}$$

with a loss of damper effectiveness of 24 %.

For one suspension of a ground effect racing car, take the data to be

$$m = 170\,\text{kg}$$

$$K_S = 200\,\text{kN/m}$$

$$K_T = 200\,\text{kN/m}$$

$$C_S = 8\,kN\,s/m$$

$$K_R = 100\,\text{kN/m}$$

$$\omega_N = 24.25\,\text{rad/s}$$

$$K_{ST} = 400\,\text{kN/m}$$

At the undamped natural frequency,

$$C_{S,opt} = \frac{K_{ST}}{\omega_N} = 16.49\,\text{kNs/m}$$

$$\frac{C_S}{C_{S,opt}} = 0.485$$

$$\frac{C_E}{C_S} = 0.202$$

$$C_E = 1.62\,\text{kNs/m}$$

In this case, then, the tyre and suspension compliance are of great importance, making it difficult to achieve good damping, and making changes of damping ineffective. Indeed, in many cases, the damping coefficient applied may be in excess of C_M. If the attempted damping is too great, the vehicle simply vibrates instead on the series spring which is an undamped motion. The highest damping that can be achieved in this case is with

$$C_S = C_{S,opt} = \frac{K_{ST}}{\omega_N} = 16.5\,\text{kNs/m}$$

$$\frac{C_E}{C_S} = 0.125$$

$$C_E = 2.06\,\text{kNs/m}$$

The best effective damping ratio achievable in this case is then about 0.250.

2.9 Free Vibration Undamped (2-dof)

A 2-dof system is one requiring two variables to describe its position. This could be a single 'point' mass moving in (x,y) coordinates, for example, controlled by two nonparallel linear springs. Such systems are found to have two 'modes' of vibration. A mode is a vibration combining the (x,y) movements in a simple way. For two degrees of freedom there will be two modes, each with its own natural frequency. In the example cited each mode will be sinusoidal motion to-and-fro along a straight line at a particular angle, combining the x and y coordinates in a particular ratio, e.g. $x = 2y$ for one mode and $x = -0.5y$ for the other mode. The ratio of the coordinates is called the mode shape, with values 2 and -0.5 in this case. An important aspect of a mode, in undamped vibration at least, is that the system can vibrate in one mode without causing any motion in the other mode—the modes are said to be decoupled. The coordinate system in use must be clearly defined, because a physical mode will appear to have a different mode shape in each different coordinate system in which it is viewed, e.g. rotating the (x, y) axes to a new position.

In the case of a motor vehicle, basic analysis of the ride quality may be done by studying the two-degree-of-freedom heave-and-pitch motion of the vehicle body. Heave is a simple up-and-down motion without any angular motion. Pitch is pure angular motion about the centre of mass. The term heave is also used for ships and aircraft, although in the particular context of ground vehicles heave is also known as bounce. The undamped body is found to have two modes, one mainly a heave action, one mainly a pitch action. The particular ratio of heave and pitch angle in a mode results in a modal node—a point at which the body does not have any vertical displacement. The nodal position is different for each mode, of course. The mode shape, expressed in the basic coordinate system of the vehicle centred at G, is the position of the node behind the centre of mass. For good ride quality, the designer must position these nodes correctly. The position of the body can be expressed as a sum of angular modal displacements θ_1 and θ_2 about the nodes. In this special coordinate system (θ_1, θ_2) the mode shapes are simply (1,0) and (0,1), i.e. each mode is simply a motion in its own modal coordinate.

2.10 Free Vibration Damped (2-dof)

When damping is added to a 2-dof system, it is very unlikely that the modes will remain accurately decoupled. The damper forces will transmit energy from one mode to another. This greatly complicates the analysis. The mode shape was previously just a ratio of displacements, but with damping in the system the solution gives a complex number. The best interpretation of this is to think of it in terms of its magnitude and angle. The magnitude is the ratio of amplitudes, as before, but now the angle of the mode shape adds a phase angle to the relative displacement in the two coordinates. In the example cited, without damping, if the modal frequencies are in a simple ratio, the point will move in a Lissajous figure. With damping, the result will be a spiral, possibly erratic, inwards to the origin, depending on the frequency ratio.

For a motor vehicle, with a damped analysis it is no longer possible in general to deduce a fixed nodal point for the mode. There are several ways to deal with this. The simplest, and possibly still the best for vehicles, is to analyse the undamped mode shape and undamped nodal positions, and then analyse each of these as a 1-dof system with damping to obtain a modal damping ratio.

In general vibration analysis, a common approach is to 'cheat' on the damping. This involves modelling the damping in a particular way, Rayleigh damping, such that the modes remain decoupled. In more elaborate (higher dof) vibration analysis, e.g. using Finite Element analysis, the addition of damping as a completely independent term can greatly increase the computational effort required. This is in contrast to normal analytical solutions, in which the linear damping is easily incorporated. In the interests of efficient calculation, it is often useful to express the damping somewhat artificially as a term directly proportional to either the mass matrix or the stiffness matrix. In fact a sum of these may also be used with little extra effort, with $C = \alpha M + \beta K$, α and β being the constants of proportionality. This method is known as Rayleigh damping or as Proportional Damping. The damping force is then proportional to the velocity, and opposes it, but the effective constant of proportionality is limited to a matrix that is a sum of constants times the stiffness and mass matrices. This keeps the theoretical vibration modes decoupled. In the 2-dof heave-and-pitch analysis of a vehicle, the mode shape will be real if the dampers agree with the springs about the nodal positions. This will occur for $C_f/C_r = K_f/K_r$. This is equivalent to making the damping matrix proportional to the stiffness matrix.

In the simple example given earlier of the 'point' mass moving in (x,y) coordinates, when damping is added then the modal motions are no longer simply separable. The coordinates x and y are no longer combined in a constant real ratio, but in a complex one. With the mode shape expressed not as real and imaginary parts, but as a magnitude and a phase angle, it may be seen that the ratio y/x has a constant magnitude, and that the y-motion has a constant phase angle relative to the x-motion.

2.11 The Resonant Absorber

The so-called Resonant Absorber or Dynamic Vibration Absorber, due to Frahm, is not really a vibration absorber or damper at all, but rather a preventer, although in operation it gives some impression of being an absorber. It is an interesting example of an invention resulting directly from theoretical analysis, in this case the theory of 2-dof vibration. Figure 2.11.1 shows a basic model to illustrate the principle.

The problem is to reduce the response of the main mass m when it is forced to vibrate. Application of the basic concept of the absorber is appropriate when it is desired to reduce the response at a particular frequency, either a fixed forcing frequency or at the resonant natural frequency of the main mass. This is achieved by the 'absorber' mass and stiffness, m_A and K_A, which are chosen to have a resonant frequency at the frequency of vibration in m to be prevented or reduced. For the basic system there should be only small damping in the absorber system.

If the main mass m makes even a small vibration at the absorber resonant frequency, the absorber will vibrate in resonance at large amplitude, because it has low damping. It does so in a way that generates a force in spring K_A which strongly opposes the vibration of m. The mode shape z_A/z is large and negative. Hence, at this frequency m is held almost stationary whilst m_A vibrates strongly. The effect is very striking when observed in practice, and can be highly effective. It has been used for many purposes, including machinery, motor vehicles and even for reducing lateral oscillatory motion of very tall buildings in wind using supplementary masses up to 1000 tonnes on rubber shear mountings.

A serious limitation of the basic principle is that it is operative at only one frequency, and so is mainly applicable to a well-known natural or forced frequency. The combined two-degree-of-freedom system actually acquires two new resonant frequencies, one on each side of the original. Also, for low frequency the system requires rather a large absorber mass m_A.

A more general version of the concept includes damping. This compromises the performance at the central frequency whilst controlling the new resonant frequencies. The damping is selected to minimise the worst magnification ratio or transmissibility, as appropriate, across the relevant frequency range, and typically requires a damping ratio of about 0.15, and almost always in the range 0.1 to 0.2.

This method is impractical for use in general on a vehicle body, although the engine can be tuned on its mounting to have a desired effect. One case of a successful application was to reduce torsional body vibrations in the early days when chassis torsional stiffness was poor, Figure 2.11.2. It has also been used to discourage wheel hop vibrations with some success on at least one production vehicle, but is now out of use. A full theoretical analysis is given in many good vibration text books (e.g. Den Hartog, 1985, Hunt, 1979).

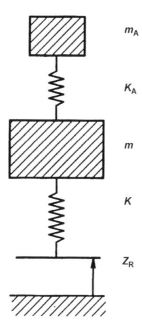

Figure 2.11.1 'Resonant vibration absorber' system (H. Frahm, 1909).

2.12 Damper Models in Ride and Handling

As will be seen in later sections, actual damper $F(V)$ characteristics can be quite complex. In a computer simulation this can be represented to any required degree of accuracy by functions such as high-order polynomials or tabulations. However, the representations used are usually not so complicated because such representations do not greatly affect the behaviour, and simple representations with fewer parameters are more comprehensible. For analytical work, as in the subsequent sections of this chapter, it is almost essential to use a simple linear model.

The models normally used may be classified as:

(1) linear;
(2) bilinear;
(3) complex.

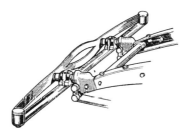

Figure 2.11.2 An early Frahm-type torsional vibration absorber. The bumper vibrated in 'roll' on the rubber mountings, reducing scuttle shake (Wilmot Breeden, 1938).

A linear model is one in which the damper is represented as having a damping force proportional to velocity with a constant coefficient

$$F_D = C_D V_D$$

A bilinear model is one similar to the above, but with different coefficients for the two directions, extension and compression:

$$F_{DE} = C_{DE} V_{DE} \quad \text{(extension, } V_{DE} > 0\text{)}$$
$$F_{DC} = C_{DC} V_{DC} \quad \text{(compression, } V_{DC} > 0\text{)}$$

This is quite a useful enhancement in some cases because asymmetry is an important feature of most real suspension dampers.

The 'complex' model heading covers all others, including nonlinear behaviour, the addition of some Coulomb friction, etc. Typically an empirical index model will be used over the speed range of interest, e.g.

$$F = C_1 V^n$$

where the index n will be less than 1 to capture the knee shape of the $F(V)$ curve found in practice over the main part of the speed range. This is also easily made asymmetrical.

For more complex curve shapes, it is usual to have a Tabulation Model - a look-up table for $F(V)$ with some sort of interpolation.

With the great speed of modern computation, it has become possible to incorporate detailed phenomenological models of the damper in ride and handling simulations. Obviously such models are of great interest to the damper designer. Whether the extra complexity of the modelling really adds any accuracy to the ride or handling prediction ability as perceived by the passengers may remain somewhat doubtful.

Although a linear $F(V)$ characteristic can be realised by viscous flow in a real damper, in automotive dampers the restriction is primarily inertial in action, depending mainly on the oil density, and valves with variable effective flow area are used, so various nonlinear characteristics are actually found in practice, discussed later (e.g. Chapter 8).

2.13 End Frequencies

Because of the small mass of the wheels compared with the body, and the relatively large vertical stiffness of the tyres compared with the suspension wheel rate, when analysing the body motion the mass of the wheels can usually be neglected, and the so-called ride stiffness K_R combining suspension (wheel rate) and tire stiffness in series is used, given by

$$\frac{1}{K_R} = \frac{1}{K_W} + \frac{1}{K_T}$$

This has some effect on the numerical values of body frequency, but no effect on the principles of analysis.

The total sprung mass (mass of the body) m_S may be divided into front and rear end masses. For a wheelbase L, and a centre of mass positioned at a behind the front axle, b in front of the rear axle,

$$L = a + b$$

The end masses are

$$m_f = \frac{b}{L}m_S$$

$$m_r = \frac{a}{L}m_S$$

However, note that if these end masses are considered to be situated at the axle positions they do not give the correct pitch inertia, in general.

Consider the individual suspensions to have stiffnesses at the wheels K_f and K_r, i.e. two front suspension units each of effective stiffness K_f, two rear each of K_r. If each end could vibrate independently, the front and rear radian natural frequencies, known as the end frequencies, would be

$$\omega_{Nf} = \sqrt{\frac{2K_f}{m_f}}$$

$$\omega_{Nr} = \sqrt{\frac{2K_r}{m_r}}$$

Also, in hertz

$$f_{Nf} = \frac{\omega_{Nf}}{2\pi}$$

$$f_{Nr} = \frac{\omega_{Nr}}{2\pi}$$

In order to obtain good ride behaviour, the rear frequency is made somewhat higher than that of the front. Considering a single isolated bump which stimulates the rear after the front, the result of the higher rear frequency is that the rear oscillation tends to catch up with that of the front, so that the total response is biased more towards a heave motion and less to pitching, the latter being more objectionable. The frequency used is typically 20–30% higher at the rear, say 1.6 against 1.30 Hz. The rear/front mass ratio is

$$R_M = \frac{m_r}{m_f} = \frac{a}{b}$$

The rear/front frequency ratio is

$$R_F = \frac{f_{Nr}}{f_{Nf}} = \frac{\omega_{Nr}}{\omega_{Nf}} = \sqrt{\frac{K_r b}{K_f a}}$$

The rear/front suspension stiffness ratio required is

$$R_K = \frac{K_r}{K_f} = \frac{m_r \omega_{Nr}^2}{m_f \omega_{Nf}^2} = \frac{m_r}{m_f} R_F^2 = \frac{a}{b} R_F^2$$

Note that the rear/front mass ratio is often less than 1.0 whereas the rear/front frequency ratio and stiffness ratio R_F and R_K are both normally greater than 1.0. By way of example values, for a mid-wheelbase centre of mass, $a = b$ and frequency ratio 1.2, $R_K = R_F^2 = 1.44$. On the other hand, with a forward centre of mass position giving $b/a = 1.44$, the mass ratio is 0.694 and the suspension stiffnesses can be equal.

2.14 Heave and Pitch Undamped 1-dof

The two ends of the vehicle are connected, and can not really vibrate independently. However, consider the case when the two end frequencies are the same. The ends can move in synchronisation, and the body can perform a simple heave motion with no pitch. With the ends synchronised in opposed motion there is pitch but no heave. In other words there will then be one pure heave mode and one pure pitch mode, which can therefore be analysed separately as 1-dof motions. This would occur for a vehicle with longitudinal symmetry, as in Figure 2.14.1, with a mid-wheelbase centre of mass point and equal front and rear suspension parameters.

The heave motion depends upon the total heave stiffness

$$K_H = 4K$$

giving a heave radian natural frequency

$$\omega_{NH} = \sqrt{\frac{K_H}{m_S}}$$

The pitch motion depends on the pitch stiffness and pitch inertia. Considering a pitch angle θ about the centre of mass, expressed in radians, the suspension deflections are

$$z_{Sf} = \tfrac{1}{2}L\theta \qquad z_{Sr} = -\tfrac{1}{2}L\theta$$

giving a restoring pitch moment from the two front springs of

$$M_f = 2K\left(\tfrac{1}{2}L\theta\right)\left(\tfrac{1}{2}L\right) = \tfrac{1}{2}KL^2\theta$$

The same value occurs at the rear. The total front and rear restoring pitch moment is then

$$M_P = KL^2\theta$$

with a pitch stiffness, for pitch rotation about the centre of mass,

$$K_P = \frac{M_P}{\theta} = KL^2$$

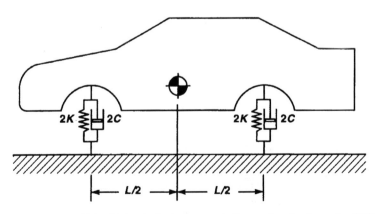

Figure 2.14.1 Vehicle with longitudinal symmetry (with zero damping for section 2.14).

Vibration Theory

The pitch inertia (second moment of mass) is

$$I_P = m_S k_P^2$$

where m_S is the sprung mass and k_P is the pitch radius of gyration for that mass. This is related to the wheelbase by the pitch dynamic index i_P, defined as

$$i_P = \frac{k_P}{\sqrt{ab}} = \frac{k_P}{\frac{1}{2}L}$$

in this case, so

$$I_P = m_S \tfrac{1}{4} L^2 i_P^2$$

A value of i_P equal to 1.0 means a radius of gyration equal to half of the wheelbase.

The pitch natural radian frequency is then

$$\omega_{NP} = \sqrt{\frac{K_P}{I_P}} = \sqrt{\frac{4 K_P}{m_S L^2 i_P^2}} = \sqrt{\frac{4K}{m_S i_P^2}}$$

The ratio of pitch frequency to heave frequency is

$$\frac{\omega_{NP}}{\omega_{NH}} = \sqrt{\frac{K_P}{I_P} \frac{m_S}{K_H}}$$

Using

$$K_P = KL^2$$

and

$$K_H = 4K$$

then

$$\frac{\omega_{NP}}{\omega_{NH}} = \frac{1}{i_P}$$

Hence, for the simple symmetrical vehicle model, the pitch frequency is close to the heave frequency, depending on the pitch dynamic index.

2.15 Heave and Pitch Damped 1-dof

Continuing with the simple symmetrical vehicle of Figure 2.14.1, damping is now included in the 1-dof analysis of heave and pitch independently. The equation of heave motion is

$$m\ddot{z} + 4C\dot{z} + 4Kz = 0$$

The radian undamped natural frequency in heave ω_{NH} is unchanged at

$$\omega_{NH} = \sqrt{\frac{4K}{m_S}}$$

The damping ratio in heave, ζ_H, is given by

$$\zeta_H = \frac{4C}{2m_S}\frac{1}{\omega_{NH}} = \frac{C}{\sqrt{m_S K}}$$

In pitch, again the motion is 1-dof By considering a pitch angle θ about the centre of mass, and the corresponding restoring moment, the pitch stiffness is still

$$K_P = KL^2$$

Considering a pitch velocity $\dot{\theta}$ and the corresponding damper moment, the pitch damping coefficient is

$$C_P = CL^2$$

The 1-dof pitch equation of motion is then

$$I\ddot{\theta} + CL^2\dot{\theta} + KL^2\theta = 0$$

Writing

$$I = m_S k_P^2 = \tfrac{1}{4} m_S L^2 i_P^2$$

where i_P is the pitch dynamic index, gives

$$m_S i_P^2 \ddot{\theta} + 4C\dot{\theta} + 4K\theta = 0$$

The undamped natural frequency in pitch is therefore, as before

$$\omega_{NP} = \sqrt{\frac{4K}{m_S i_P^2}}$$

The damping ratio in basic pitch is

$$\zeta_P = \frac{C}{i_P \sqrt{m_S K}}$$

Comparing the pure pitch and heave modes, the ratio of undamped natural frequencies is, as before,

$$\frac{f_{NP}}{f_{NH}} = \frac{\omega_{NP}}{\omega_{NH}} = \frac{1}{i_P}$$

The ratio of the pitch and heave damping ratios is

$$\frac{\zeta_P}{\zeta_H} = \frac{1}{i_P}$$

The pitch dynamic index is around 1.0 for a passenger car, often more for a large car, less for small cars which tend to have less overhang, and substantially less than 1.0 for formula racing cars (e.g. 0.6). This indicates that for ordinary cars the undamped pitch frequency is similar to the heave frequency, as seen before, and that the pitch damping ratio is similar to the heave damping ratio, perhaps a little higher. Hence, the provision of suitable heave damping will probably also be suitable for pitch.

Some cars have suspensions interconnected front-to-rear to reduce pitch stiffness, to lower the pitch frequency and to improve the ride. If the effective damping coefficients are the same for pitch as for heave, then this will increase the pitch damping ratio.

The damped radian natural frequency in heave (i.e. the actual frequency with damping present) is

$$\omega_{DNH} = \omega_{NH}\sqrt{1 - \zeta_H^2}$$

This goes to zero for critical damping, but for a mean damping ratio of 0.3 is only 5% below the undamped frequency. This also applies to pitch. For a damping ratio of 0.6 the frequency reduction is 20%. The reduction of frequency by damping may well, therefore, need to be taken into account.

2.16 Roll Vibration Undamped

The roll mode of vibration is largely independent of pitch or heave, and can be approximately assessed as an independent 1-dof motion. The roll stiffness K_{RS}, for independent suspension, springs only, without anti-roll bars, and with equal tracks and stiffnesses, where K is the spring rate at each wheel, is

$$K_{RS} = KT^2$$

The sprung-mass roll inertia, about the roll axis, is

$$I_R = m_S k_R^2 = \tfrac{1}{4} m_S T^2 i_R^2$$

where k_R is the roll radius of gyration and i_R is the roll dynamic index, defined as

$$i_R = \frac{k_R}{\tfrac{1}{2}T}$$

The equation of free roll motion is

$$I_R \ddot{\phi} + K_R \phi = 0$$

The roll natural frequency ω_{NR} is

$$\omega_{NR} = \sqrt{\frac{K_R}{I_R}} = \sqrt{\frac{4K}{m_S i_R^2}}$$

Because the roll dynamic index tends to be less than 1, around 0.85, the roll natural frequency tends to be a little higher than for pitch or heave. Also, the roll stiffness is frequently increased by anti-roll bars, to improve the handling.

Comparing ω_{NR} with the natural heave frequency,

$$\frac{\omega_{NR}}{\omega_{NH}} = \frac{1}{i_R}$$

for no anti-roll bars. With an anti-roll bar roll stiffness factor defined by

$$f_{ARB} = \frac{k_{\phi ARB}}{K_{RS}}$$

the roll stiffness is multiplied by the factor $1 + f_{ARB}$ which may in some cases take a value of 2:

$$K_R = K_{RS}(1 + f_{ARB})$$

and the frequency ratio becomes

$$\frac{\omega_{NR}}{\omega_{NH}} = \frac{1}{i_R}\sqrt{1 + f_{ARB}}$$

In the case of a solid axle, the roll stiffness arising from the springs depends on their actual spacing squared, which gives considerably reduced roll stiffness. This may be compensated to a limited extent by the inherent roll stiffness within leaf springs.

2.17 Roll Vibration Damped

For independent suspension, considering equal stiffness and damping as seen at the wheels, the inertia, stiffness and damping for roll are:

$$I_R = m_S k_R^2 = m_S \tfrac{1}{4} T^2 i_R^2$$
$$K_R = KT^2(1 + f_{ARB})$$
$$C_R = CT^2$$

The equation of free roll motion is

$$I_R \ddot{\phi} + C_R \dot{\phi} + K_R \phi = 0$$

The undamped natural frequency is

$$\omega_{NR} = \sqrt{\frac{K_R}{I_R}} = \sqrt{\frac{4K(1 + f_{ARB})}{m_S i_R^2}}$$

The damping factor is

$$\alpha_R = -\frac{C_R}{2I_R} = -\frac{2C}{m_S i_R^2}$$

The damping ratio is

$$\zeta_R = -\frac{\alpha_R}{\omega_{NR}} = \frac{C}{i_R\sqrt{m_s K(1+f_{ARB})}}$$

The ratio of roll damping ratio to heave damping ratio is

$$\frac{\zeta_R}{\zeta_H} = \frac{1}{i_R\sqrt{1+f_{ARB}}}$$

The roll damping ratio is therefore basically similar to that for 1-dof heave, but dependent upon the roll dynamic index, which is likely to be about 0.8–0.85, enhancing roll damping. On the other hand, the use of anti-roll bars will reduce the roll damping ratio.

2.18 Heave-and-Pitch Undamped 2-dof

Consider now a heave-and-pitch analysis for a more realistic vehicle model with longitudinal asymmetry, i.e. $a \neq b, K_f \neq K_r$, and/or $C_f \neq C_r$, as in Figure 2.18.1.

For an undamped 2-dof analysis of heave and pitch, consider the vehicle body with heave position z (at the centre of mass), positive upwards, and pitch angle θ, positive pitch nose up, Figure 2.18.2. The front and rear body ride positions are

$$z_f = z + a\theta$$
$$z_r = z - b\theta$$

The front and rear suspension forces on the body, positive upwards, are

$$F_f = -2K_f z_f = -2K_f(z + a\theta)$$
$$F_r = -2K_r z_r = -2K_r(z - b\theta)$$

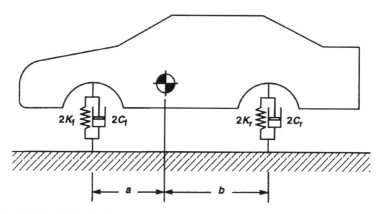

Figure 2.18.1 Vehicle with longitudinal asymmetry, i.e. centre of mass not at the mid-wheelbase, and/or different front and rear suspension parameters (with zero damping in Section 2.18).

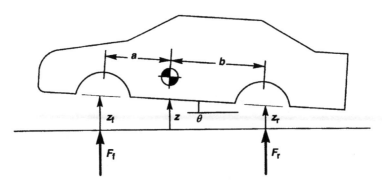

Figure 2.18.2 Vehicle body position with heave and pitch deflections.

The equations of motion are

$$m\ddot{z} = F_f + F_r$$
$$= -2K_f(z + a\theta) - 2K_r(z - b\theta)$$
$$= -(2K_f + 2K_r)z - (2aK_f - 2bK_r)\theta$$

$$I\ddot{\theta} = aF_f - bF_r$$
$$= -2aK_f(z + a\theta) + 2bK_r(z - b\theta)$$
$$= -(2aK_f - 2bK_r)z - (2a^2K_f + 2b^2K_r)\theta$$

Using Heaviside D-operator notation and collecting terms these become

$$\{mD^2 + (2K_f + 2K_r)\}z + \{(2aK_f - 2bK_r)\}\theta = 0$$
$$\{(2aK_f - 2bK_r)\}z + \{ID^2 + (2a^2K_f + 2b^2K_r)\}\theta = 0$$

These are the simultaneous differential equations of motion for undamped pitch and heave.

These equations can be expressed more concisely by considering the constant terms as zeroth, first and second moment vehicle ride stiffness coefficients:

$$C_{K0} = 2K_f + 2K_r \qquad [\text{N/m}]$$
$$C_{K1} = 2aK_f - 2bK_r \qquad [\text{N/rad}]$$
$$C_{K2} = 2a^2K_f + 2b^2K_r \qquad [\text{Nm/rad}]$$

The first of these is in fact the total stiffness in 1-dof heave, the last is the pitch stiffness, in 1-dof motion. The second one, C_{K1}, is a coupling coefficient between simple heave and pitch in the coordinates being used for the analysis, namely z and θ as seen at the centre of mass. Substituting the coefficients gives the standard form

$$\{mD^2 + C_{K0}\}z + \{C_{K1}\}\theta = 0$$
$$\{C_{K1}\}z + \{ID^2 + C_{K2}\}\theta = 0$$

Vibration Theory

The characteristic equation for 2-dof heave and pitch motion is obtained by eliminating either z or θ, the remaining variable then dividing out. The result is

$$(mD^2 + C_{K0})(ID^2 + C_{K2}) - C_{K1}^2 = 0$$

giving the characteristic equation

$$(mI)D^4 + (mC_{K2} + IC_{K0})D^2 + (C_{K0}C_{K2} - C_{K1}^2) = 0$$

The solutions required for D are the imaginary values associated with the frequency of undamped natural vibration, i.e.

$$D = +i\omega_N$$
$$D^2 = -\omega_N^2$$
$$D^4 = +\omega_N^4$$

so the undamped modal radian natural frequencies are the solutions to

$$(mI)\omega_N^4 - (mC_{K2} + IC_{K0})\omega_N^2 + (C_{K0}C_{K2} - C_{K1}^2) = 0$$

This characteristic equation is a quartic, but has no odd powers, so it can be considered to be a quadratic in $E = \omega_N^2$. Also, to condense the equations further, write vehicle ride stiffness coefficients

$$C_{K3} = \frac{C_{K2}}{I} + \frac{C_{K0}}{m} \quad [\text{s}^{-2}]$$

$$C_{K4} = \frac{C_{K0}C_{K2} - C_{K1}^2}{mI} \quad [\text{s}^{-4}]$$

This all gives the characteristic equation in the compact form

$$\omega_N^4 - C_{K3}\omega_N^2 + C_{K4} = 0$$

with the standard solution

$$\omega_N^2 = \tfrac{1}{2}C_{K3} \pm \sqrt{\tfrac{1}{4}C_{K3}^2 - C_{K4}}$$

The two modal natural frequencies are therefore, taking the lower one first,

$$\omega_{M1} = \sqrt{\tfrac{1}{2}C_{K3} - \sqrt{\tfrac{1}{4}C_{K3}^2 - C_{K4}}}$$

$$\omega_{M2} = \sqrt{\tfrac{1}{2}C_{K3} + \sqrt{\tfrac{1}{4}C_{K3}^2 - C_{K4}}}$$

which are easily computed. These frequencies depend on the physical properties of the system, the inertias and stiffnesses, not on the initial conditions of the particular vibration. These frequency values are around 9 rad/s (1.4 Hz) for a passenger car.

In each of the two modes of vibration, considered separately, the proportions of heave and pitch are defined by the mode shape factor, which is the quotient of heave over pitch in the mode. To find this, consider the solution, which for an undamped linear system will be sinusoidal in time, to be

$$z = Z \sin \omega t$$
$$\theta = \Theta \sin \omega t$$

where Z and Θ are to be determined. Differentiating twice gives

$$\ddot{z} = -\omega^2 Z \sin \omega t$$
$$\ddot{\theta} = -\omega^2 \Theta \sin \omega t$$

Substituting these into the standard form of the differential equations,

$$-m\omega^2 Z \sin \omega t + C_{K0} Z \sin \omega t + C_{K1} \Theta \sin \omega t = 0$$
$$C_{K1} Z \sin \omega t - I\omega^2 \Theta \sin \omega t + C_{K2} \Theta \sin \omega t = 0$$

From the first of these equations

$$Z(C_{K0} - m\omega^2) + C_{K1}\Theta = 0$$

and the mode shape S is, by definition,

$$S = \frac{Z}{\Theta} = \frac{C_{K1}}{(m\omega^2 - C_{K0})}$$

From the second equation

$$C_{K1} Z + (C_{K2} - I\omega^2)\Theta = 0$$
$$S = \frac{Z}{\Theta} = \frac{(I\omega^2 - C_{K2})}{C_{K1}}$$

These two equations for S must give the same value for the mode shape, of course, so either may be used. They are to be evaluated at the modal natural frequencies ω_{M1} and ω_{M2} found earlier, to give the two mode shape values. They only agree in value at the modal frequencies. This necessary agreement is an alternative way to derive the frequencies.

The two mode shapes, in full notation, are

$$S_1 = \left(\frac{Z}{\Theta}\right)_1 = \frac{C_{K1}}{(m_s\omega_{M1}^2 - C_{K0})}$$
$$S_2 = \left(\frac{Z}{\Theta}\right)_2 = \frac{C_{K1}}{(m_s\omega_{M2}^2 - C_{K0})}$$

The unit of each of these mode shapes is the metre (m). This is the quotient of heave over pitch, and represents the distance behind the centre of mass at which the effective nodal 'pivot' point will be found, as seen in Figure 2.18.3. One of the mode shapes is negative, indicating a nodal point forward of the centre of mass for that mode.

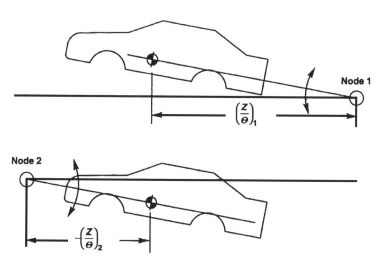

Figure 2.18.3 Nodal points and body modal displacements.

As with 1-dof vibration, the amplitude is arbitrary, depending upon initial conditions. In general of course, an initial deflection would be some mixture of the two mode shapes, and the two modes would then proceed in superposition, each at its own frequency.

One of the mode shapes has a greater magnitude than the other; this is the predominantly heave mode, the other being the predominant pitch mode.

For practical values of end frequencies, 20–30% higher at the rear, the node of the heave mode is approximately one wheelbase behind the centre of mass. The pitch node is about one quarter of the wheelbase in front of the centre of mass, round about midway between the centre of mass and the front axle. Hence the pitch mode gives a much greater motion at the rear suspension than at the front. Damping of the pitch mode is therefore largely dependent on the rear dampers.

The importance of these modes is that they are uncoupled—the vehicle can vibrate in one mode without any motion in the other mode. In the equations of motion it was seen that there was a coupling term C_{K1}. In the original coordinates (z, θ) at the centre of mass, there is coupling—the vehicle can not vibrate in simple heave or pitch about the centre of mass. By changing to coordinates to angular motion in the two modes (θ_1, θ_2), decoupling is achieved.

The relationship between heave and pitch positions and the modal positions may be found as follows. For a vehicle body position specified by heave position z and pitch angle θ, with corresponding modal positions θ_1 and θ_2, rotating about the corresponding node, positive angle for pitch up,

$$z = S_1 \theta_1 + S_2 \theta_2$$
$$\theta = \theta_1 + \theta_2$$

Simultaneous solution of these gives

$$\theta_1 = \frac{z - S_2 \theta}{S_1 - S_2}$$

$$\theta_2 = \frac{S_1 \theta - z}{S_1 - S_2}$$

Corresponding equations apply for velocities and accelerations.

If the initial vehicle position is expressed by z_f and z_r, to find the corresponding modal positions θ_1 and θ_2:

$$z_f = (S_1 + a)\theta_1 + (S_2 + a)\theta_2$$
$$z_r = (S_1 - b)\theta_1 + (S_2 - b)\theta_2$$

The solution to these simultaneous equations is

$$d = (S_1 + a)(S_2 - b) - (S_2 + a)(S_1 - b)$$
$$\theta_1 = \frac{z_f(S_2 - b) - z_r(S_2 + a)}{d}$$
$$\theta_2 = \frac{-z_f(S_1 - b) + z_r(S_1 + a)}{d}$$

Any initial body position can then be represented by the sum of two modal positions. If the body is then released in free vibration, the modes proceed independently of each other, each at its own frequency. The body position at time t in terms of z and θ may then be found by superposition of the modal positions.

2.19 Heave-and-Pitch Damped 2-dof Simplified

The next stage of modelling complexity is to consider separately a damped 1-dof motion of each of the mode shapes as found by undamped 2-dof analysis. This is a very effective method. In contrast with more elaborate analysis, the highly meaningful undamped mode shapes (nodal positions) are retained and, although imperfect, have a clear interpretation.

The mode shapes, giving the nodal positions for each of the two modes, are found by undamped analysis as shown earlier:

$$\omega_{M1} = \sqrt{\tfrac{1}{2}C_{K3} - \sqrt{\tfrac{1}{4}C_{K3}^2 - C_{K4}}}$$

$$\omega_{M2} = \sqrt{\tfrac{1}{2}C_{K3} + \sqrt{\tfrac{1}{4}C_{K3}^2 - C_{K4}}}$$

$$S_1 = \left(\frac{Z}{\Theta}\right)_1 = \frac{C_{K1}}{(m_S \omega_{M1}^2 - C_{K0})}$$

$$S_2 = \left(\frac{Z}{\Theta}\right)_2 = \frac{C_{K1}}{(m_S \omega_{M2}^2 - C_{K0})}$$

Consider now 1-dof vibration about the node for mode 1. Imagine the vehicle pinned at the nodal point, so that despite any effect of the damping the motion is still pitching about that nodal point. This is not exactly the real motion, which depends on the distribution of damping compared with the distribution of stiffness, but it is generally an adequate approximation. By the Parallel Axis Theorem, the modal inertias are

$$I_{M1} = I_P + mS_1^2$$
$$I_{M2} = I_P + mS_2^2$$

By considering the restoring moment from a modal deflection θ, the modal stiffnesses are

$$K_{M1} = 2K_f(S_1 + a)^2 + 2K_r(S_1 - b)^2$$
$$K_{M2} = 2K_f(S_2 + a)^2 + 2K_r(S_2 - b)^2$$

By considering a modal (angular) velocity, the modal damping coefficients C_{M1} and C_{M2} are

$$C_{M1} = 2C_f(S_1 + a)^2 + 2C_r(S_1 - b)^2$$
$$C_{M2} = 2C_f(S_2 + a)^2 + 2C_r(S_2 - b)^2$$

Considering the modal motion as 1-dof, the equation of motion is simply

$$I_M \ddot{\theta}_M + C_M \dot{\theta}_M + K_M \theta_M = 0$$

The radian natural frequency is as usual the square root of the stiffness over the inertia:

$$\omega_{M1} = \sqrt{\frac{K_{M1}}{I_{M1}}}$$

$$\omega_{M2} = \sqrt{\frac{K_{M2}}{I_{M2}}}$$

Of course, these frequencies are already known, having been found during the evaluation of the modal node positions, but these new frequency expressions should be in agreement with the earlier results, and provide a useful check. The modal damping factors (the real part of the complex root), negative for practical cases, are

$$\alpha_{M1} = -\frac{C_{M1}}{2I_{M1}}$$

$$\alpha_{M2} = -\frac{C_{M2}}{2I_{M2}}$$

The modal damping ratios are then

$$\zeta_{M1} = -\frac{\alpha_{M1}}{\omega_{M1}}$$

$$\zeta_{M2} = -\frac{\alpha_{M2}}{\omega_{M2}}$$

The damped natural frequencies ω_{D1} and ω_{D2} in rad/s may then be calculated in the usual way from

$$\omega_{D1} = \omega_{M1}\sqrt{1 - \zeta_{M1}^2}$$

$$\omega_{D2} = \omega_{M2}\sqrt{1 - \zeta_{M2}^2}$$

In fact the presence of damping in general strictly invalidates the assumption of independent motion in the two modes, by introducing some modal coupling, but the approximation is usually a reasonably good one. Therefore the above calculation gives a very simple way to obtain an estimate of the value of the damping ratio for each of the two modes. A valuable aspect of the above analysis is that the idea of a fixed modal node is retained, allowing easy interpretation of the mode shapes and motion, and the calculations are easily performed by hand.

2.20 Heave-and-Pitch Damped 2-dof Full Analysis

With computational facilities, a full 2-dof analysis with damping can be performed. Solution of a quartic equation is required, which is rather time consuming for manual solution. The results are usually close to those found by the simplified method above.

Consider the vehicle of Figure 2.18.1 with a 2-dof position defined by a heave value z at the centre of mass, plus a pitch-up angle θ (in radians). The body positions z_f and z_r and the corresponding vertical velocities are

$$z_f = z + a\theta$$
$$z_r = z - b\theta$$
$$\dot{z}_f = \dot{z} + a\dot{\theta}$$
$$\dot{z}_r = \dot{z} - b\dot{\theta}$$

The total suspension forces F_f and F_r on the body are

$$F_f = -2K_f(z + a\theta) - 2C_f(\dot{z} + a\dot{\theta})$$
$$F_r = -2K_r(z + a\theta) - 2C_r(\dot{z} + a\dot{\theta})$$

The equations of motion are

$$m\ddot{z} = -2K_f(z + a\theta) - 2C_f(\dot{z} + a\dot{\theta}) - 2K_r(z - b\theta) - 2C_r(\dot{z} - b\dot{\theta})$$
$$I\ddot{\theta} = -2aK_f(z + a\theta) - 2aC_f(\dot{z} + a\dot{\theta}) + 2bK_r(z - b\theta) + 2bC_r(\dot{z} - b\dot{\theta})$$

Using the vehicle ride stiffness coefficients C_{K0}, C_{K1} and C_{K2} and vehicle ride damping coefficients C_{D0}, C_{D1} and C_{D2}, which are as follows:

$$C_{K0} = 2K_f + 2K_r$$
$$C_{K1} = 2aK_f - 2bK_r$$
$$C_{K2} = 2a^2 K_f + 2b^2 K_r$$
$$C_{D0} = 2C_f + 2C_r$$
$$C_{D1} = 2aC_f - 2bC_r$$
$$C_{D2} = 2a^2 C_f + 2b^2 C_r$$

and collecting terms, with Heaviside D-operator notation for d/dt, the condensed result is:

$$\{mD^2 + C_{D0}D + C_{K0}\}z + \{C_{D1}D + C_{K1}\}\theta = 0$$
$$\{C_{D1}D + C_{K1}\}z + \{ID^2 + C_{D2}D + C_{K2}\}\theta = 0$$

These are the simultaneous differential equations of motion for 2-dof heave and pitch including damping. The above equations may be expressed more briefly as

$$\{A\}z + \{B\}\theta = 0$$
$$\{B\}z + \{C\}\theta = 0$$

where

$$A = mD^2 + C_{D0}D + C_{K0}$$
$$B = C_{D1}D + C_{K1}$$
$$C = ID^2 + C_{D2}D + C_{K2}$$

The term B is the coupling term for analysis in the coordinates (z,θ) at the centre of mass; if the coupling is zero then heave and pitch about G are independent, which is not normally the case. This would require that C_{D1} be zero (zero damper coupling) in addition to zero stiffness coupling C_{K1}. In the undamped model, decoupling was achieved by changing to modal coordinates, representing the position by modal angles (θ_1,θ_2) rotating about the nodal points. In general, this is not possible in the damped case because the dampers do not agree with the springs about where the nodes should be. It would only occur if $C_f/C_r = K_f/K_r$. Therefore, in general the mode shape is no longer a real number.

The characteristic equation is obtained from the two simultaneous equations of motion in the usual way, by eliminating either z or θ, the other then dividing out, giving

$$AC - B^2 = 0$$

The result is

$$\{mD^2 + C_{D0}D + C_{K0}\}\{ID^2 + C_{D2}D + C_{K2}\} - \{C_{D1}D + C_{K1}\}^2 = 0$$

Expanding and collecting terms gives a quartic equation:

$$D^4\{mI\}$$
$$+ D^3\{mC_{D2} + IC_{D0}\}$$
$$+ D^2\{mC_{K2} + C_{D0}C_{D2} + IC_{K0} - C_{D1}^2\}$$
$$+ D^1\{C_{D0}C_{K2} + C_{D2}C_{K0} - 2C_{D1}C_{K1}\}$$
$$+ D^0\{C_{K0}C_{K2} - C_{K1}^2\} = 0$$

This is the characteristic equation in pitch and heave. Setting damping terms to zero eliminates the odd-powered terms, giving the quadratic in D^2 as shown earlier for undamped motion.

The above equation is a quartic with all terms present, so a general analytical solution is impractical. For a particular design, the numerical results may be obtained by hand computation; this is cumbersome and prone to error, but possible. The coefficients in the above equation are evaluated, to give

$$aD^4 + bD^3 + cD^2 + dD + e = 0$$

and standard methods of quartic solution are employed. This is ideal for computer numerical solution, revealing the modal damping ratios and the complex mode shapes. A method for solution of quartic equations is given in Appendix E.

A quartic equation has four roots (solutions). In general these are complex rather than just real values, even though the quartic coefficients above are all real. A practical passenger car will have heave and pitch motions which are damped oscillatory, so the roots will be complex, say z_1 to z_4 in conjugate pairs. The quartic therefore factors into

$$(D - z_1)(D - z_2)(D - z_3)(D - z_4) = 0$$

There are three ways to combine these factors into a pair of quadratic factors for the quartic. When paired correctly, one quadratic corresponds to the primarily heave mode, the other to the primarily pitch mode. Because the coefficients of the quartic are real, the complex roots are in conjugate pairs (the same real part, equal magnitude imaginary parts of opposite sign), each pair from one quadratic. Hence the correct factorisation of the quartic produces two quadratic equations with coefficients which are all real. The quadratic factor equations give the modal frequencies and damping ratios in the usual way, one quadratic representing each mode. The fact that the quartic factors into quadratics with real coefficients does not imply that the modes are uncoupled.

3
Ride and Handling

3.1 Introduction

The purpose of this chapter is to review some principles of vehicle ride and handling, so that the rôle of the damper can be better understood. This shows why vehicle damping coefficients are given the values that they are, and illustrates the function of the damper in the ride–handling compromise.

In practice, most vehicle designs are evolutionary. Prototypes will be assembled using the springs and dampers expected to be suitable, based on prior experience with similar vehicles. Small, unimportant, changes may be made so that advertisements can say that the suspension is 'improved'. Road testing by professional test drivers is used to tune the values before final specification for production. It is relatively rare for a new vehicle to be a substantial extrapolation of design. One interesting example when this was certainly the case was the Apollo Moon Rover vehicle, required to operate in an abnormally low gravitational field over terrain that was rough and rocky in some areas but soft and dusty in others.

The significance of the damper in Ride and Handling relates closely to the free vibrations and forced vibrations that occur because of road roughness and control inputs. The passenger on the seat cushion, the sprung mass on the suspension springs and dampers, and the unsprung mass (wheels) on the tyre stiffness and tyre damping are all involved.

The damping forces seen at the wheel can be expressed in various ways. For example, dividing the vehicle weight ($W = mg$ (N), not mass m (kg)) by the total damping coefficient at the wheels in bump gives a characteristic speed, of order 10 m/s, representing a notional steady sink rate at which the car would settle on its suspension when resisted by its dampers without spring support:

$$V_{ZS} = \frac{mg}{\sum C_W}$$

However, the optimum damping depends not just on the vehicle mass or weight, but on the spring stiffness too, so it is better to refer to the classic vibration concept of damping ratio. This may not be a good approximation to apply when damping coefficients for the two motion directions (bump and droop) are widely different, but it is useful in a qualitative way. Manufacturers' opinions regarding the optimum damping ratio seem to vary over a range of at least 2:1 for a given type of vehicle, which is not surprising in view of the subjective nature of ride assessment.

3.2 Modelling the Road

Any particular track along a road has a longitudinal profile representing its vertical sectional shape for ride purposes. This neglects lateral curvature (cornering). For advanced analysis, two parallel

---

The Shock Absorber Handbook/Second Edition John C. Dixon
© 2007 John Wiley & Sons, Ltd

tracks may be considered, for left and right wheels, or even four road tracks if the vehicle axle tracks (US treads) are different. This road profile may be analysed and used in various ways. Typically it is subject to a Fourier analysis to reveal its characteristics as a function of spatial frequency.

In contrast, in simple analytical ride studies the road is considered to have only one spatial frequency of bumps at a time. This is known as a sinusoidal road. All frequencies may be considered, but only one at a time. Considering a sinusoidal road shape of wavelength λ_R, the spatial frequency of the road n_{SR} in cycles per metre is

$$n_{SR} = \frac{1}{\lambda_R}$$

The radian spatial frequency ω_{SR} in radians per metre is

$$\omega_{SR} = 2\pi n_{SR} = \frac{2\pi}{\lambda_R}$$

At a vehicle longitudinal speed V, the frequency of vibration stimulus is f_R (Hz) with corresponding radian frequency ω_R (rad/s) and period T_R (in seconds):

$$T_R = \frac{\lambda_R}{V}$$

$$f_R = \frac{1}{T_R} = \frac{V}{\lambda_R} = V n_{SR}$$

$$\omega_R = V \omega_{SR}$$

It is therefore easy to find the speed or wavelength which will stimulate vehicle resonances at a natural frequency f_N:

$$V = \lambda f_N$$

A heave or pitch resonance at 1.6 Hz occurs for a wavelength of 10 m at a speed of $10 \times 1.6 = 16$ m/s. At a typical speed of 20 m/s and $f_N = 1.6$ Hz the resonant wavelength is 12.5 m. A wheel hop resonance at 10 Hz will be stimulated at a speed of 20 m/s by a wavelength of $20/10 = 2.0$ m. At a wavelength of 0.1 m, as found with cobblestones, wheel hop is stimulated at a speed of around 1 m/s.

Real roads, subject to Fourier analysis, are found to have a spectral distribution of roughness declining rapidly with spatial frequency, Figure 3.2.1 giving two examples. A commonly used road model has been that in Figure 3.2.2, with power spectral density

$$S = S_{ref} \left(\frac{n_S}{n_{S,ref}} \right)^{-W}$$

with W usually equal to 2.5 or 2.7. More complex models are also used with two gradients, and various cut-off methods, and sometimes continuous curves.

The ISO road surface model has two linear sections as in Figure 3.2.3, with an average road specified by a break point or reference point at a spatial frequency of 1 rad/m, which is a wavelength of 2π metres. In cycles/metre this is

$$n_{S,ref} = \frac{1}{2\pi} \, c/m = 0.1592 \, c/m$$

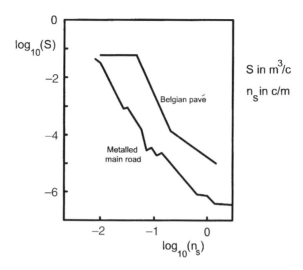

Figure 3.2.1 The spectral analysis of two example roads (spectral density versus spatial frequency).

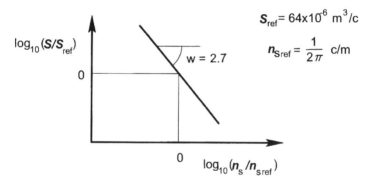

Figure 3.2.2 The basic spectral model of a representative road surface.

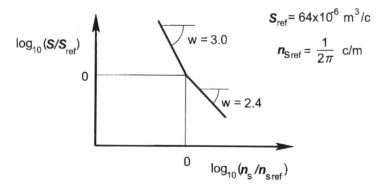

Figure 3.2.3 The ISO spectral road model.

At a speed of 20 m/s, this break point at a wavelength of 2π metres corresponds to a frequency of 3.18 Hz, in between the body natural frequency and the wheel hop frequency. The spectral density of the ISO standard road at this frequency is

$$S_{\text{ref}} = 64 \times 10^{-6} \, \text{m}^3/\text{c} = 64 \, \text{cm}^3/\text{c}$$

The line negative gradients are

$$W_1 = 3.0; \quad W_2 = 2.4$$

The particular metalled main road in Figure 3.2.1 can be seen to have two gradients, which are about -3.0 and -1.5 with a break point at a wavelength of 10 m. Data published by Verschoore *et al.* (1996) show a 'perfect asphalt' road to have a single gradient of -2.2 from 100 m to 0.4 m wavelength. Therefore all roads are not ISO roads.

Evidently roads vary considerably in quality. It is useful to consider a grading of roads for simulation purposes. The road model is assumed to have one or two gradients with a reference point at a spatial frequency of $1/2\pi$ c/m. The spectral density at this reference point will then be as in Table 3.2.1. The spectral density is 2^R cm^3/cycle, where R is the road profile roughness rating. The table covers a range from a very good (motorway) quality to a poor minor road. The 'medium' grade, $R = 6$, is the same as that of the ISO standard road. Road types may actually vary by a factor of up to 8 in the reference spectral density, e.g. motorways may vary from 1 to 64 according to the particular road and maintenance state. However, the table is a useful guide to values.

Displacement can be scaled after making the basic profile shape, so one can make the road shape for 1 cm^3/c or 64 cm^3/c and adjust for road quality later. If the vehicle model is linear, the response is proportional to the stimulus, and the transmissibility is unchanged by the roughness grade of the road, which is important only for a non-linear vehicle model or when considering human vibration tolerance.

With a negative gradient of 2.5 in a spectral model, doubling the speed of travel increases the roughness at a given frequency, e.g. the body heave natural frequency, by a factor of $2^{2.5} = 5.7$. The provision of high-quality surfaces for high-speed roads is therefore not just a matter of ride quality and comfort, but also one of safety.

Given a spectral distribution for a track, the numerical track profile may be built. This is a process of inverse Fourier transform. It may be done by an inverse FFT (Fast Fourier Transform), but this is not essential. A 'slow FT' may be used because the road generation need only be done once, the road profile then being stored and used many times. A series of spatial frequencies is chosen, and the spectral distribution condensed down into these frequencies, producing specific amplitudes for each one. In effect, the spectral distribution is converted into a histogram, with one column for each spatial frequency. The area under the spectral distribution within the width of the column, which depends on

Table 3.2.1 Spectral density of major roads at spatial frequency $1/2\pi$ c/m

Rating	Description	Spectral density (cm^3/cycle)
3	Very good	8
4	Good	16
5	Medium-good	32
6	Medium	64
7	Medium-bad	128
8	Bad	256
9	Very bad	512

the adjacent frequencies, gives the column height. The column height is an integral of the spectral density, so having units m². The square root of twice this is the amplitude for that frequency. The wavelengths must all fit exactly into the length of road. The phase angle for each frequency is chosen at random in the range $0-2\pi$, this giving amplitudes for the sine and cosine components.

The road is then the sum of all the displacements from the various spatial frequencies, giving N data points. In the case of an FFT, the frequencies are all predetermined, because the wavelengths are L/K where L is the length of road and K takes all integer values from 1 to $N/2$. The total number of frequencies is one-half of the number of profile data points. This is because this gives a total of N sine and cosine coefficients. In effect, a Fourier analysis gives an alternative representation of the N profile points, and solves N simultaneous equations for the coefficients, which is easily done because the equations are decoupled. For efficient computation, it is usual to take the number of data points as 2^M, say $M = 10$ with 1024 points at one metre spacing.

In more advanced ride models, using a vehicle model with two wheels on each axle at the track (tread) lateral spacing, the correlation of the road profile at adjacent positions in the two wheel tracks is of interest. This obviously depends on the method of road manufacture and on the types of machine used for finishing, and there could easily be very high coherence at some short wavelengths. However, Figure 3.2.4 shows the coherence found in one study, track value not specified, but presumably that of a typical road vehicle. As would be expected, the coherence is unity at long wavelengths, falling to zero at short ones. At a vehicle speed of 20 m/s the resonant wavelength is about 14 m, 0.07 cycles/m, where the coherence may be seen to be quite high, 0.4 to 1.0 for the wide range of types of the three roads reported. At a wheel hop frequency of about 10 Hz, the wavelength is 2 m (depending on the speed), which is a spatial frequency of 0.5 cycles/m, requiring extrapolation of an undulating curve, with a coherence possibly anywhere in the range 0–0.4, perhaps even more in some cases.

The real road is manufactured in a highly directional manner, so there is little reason, *a priori*, to expect the road to be isotropic (properties independent of direction), and in some cases clearly it is not. For example, in soft ground when there are ruts from the passage of the wheels the transverse section has a very strong spectral density at a particular wavelength. Studies of metalled roads do indicate lack of isotropy, but in a highly unsystematic way, so isotropy may still be a reasonable way to anticipate the relationship between adjacent tracks. When there is no actual data, and a road of two or more tracks is to be generated, isotropy is the logical assumption. Figure 3.2.5 shows the coherence between tracks calculated on this basis.

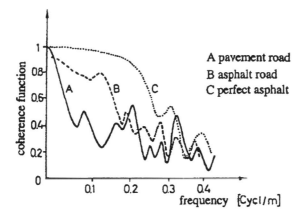

Figure 3.2.4 Coherence between two adjacent tracks. Reproduced with permission from Verschoore, Duquesne and Kermis (1996) Determination of the vibration comfort of vehicles by means of simulation and measurement, *Eur. J. Mech. Eng.*, 41(3), pp.137–143.

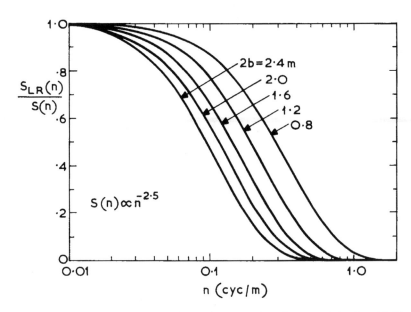

Figure 3.2.5 Coherence between adjacent tracks, on the assumption of a spectral density with $W = 2.5$ and isotropy. The track spacing is $2b$. Reproduced from Robson (1979) Road surface description and vehicle response, *IJVD*, 1(1), pp.25–35.

A numerical road of two adjacent tracks is then to be created from a spectral density and a coherence graph. For each spatial frequency in turn, one track is made as for a single track. The second track has the same amplitude. The phase difference between the tracks depends on the coherence for that wavelength. However, the coherence is not explicit, and is only a guide to the range of randomness, and actually the distribution is not specified, only the effective mean. For a given wavelength and coherence from the graph, the phase difference ϕ may be taken with reasonable realism as

$$\phi = \pm \pi (1 - C) \text{ Rand}$$

with equal probability for either sign, and Rand is a 0–1 uniform random deviate. Alternatively one can just use a phase difference of

$$\phi = \pm 2\pi \, n_s T \text{ Rand} \quad \text{or} \quad \phi = \pi \, n_s T \{2(\text{Rand} + \text{Rand} + \text{Rand} - 1.5)\}$$

As a matter of interest, Figure 3.2.6 shows a contour map of a poor-quality road surface.

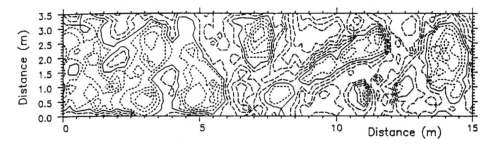

Figure 3.2.6 Contour map of MIRA Belgian Pavé test track. This is similar to an isotropic surface. Long dashes are contours below datum, short ones above. Reproduced from Cebon and Newland (1984) The artifical generation of road surface topography by the inverse FFT method, *Proc. 8th IAVSD-IUTAM Symp.*, Tayor & Francis, 1984, pp.29–42.

3.3 Ride

The ride of a vehicle is the heaving, pitching and rolling motion in forced vibration caused by road roughness. The purpose of the suspension in this context is to minimise the discomfort of the passengers, which obviously involves a minimisation of some measure of the vehicle body motion, by choice of the springs and dampers, Figure 3.3.1. The optimum values will of course be dependent upon the quality of roads over which a vehicle is expected to operate; the best suspension parameters for high-speed cruising on good-quality roads may be quite different from the best over cobblestones. Another important factor to bear in mind is that, although the ride motions can be measured accurately, ride quality is really a subjective opinion, and as such depends very much upon the individual. Ultimately, therefore, there is significant variation of opinion on optimum suspension design, for both stiffness and damping, but especially of the latter.

Ride behaviour may be analysed in the time domain or in the frequency domain. Time-domain analysis predicts positions, velocities and accelerations as functions of time, with results often displayed as such graphs, e.g. Figure 2.3.1 (in Chapter 2). Frequency domain analysis predicts the characteristics as functions of frequency, e.g. the transmissibility of Figure 2.4.1, hence revealing resonances, etc.

For ride analysis, a fairly complex model would have eight degrees of freedom (seat height, body heave, pitch and roll, four wheel heights). A simpler model would have only four degrees of freedom (body heave and pitch plus front and rear wheel heights). The simplest model, the heave model (so-called quarter-car model), simply has a body mass and wheel mass with associated suspension and tyre stiffnesses and damping. Only heave motion is allowed. For racing cars, the suspension structure compliance may also be significant. The passenger on a seat may be added. Even such a very simple model can reveal a great deal about ride behaviour, and has been widely studied in the research literature. Simple models can be solved analytically, and give useful insight into trends, whereas more complex models may be more accurate in principle but require numerical solution. To illustrate the principle, Figure 3.3.2 shows such a heave ride model. The terminology 'quarter-car model' arises because it appears to have one suspension only. However, this leaves a problem with regard to the mass of the passenger (one quarter of a passenger?). It is better to deem it to be a heave model with a mass equal to the whole vehicle body, suspension comprising the total stiffness of four corners, damping comprising the total damping coefficients, and one complete passenger on one seat cushion.

The parameters of this model are as follows, as seen in Figure 3.3.2:

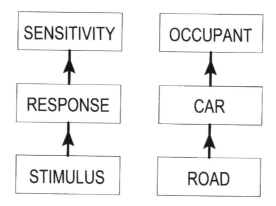

Figure 3.3.1 Ride system analysis.

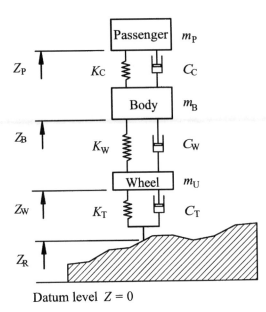

Figure 3.3.2 Heave ride model including passenger, defining inertia, stiffness and damping coefficient notation and ride positions.

(1) passengers (or driver or load) mass m_P (usually one);
(2) cushion (seat) vertical stiffness K_C (one);
(3) cushion (seat) vertical damping coefficient C_C (one);
(4) vehicle body (sprung mass) mass m_B (whole car);
(5) suspension vertical stiffness wheel rate K_W (total);
(6) suspension vertical damping coefficient at the wheel C_W (total);
(7) wheel mass (i.e. unsprung mass) m_U (total);
(8) tyre vertical stiffness K_T (total);
(9) tyre vertical damping coefficient C_T (total, usually small).

The tyre damping is frequently neglected. The seat damping in many cases depends upon rubbing of the internal structure. However in some cases, especially in large commercial vehicles, the seat may have a sprung mechanism with a hydraulic damper.

Above the datum level at $Z = 0$ there are the following vertical positions of defining points on the masses:

(1) road height (from 'roughness') Z_R;
(2) wheel centre height Z_W;
(3) body ride height Z_B;
(4) passenger ride height Z_P.

The variations of these from the static values are the ride displacements, e.g.

$$z_B = Z_B - Z_{B0}$$

and are as follows:

(1) passenger ride displacement z_P;
(2) body ride displacement z_B;
(3) wheel ride displacement z_W.

There are also:

(1) tyre ride deflection $z_T = z_R - z_W$;
(2) suspension ride deflection (bump) $z_S = z_W - z_B$;
(3) cushion (seat) ride deflection $z_C = z_B - z_P$.

3.4 Time-Domain Ride Analysis

Time-domain analysis considers the vehicle behaviour in terms of its motion as time passes. In time stepping, the velocity of each component follows from position or displacement changes divided by the time step or from acceleration multiplied by the time step, giving, *inter alia*, $\dot{z}_T, \dot{z}_S, \dot{z}_C$. In a time-stepping simulation, of course, it is practical to have nonlinear springs and nonlinear damping.

Given the positions and velocities at a given time, all forces may be deduced.

Assuming a linear model for springs and dampers (or more complex if desired),

$$F_{KC} = K_C z_C$$
$$F_{CC} = C_C \dot{z}_C$$
$$F_{KW} = K_W z_S$$
$$F_{CW} = C_W \dot{z}_S$$
$$F_{KT} = K_T z_T$$
$$F_{CT} = C_T \dot{z}_T$$

The consequent passenger, body and unsprung-mass accelerations A_P, A_B and A_U are

$$A_P = \frac{F_{KC} + F_{CC}}{m_P}$$
$$A_B = \frac{F_{KW} + F_{CW} - F_{KC} - F_{CC}}{m_B}$$
$$A_U = \frac{F_{KT} + F_{CT} - F_{KW} - F_{CW}}{m_U}$$

The new time-stepped velocities and positions then follow, e.g.

$$\dot{Z}_B = \dot{Z}_{B,\text{last}} + A_B \Delta_t$$
$$Z_B = Z_{B,\text{last}} + \dot{Z}_B \Delta_t$$

Hence, given some particular road profile, the time stepping will predict a position for the wheel, body and passenger over time. This final step may be elaborated by the use of more complex differential equation solutions, e.g. by Runge–Kutta methods. However, although such more elaborate methods are often necessary in the accurate solution of differential equations by numerical means, there seems to be little if any benefit for road vehicles at ride frequencies. This is due to the nature of the problem. Ride quality analysis does not depend upon distant extrapolation of a motion by a differential equation from an initial condition, as does, say, a rocket trajectory, so sophisticated methods, although essential in some problems, are not necessarily helpful in ride analysis.

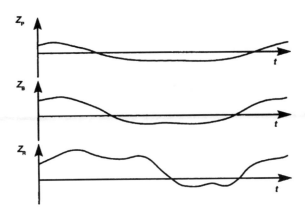

Figure 3.4.1 Passenger and vehicle body ride positions in response to road position, as functions of time.

Figure 3.4.1 illustrates a resulting graph, which may alternatively be obtained by a data logger with instrumentation on an actual vehicle. There are positions for the wheel, body and passenger, with corresponding deflections of the tyre, suspension and seat cushion.

The road stimulus considered may be a simple sinusoid to investigate the vehicle behaviour at a given frequency. This may also be done analytically, but with computer time stepping it is feasible to have nonlinear components. In this case the resulting positions and displacements vary sinusoidally (for linear components) or approximately sinusoidally (for nonlinear) at the forcing frequency, and the ratio of the response amplitude to the road amplitude gives a transmissibility T to each of the wheel, body and passenger for that frequency.

Alternatively, the simulated road may be given a white noise characteristic, at least over a frequency range of interest, say up to 15 Hz; the response of each mass then has a semi-random nature which can be statistically analysed. There will be an overall root-mean-square displacement, which can be divided by the road r.m.s. displacement to give an r.m.s. transmissibility for these conditions. Spectral analysis will also reveal certain resonances and responses, e.g. there will be an increased body response at the body resonant and wheel hop resonant frequencies.

A third option for the model road, frequently favoured, is to give the road a random nature based statistically on the character of real roads, in which case the responses can be regarded as more realistic. The random response of the bodies can then be analysed. The most important results are those for:

(1) the passenger response (for passenger comfort, i.e. ride);
(2) the tyre deflection r.m.s. variation (for tyre comfort, i.e. handling);
(3) the suspension r.m.s. displacement (suspension workspace).

The job of the suspension design engineer is to juggle the available variables (springs, dampers) to obtain the best overall behaviour for the type of vehicle, considering these conflicting requirements.

The assessment of passenger ride perception is called the 'passenger discomfort' D_P. This may be defined as the root-mean-square acceleration experienced by the passenger, normalised by dividing by g, standard gravity (9.80665 m/s^2, 32.174 ft/s^2):

$$D_P = \left[\frac{A_{rms}}{g}\right]$$

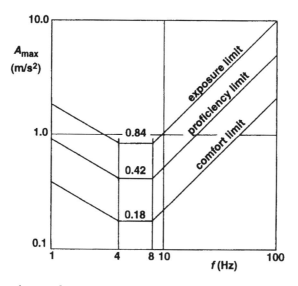

Figure 3.4.2 Average tolerance of a seated human to vertical vibration at a single frequency, variation with frequency, 8 hour exposure (ISO 2631, 1974).

Multiplying by the passenger mass, it is seen that this is also equal to the rms variation of vertical force supporting the passenger divided by the mean vertical force.

Often, the passenger discomfort calculation is adjusted by a frequency-dependent loading function S_P:

$$D_P = \left[\frac{S_P(f)A(f)}{g}\right]_{rms}$$

on the basis that the human body is not uniformly tolerant of different frequencies, Figure 3.4.2.

For 8 h exposure, the ISO 2631 (1974) graph for the seated human vertical motion comfort limit is as follows:

$$\begin{array}{lll} f < 4\,\text{Hz} & a_C = 0.180(f/4)^{-0.50} & \text{m/s}^2 \\ 4\,\text{Hz} < f < 8\,\text{Hz} & a_C = 0.180 & \text{m/s}^2 \\ 8\,\text{Hz} < f & a_C = 0.180\,(f/8) & \text{m/s}^2 \end{array}$$

The proficiency limit at any frequency is

$$a_P = 2.333 a_C$$

and the exposure limit (i.e. damage limit) is

$$a_L = 4.667 a_C = 2 a_P$$

The use of frequency loading is a contentious point, however. Although extensive research has confirmed that human tolerance varies with frequency, there are some objections to its application in this case:

(1) There is very poor agreement on just how the tolerance varies with frequency. This may be due to individual differences in humans (biodiversity).

(2) Research on human tolerance has frequently neglected the effect of seat isolation, giving incorrect results.
(3) Other factors may be highly influential and confuse the results, e.g. noise levels.
(4) There is some doubt that using a frequency-dependent loading function actually improves the accuracy of the ride quality prediction.

The ride simulation also gives a figure for the r.m.s. variation of tyre vertical force. This is the tyre discomfort:

$$D_T = \frac{(\Delta_{FVT})_{rms}}{F_{VT,mean}}$$

Here F_{VT} is the vertical force exerted by the road on the tyre. Hence the tyre discomfort is the r.m.s value of the vertical force fluctuations, divided by the mean vertical force. By these definitions, both the passenger discomfort and the tyre discomfort are non-dimensional values given by the rms variation of vertical force divided by the mean vertical force. This raises the possibility that tyres are not equally sensitive to all frequencies (e.g. carcase resonances), so possibly tyre discomfort should have a frequency loading function.

The basic definition of passenger and tyre discomfort corresponds to the statistician's coefficient of variation, or the normalised standard deviation, the SD divided by the mean. The r.m.s value of a Gaussian distribution with zero mean is one standard deviation. The r.m.q (root mean quartic) is $1.316 = 3^{0.25}$ standard deviations.

Zero tyre discomfort is then a constant vertical force at the tyre contact patch, which is best for maximum tyre shear forces (cornering, braking and acceleration forces) and hence gives best handling. A high tyre discomfort value (i.e. a numerical value significant relative to 1.0) causes significant deterioration of the tyre mean shear force capability and therefore of the vehicle cornering and handling capability.

In Figure 3.3.2 three dampers were shown. However the tyre damping is generally very small, to the point of being negligible for passenger cars. Far and away the dominant damping comes from the suspension damper.

It has been suggested that the RMS (root mean square) measure of body acceleration undervalues the discomfort of the occasional more severe bump, and that a better mean value is the RMQ (root mean quartic):

$$A_{RMQ} = \left\{ \frac{1}{T} \int_0^T A^4(t) dt \right\}^{1/4}$$

Again, this draws attention to the point that passenger discomfort is subjective, and opinions vary.

The above brief review of ride analysis gives a small indication of the difficult task facing the ride and suspension engineer, and of the important rôle of the suspension damper. In time-domain analysis, the function of the damper can be seen to be to damp out the free response to a given bump, which would otherwise result in a persistent oscillation. In time-domain response to forced vibration, the job of the damper is to limit the response when the forcing frequency is equal to or near to the resonant frequency. In the case of random stimulation, the response can be thought of either as continually forced, or as the sum of the remnant free response from all the previous stimuli. In the latter case the damper is clearly limiting a build-up of response by disposing of the older responses.

In time-domain analysis, the effect of the damper is to influence the responses in a broad way, and the optimisation of a damper will be performed by investigating the vehicle behaviour for various damper values (including probably asymmetries and possibly nonlinearities) and studying the resulting passenger discomfort and tyre discomfort values, preferably as graphs against damper parameters. Ultimately, the dampers are specified by road testing, but computer simulations are relatively quick and easy to perform, and permit ready variation of the damper, and other, specifications, so the suspension engineer can learn a great deal about the influence of the damper on the vehicle behaviour — quickly and in relative comfort! Such theoretical results are, of course, always subject to confirmation by road testing.

3.5 Frequency-Domain Ride Analysis

Frequency-domain analysis considers the behaviour of the vehicle in terms of its response at any given frequency of stimulus. Such analysis therefore produces results such as graphs of the transmissibility against frequency.

For a simple linear system the behaviour at a given frequency may be obtained analytically. For more complex models, or with nonlinear components, it may be obtained by time-stepping analysis at each of a series of frequencies of sinusoidal stimulus or, more efficiently in general, by a broad spectrum stimulus, with the response being spectrally analysed, the transmissibility then being the amplitude ratio of response to stimulus, for each frequency of interest through the spectrum. This analysis may be applied to a simple model such as that in Figure 3.3.2, or to more complex ones.

The stimulating road, then, may be a sinusoidal one, with various spatial frequencies taken in turn, or a stochastic one, being either an actual road profile trace, or, more likely, a profile generated by inverse Fourier transform from a spectral density definition such as the standard ISO road.

The vibration review of Chapter 2 showed some results for simple systems, such as the forced vibration of the 1-dof system as in Figure 2.4.1 (undamped) and 2.5.1 (damped). These illustrate the nature of such curves: the enhanced responses at certain natural frequencies — the resonances.

Figure 3.5.1 shows a possible passenger transmissibility T_P (i.e. for the road to the passenger) in the model of Figure 3.3.2. There is a resonance of the vehicle body on the suspension at about 1–1.5 Hz. There is also a resonance for the passenger on the seat at about 3 Hz, rather exaggerated in the figure. There would also be another one for wheel hop at around 10 Hz. These enhanced resonant responses are a regrettable side effect of using springs to obtain a desirable small transmissibility at high frequencies. As seen in Chapter 2, the function of the damper in this context is to limit the resonant response, although this tends to make the transmissibility greater at higher frequencies, beyond $\sqrt{2}$ times the particular natural frequency.

For a fairly simple model, provided that the individual resonant modes are fairly independent, then they can be analysed separately. This is often possible for a vehicle. The vibration of the passenger on the seat, of the body on the suspension and of wheel hop can be analysed with useful accuracy quite separately. The basic analysis of resonant frequencies and peak responses can therefore be made quite well with 1-dof analysis as reviewed in Chapter 2, (1) for the vehicle body on four springs and dampers, (2) for the passenger on the seat cushion on a given vehicle body motion, and (3) for wheel hop with the

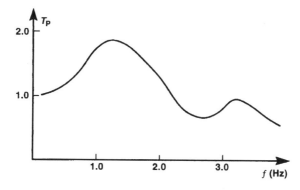

Figure 3.5.1 Hypothetical transmissibility factor from road to passenger, showing frequency dependence including resonances.

vehicle body stationary or in predetermined motion. The passenger resonant frequency on the seat is analysed in the next section. The wheel hop frequency and peak transmissibility are considered in a subsequent section.

The basic character of the passenger transmissibility–frequency curve follows from these natural frequencies and their peak values.

In frequency-domain ride analysis then, the damper is provided primarily to control the amplitude of resonant responses, and the value of damping coefficient for optimum ride is based primarily on this consideration. The suspension dampers are sized primarily to control the motion of the body (sprung mass), which can generally be considered in isolation of the passenger-on-seat or wheel hop behaviour.

3.6 Passenger on Seat

In a normal passenger car, the seat suspension is provided by the structure of the seat cushion. The effective stiffness and damping are then not readily analysed, because they derive from a complex seat structure, not from simple direct-acting components. In some commercial and other large vehicles, the

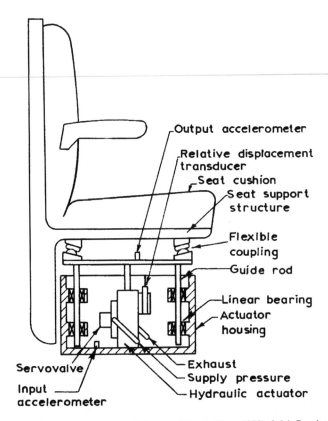

Figure 3.6.1 Advanced vibration-isolating seat (Calcaterra,1972, in Hunt, 1979). 3.6.1. Reprinted from *Journal of Sound and Vibration, 6,* Calcaterra, Active vibration isolation for aircraft seating, pp.18–23, Copyright 1972, with permission from Elsevier.

seat has a support mechanism with a spring and damper. In that case the parameter values are clearer, but note that a significant seat mass must then be added to that of the passenger. In some commercial vehicles, and also in some aircraft, a more complex and expensive seat arrangement is justified, with springs and dampers, as in Figure 3.6.1.

Because the mass of the passenger is much less than that of the vehicle body, this vibration can be analysed by the approximation that the vehicle body is fixed in position or moving in a predetermined way. Really the passenger-on-seat mode of vibration would involve some motion of the rest of the vehicle, having some, usually small, effect on the natural frequency and damping of the mode. One driver or passenger is about 70 kg, some 4–5% of the vehicle body mass, so it is understandable that the body motion would be relatively small i.e. the vehicle body is seismic in this analysis. If four passengers, of total mass around 20% of that of the body, were to vibrate synchronously then the approximation may not be very good. With a fixed body, the passenger-on-seat vertical vibration becomes a simple 1-dof analysis as in Chapter 2. Hence the required seat stiffness for a given natural frequency can be calculated, and the necessary minimum damping coefficient to control resonance (damping ratio around 0.2) can be evaluated from:

$$f_{NP} = \frac{1}{2\pi}\sqrt{\frac{K_C}{m_P}}$$

$$\zeta_P = \frac{C_C}{2\sqrt{m_P K_C}}$$

Alternatively, when a value for the passenger mass is known, and the seat stiffness and damping coefficient are known, then the natural frequency and peak response follow easily by a 1-dof analysis as in Chapter 2. Adult passenger masses from 50–100 kg (or even more, especially US) give a frequency ratio of 0.7 over the mass range on a given seat stiffness.

3.7 Wheel Hop

The unsprung mass is generally substantially less than that of the body, so a useful wheel hop analysis for independent suspension may be made by considering the body to be fixed. For a wheel displacement, the total restoring stiffness is $K_T + K_W$, where K_T is usually much larger than K_W (although not for ground-effect racing cars). Hence the approximate equation of motion for free vibration of the wheel is

$$m_U \ddot{z}_W + (C_W + C_T)\dot{z}_W + (K_W + K_T)z_W = 0$$

The undamped wheel hop natural frequency is then

$$f_{NWH} = \frac{1}{2\pi}\sqrt{\frac{K_W + K_T}{m_U}}$$

For a passenger car, example values for one suspension might be

$$K_W = 28 \text{ kN/m}$$
$$K_T = 200 \text{ kN/m}$$
$$m_U = 50 \text{ kg}$$
$$f_{NWH} = 10.7 \text{ Hz}$$

Hence, the wheel hop natural frequency is about eight times the body heave natural frequency, largely governed by the tyre vertical stiffness, unless the wheel leaves the ground, in which case the natural frequency drops dramatically. The wheel hop damping ratio ζ_{WH} is given by

$$\zeta_{WH} = \frac{C_W + C_T}{2\sqrt{m_U(K_W + K_T)}}$$

Example values might be

$$C_W = 1200 \, \text{Ns/m}$$
$$C_T = 100 \, \text{Ns/m}$$
$$\zeta_{WH} = 0.19$$

Making a direct comparison of the wheel hop and body heave damping ratios,

$$\frac{\zeta_{WH}}{\zeta_H} = \left(\frac{C_W + C_T}{C_W}\right)\left(\frac{K_W}{K_W + K_T}\right)^{1/2}\left(\frac{m_B}{m_U}\right)^{1/2} \approx \left(\frac{K_W m_B}{K_T m_U}\right)^{1/2}$$

The mass ratio m_B/m_U is around 4–6, depending upon the type of suspension. The stiffness ratio K_W/K_T is around 0.1 for a normal car (much higher for some racing cars). Hence, for a passenger car $\zeta_{WH}/\zeta_H \approx 0.6$, and the wheel hop is less well damped than the body, and a little more prone to resonance. Fortunately the natural frequency is higher, so, from the spectral analysis of real roads, the stimulus at the wheel hop frequency is normally much less. For a racing car the ratio of modal damping factors tends to be higher, perhaps 1.4.

One other possible stimulus of wheel hop is an unbalanced wheel. For wheel diameter D the stimulus frequency is $f = V/\pi D$. At $V = 20 \, \text{m/s}$ and a wheel diameter of 0.6 m the frequency is 10.6 Hz, so the resonance will be struck at a normal speed of travel. The temporary fix is to adjust the speed.

3.8 Handling

Handling is the quality of a vehicle enabling it to be controlled by the driver in a safe and predictable manner, so that it is easy to maintain a desired course, and easy to control the vehicle at high longitudinal and lateral accelerations, should this be needed. Dampers undoubtedly play an important rôle in this, partly because badly controlled ride motions cause severe problems during cornering or braking, but also because the variations of acceleration cause pitch and roll angles to develop, and this must occur in a controlled way.

The discussion of handling is divided into three sections:

(1) straight-line handling;
(2) pitch vibration;
(3) roll vibration.

Straight-Line Handling

At high forward speeds, even straight-line handling may become problematic due to ride motions of the body, particularly at its natural frequency, stimulated by the road. At high speed, the resonant body-on-suspension frequencies correspond to longer road wavelengths and lower spatial frequencies n_S, because at vehicle speed V the stimulus frequency is

$$f = \frac{V}{\lambda} = Vn_S$$

From the road spectral density of Figures 3.2.1–3.2.3, these lower spatial frequencies have a much higher spectral density. For example, with the standard road spectral model gradient exponent -2.7, a doubling of the speed gives 6.5 times the spectral density at the resonant frequency. Hence as speed increases on a given road it becomes more difficult to control the body resonances in pitch, heave and roll, and the damping coefficient needs to be greater for adequate control and safety. Also, at higher speed the body response at a given spatial frequency has a higher chronological frequency with an ω^2 factor in the resulting body vertical accelerations, with greater discomfort.

Also, for a given speed, greater suspension stiffness raises the natural frequency and raises the corresponding spatial frequency, slightly reducing the stimulus at resonance.

Pitch Vibration

On application of the accelerator or brakes, there is a longitudinal load transfer moment

$$m_{TX} = mA_X H_G$$

causing a sudden change of equilibrium pitch angle

$$\theta_{FTX} = \frac{m_{TX}}{K_P} = k_\theta A_X$$

where K_P is the pitch stiffness (Section 3.2) and k_θ is the pitch gradient (variation of 'equilibrium' pitch angle with longitudinal acceleration). In the absence of pitch damping, the result would be a pitch oscillation of amplitude θ_{FTX}, the pitch angle varying from zero to $2\theta_{FTX}$, persisting in oscillation at the pitch natural frequency. This alone would make any vehicle virtually undriveable even on a smooth road. Pitch damping must be provided to give a smooth development of pitch angle, preferably with little or no overshoot. This damping is provided exclusively by the suspension dampers. From the free motion curves of Figure 2.3.1, a pitch damping ratio of 0.7 would be desirable. This sort of value is achieved on some well-damped sports cars and racing cars, but values are smaller on most passenger cars in consideration of ride quality.

Roll Vibration

Initiation of a turn similarly creates a change of equilibrium roll angle. The lateral load transfer moment is approximately

$$M_{Ty} = mA_y H_G$$

with a change of equilibrium roll angle

$$\phi_{FTy} = \frac{M_{Ty}}{K_R} = k_\phi A_y$$

where K_R is the roll stiffness and k_ϕ is the roll angle gradient (variation of roll angle with lateral acceleration). Again, if undamped this would result in a persistent roll oscillation of amplitude

ϕ_{FTy}, with a maximum value of 2 ϕ_{FTy}, and make the vehicle undriveable. For handling purposes, the ideal roll damping ratio would be around 0.8–1.0.

A fuller understanding of roll in handling requires consideration of yaw and sideslip behaviour, because these three aspects are closely coupled. However, without more detailed analysis, it is clear that for good handling the roll motion needs to be adequately damped, and that the suspension dampers are an important factor in this.

Another aspect of the effect of roll is that of the roll velocity at corner entry or exit as the vehicle changes its roll posture. Some lateral load transfer occurs through the dampers during this process, so the front/rear distribution of lateral load transfer in corner transients can be affected by the dampers. This is considered in the setting-up of racing cars. In addition to the usual steady-state front/rear lateral load transfer moment distribution factor there is therefore a front/rear roll velocity lateral load transfer moment distribution factor.

3.9 Axle Vibrations

Live axles, and even De Dion axles, are prone to certain vibrations. These may be classified by the six individual degrees of freedom. The actual vibrations are in modes, with some combination of the simple cases, but one of the basic six degrees of freedom is typically dominant. The fundamental six single degrees of freedom are:

(1) Heave. Axle hop remaining substantially parallel to the ground, against spring and tyre stiffness and suspension damping, and therefore similar to independent suspension wheel hop, but of lower frequency because of the additional axle mass, and therefore subject to greater road stimulus.
(2) Surge. Fore-and-aft vibration without angle change against link and bush longitudinal compliance.
(3) Sway. Lateral vibration without angle change, against link and bush lateral compliance.
(4) Yaw. Angular yaw vibration against longitudinal link and bush compliance.
(5) Roll. Angular axle hop with one wheel up whilst the other is down, against suspension and tyre stiffness and damping.
(6) Pitch. Angular pitching against link and bush compliance.

For example, tramp of a live axle is predominantly a roll motion of the axle with some yaw action, stimulated by tractive forces and involving tyre characteristics.

Analysis of the above vibrations requires data on the particular stiffnesses and damping coefficients. Where the motion is against the link compliances, this requires detailed knowledge of the particular suspension layout, including rubber bush stiffnesses. Axles with link location and coil springs rather than those with location and springing from beam springs (leaf springs, e.g. Hotchkiss axle) are generally less troublesome because the location is more precise, stiffness greater and natural frequency higher so less stimulated, and rubber bushes have much more inherent damping than metal.

The suspension dampers can be used to damp axle sway vibrations by tilting them in front or rear view, at θ from the vertical, giving a damper speed $V_A \sin \theta$ for an axle sway speed V_A.

The axle pitch vibration is potentially troublesome in the case of longitudinal beam springs (leaf springs). Consider a leaf spring providing suspension spring stiffness wheel rate K_W, with the beam held rigidly on the axle, removed from the body. The independent stiffness of each end of the beam, considered symmetrical, is ½K_W. For a beam of total length l, an axle pitch angle will deflect the beam ends by ½$l\theta$, Figure 3.9.1. Each end exerts a force

$$F = \tfrac{1}{2}l\theta \tfrac{1}{2}K_W = \tfrac{1}{4}lK_W\theta$$

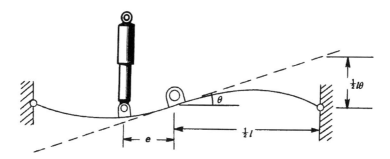

Figure 3.9.1 Axle pitch displacement with longitudinal leaf springs.

The restoring moment on the axle, for two such beam springs, one or each side, is

$$M = 4F\tfrac{1}{2}l = \tfrac{1}{2}l^2 K_W \theta$$

Hence, the axle positional pitch stiffness k_{AP} is

$$k_{AP} = \tfrac{1}{2}l^2 K_W$$

The pitch inertia of the complete axle resides mainly in the wheels and tyres because although the axle body is massive it is close to the pitching axis. For a rigid axle, the wheels are coupled to the axle body by the half-shafts and differential, which are not perfectly rigid. The springs also contribute some pitch inertia. The result is some axle pitch inertia I_{AP}.

The pitch radian natural frequency of the axle ω_{NAP} is

$$\omega_{NAP} = \sqrt{\frac{k_{AP}}{I_{AP}}}$$

Realistic values would be around 3000 Nm/rad and 3.5 kg m² giving an ω_{NAP} value of 29.3 rad/s and

$$f_{NAP} \approx 5\,\text{Hz}$$

which is a fairly low value that is likely to be activated.

The problem is exacerbated by the poor damping. One possible palliative is to offset the normal suspension dampers from the axle fore or aft by a distance e, possibly one forwards and one backwards. Then, at axle pitch angular velocity $\dot{\theta}$ relative to the body, the damper force (two dampers) is $2eC_D\dot{\theta}$ with a damping moment about the pitch axis of

$$M_D = 2e^2 C_D \dot{\theta}$$

By writing the full equation of pitch motion, the axle pitch damping ratio ζ_P may be found to be

$$\zeta_P = \frac{2e^2 C_D}{l}\sqrt{\frac{I_A}{K_W}}$$

or, for design purposes, to achieve a desired value of damping ratio, the offset required e is given by

$$e^2 = \frac{l\zeta_P}{2C_D}\sqrt{\frac{K_W}{I_A}}$$

With realistic values, it may be seen that an offset of 50 mm can be expected to be significant, and 100 mm will give considerable damping of this undesired behaviour.

3.10 Steering Vibrations

Steering vibrations are a complex and specialised area of study, highly dependent upon some esoteric tyre characteristics. Therefore no detailed analysis will be attempted here. Den Hartog (1985) gives an introduction.

For independent suspension, the road wheels simply pivot about the kingpin (steering) axis. On the assumption that the steering hand wheel is held firmly, which may in fact be difficult if a severe steering vibration develops, the main compliance is that of the steering column, and the main inertia is that of the road wheels about the kingpin axis. Friction of various kinds is present in the system including the ball joints which are pre-loaded to give a desirable small Coulomb friction to discourage such problems. The difficulties may be considered to stem from an effective negative damping in the tyre due to its motion against the road, coupled with flexibility of the kingpin axis fixture. To discourage steering vibration, some vehicles have a steering damper, often laid out horizontally behind the steering rack.

Steering vibrations were very problematic in the early days of motoring, with rigid front axles, because of interaction of the tyre characteristics with gross yaw and roll motions of the complete axle and wheel gyroscopic action. The adoption of front independent suspension improved the situation considerably. These problems have been analysed in detail, not least in the context of aircraft landing gear. With improved understanding and design, nowadays road vehicles are largely free of spontaneous steering vibration. Where hand wheel vibrations are felt, they are nearly always due to poor wheel balance. Where the two front wheels have an equal imbalance, then the slightly different effective rolling radius of the two wheels results in the vibrations beating with a cycle of ten seconds or more. The solution is not more steering damping but improved wheel balance. Some front suspension designs are much more sensitive to poor wheel balance than others. Double wishbone suspensions have a better reputation in this respect than do strut suspensions.

3.11 The Ride–Handling Compromise

The quality of ride and handling of a vehicle is certainly influenced by many factors, including the springs and dampers. In the case of ride, too soft a spring will give a very low natural frequency which leads to passenger travel sickness. Too stiff a spring with a high frequency gives too high a transmissibility of higher frequencies. For handling, a soft spring allows excessive pitch and roll angles in acceleration, whilst a very stiff spring does not permit the wheel to move adequately relative to the body and conform to the road shape, so the tyre grip becomes worse.

Considering some sort of quality rating Q_R for ride and Q_H for handling, plotting these against spring stiffness, specific stiffness, or natural frequency, gives Figure 3.11.1. The optimum stiffness for ride is less than that for optimum handling. Hence there is evidently a compromise to be made; the well-known ride–handling compromise. Practical values of stiffness will lie between K_{R1}, the best for ride, and K_{H1}, the best for handling, depending upon the type of vehicle. Plotting Q_R against Q_H gives a polar plot along the lines of Figure 3.11.2. Here it is clear that around the best ride, with stiffness K_{R1}, an increase of stiffness will give a substantial (first-order) increase of handling with little loss of ride quality (second-order). Near to optimum handling, with stiffness K_{H1}, a reduction of stiffness gives substantial improvement of ride with little deterioration of handling. Hence, for passenger vehicles the practical range of choice lies within a range somewhat less than K_{R1} to K_{H1}.

Realistic values of natural frequencies range from around 1 Hz for large passenger cars ($k_{SS} = 40$ N m^{-1}/kg) to around 1.5 Hz for small passenger cars (90 Nm^{-1}/kg), and even more for sports cars,

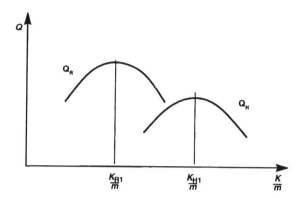

Figure 3.11.1 The influence of suspension stiffness on the quality of ride and handling.

2 Hz or even 2.5 Hz (160 to 250 Nm^{-1}/kg). Racing cars may have even higher values, especially where they have ground-effect aerodynamic downforce; such cars are very critical on ride height and need a very stiff suspension to control this, with natural heave frequencies of 5 to 6 Hz, possibly 1600 Nm^{-1}/kg ride stiffness, although at such a high ride stiffness the tyre stiffness is very important factor, and must certainly be included in frequency calculations. In such cases, at high speed the aerodynamic stiffness must be taken into account. The front wing in particular is near to the ground and small changes of ride height have a large effect on the air flow. The result may be a large negative stiffness contribution from the aerodynamics.

Required values of the damping ratio vary in a somewhat similar way to stiffness. Optimum values depend very much upon particular conditions, and especially on personal preferences, but the ride of a passenger car will generally be best at a damping ratio around 0.2, and the best handling may require an average damping ratio around 0.8. Hence values chosen in practice are likely to be in the range 0.25–0.75, Figure 3.11.3. For example, in a study of variable damping for a small passenger car, using a simple heave model, Sugasawa *et al.* (1985) found theoretically a damping ratio of 0.17 to be the ride optimum (minimum spectral energy of body heave motion) and a ratio of 0.45 the optimum for road

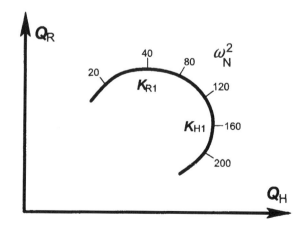

Figure 3.11.2 The ride–handling quality loop with varying stiffness.

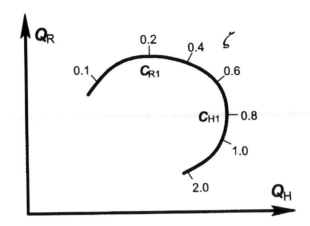

Figure 3.11.3 The ride–handling quality loop with varying damping.

holding (minimum tyre force variation). In more detail, the value found for optimum ride in ordinary driving was 0.16, a value of 0.43 to minimise 'bouncy feel', a value of 0.44 for road holding on rough roads, and a value of 0.71 for roll and pitch minimisation with control inputs. The analytical model for this did not include tyre stiffness.

Given values for the vehicle body mass and heave frequency, the total suspension heave stiffness is easily calculated, and given a desired heave damping ratio then the required total heave damping coefficient is easily obtained by 1-dof analysis as covered in Chapter 2.

The above variations of specific stiffness and damping ratio are not independent; the low damping ratio will go with the low stiffness for a vehicle having the emphasis very much on ride. The damping coefficient C is

$$C = 2\zeta\sqrt{mK}$$

so the specific damping coefficient required, c_{SD}, is

$$c_{SD} = \frac{C}{m} = 2\zeta\sqrt{\frac{K}{m}} = 2\zeta\omega_{NH} = 4\pi\zeta f_{NH}$$

For the fully ride-optimised case at say 0.8 Hz with $\zeta = 0.2$, this gives 2 s^{-1} [= N/m s^{-1}/kg]. At a handling-optimised $\zeta = 0.8$ on 1.6 Hz, the specific damping coefficient is the very much higher value of 16 s^{-1}. Hence there is a tremendous variation in the damping coefficients, arising from the product of the ranges of frequency (1:2) and damping ratio (1:4).

In practice, the asymmetry of damping may also vary. At the ride optimisation this may be distributed 20/80 bump/rebound, whereas at the handling optimum this may have shifted to a more equal 40/60.

It is of interest to try to define a suspension ride–handling parameter f_{SRH} which expresses the choice of springs and dampers on a single scale. One candidate would be the product of specific heave stiffness and specific heave damping coefficient. In fact it is better to take the cube root of this product, to produce a more convenient scale and to give it simple units of s^{-1}.

Considering the 1-dof heave equation

$$\ddot{x} + \frac{C}{m}\dot{x} + \frac{K}{m}x = 0$$

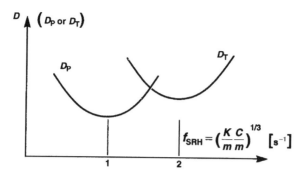

Figure 3.11.4 The effect of the ride-handling parameter on passenger and tyre discomfort.

then

$$\ddot{x} + 2\alpha\dot{x} + \omega_{NH}^2 x = 0$$

$$\ddot{x} + 2\zeta\omega_{NH}\dot{x} + \omega_{NH}^2 x = 0$$

then the suspension ride-handling parameter f_{SRH} may be expressed as

$$f_{SRH} = \sqrt[3]{\frac{KC}{m^2}} = \omega_{NH}\sqrt[3]{2\zeta}$$

On such a scale, normal vehicles fall in the range 1–2 s^{-1}. Higher values occur for racing cars, especially with ground-effect.

From a practical point of view, it is common to work not with Q, a measure of quality, but with D_P, the measure of discomfort (r.m.s. passenger acceleration, an approximate inverse of Q) nondimensionalised as A_{rms}/g. This is also the r.m.s. variation of vertical force divided by the mean supporting force for the passenger. The handling poorness may be based approximately on tyre discomfort D_T (the variation $(F_V - F_{Vm})_{rms}/F_{Vm}$). Plotting these against f_{SRH} gives curves qualitatively such as those of Figure 3.11.4. As a specific example, Figure 3.11.5 shows the variation of vehicle body acceleration and tyre deflection against damping coefficient found in one study.

The effect of the spring and damper values may be combined into one loop, the polar plot of Figure 3.11.6, the passenger–tyre discomfort loop. This shows the passenger discomfort on the y-axis, against the tyre discomfort on the x-axis, both parameters being nondimensional. As the parameter values are individually desirable but must be compromised one against the other, this is also sometimes known as a conflict diagram.

Table 3.11.1 Ride–handling parameter values

Vehicle	f_{SRH}
Ride optimised	1.0
Passenger car	1.2
Sports car	1.5
Racing car	2.0
Ground-effect racing car	3.0

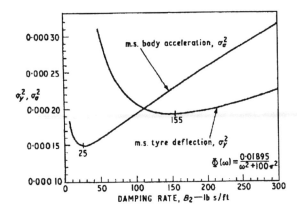

Figure 3.11.5 Body acceleration and tyre deflection variation against damping coefficient as seen at the wheel (Thompson, 1969).

The passenger and tyre discomfort values can be investigated experimentally, at significant expense, or by a time-stepping simulation, allowing some synchronised variation of stiffness and damping, from which the loop can be constructed. A basic ride simulation giving the tyre discomfort is not a complete representation of the effect of the suspension on handling, of course, because it does not include the effect of stiffer springs and suspension in reducing body pitch and roll, but it is a useful guide to trends. This aspect may be improved by incorporating into the ride simulation variations of longitudinal acceleration to stimulate gross pitch motions so that the ability of the suspension to control these is tested. For a more complex model, lateral acceleration variations may also be included.

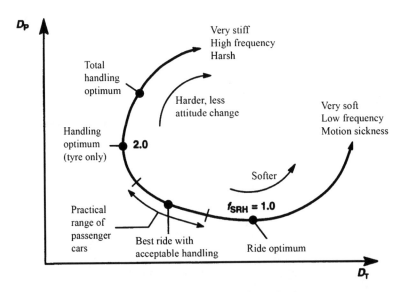

Figure 3.11.6 The passenger–tyre discomfort loop.

3.12 Damper Optimisation

The basic damper as a linear device has a damping coefficient. The approximate values for this follow from simple analysis of the whole vehicle and selected damping ratios. These desirable ratios are low for comfort, medium for straight driving on rough roads, and high for best handling. This creates the ride–handling compromise. A study by Fukushima, Hidaka and Iwata (1983), Figures 3.12.1–3.12.3, showed that having the damper force as a function of velocity only was a serious limitation, and that variation with stroke was highly desirable. Figure 3.12.2(a) shows that, with a conventional damper, passing over solitary bumps requires low damping but receives high damping whilst steering manoeuvres require high damping (e.g. in roll) but are given low damping. Figure 3.12.2(b) shows that categorising the requirements by stroke successfully separates the cases. Further detailed studies

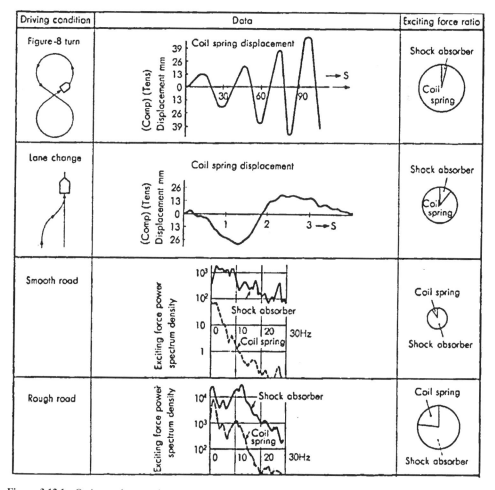

Figure 3.12.1 Optimum damper characteristics under various driving conditions and road surfaces; (a) figure-8 turns at increasing speed; (b) lane change; (c) straight driving on a smooth road; (d) straight driving on a rough road. Reproduced from Fukushima, Hidaka and Iwata (1983) Optimum characteristics of automotive shock absorbers under various driving conditions and road surfaces, *JSAE Review*, pp.62–69.

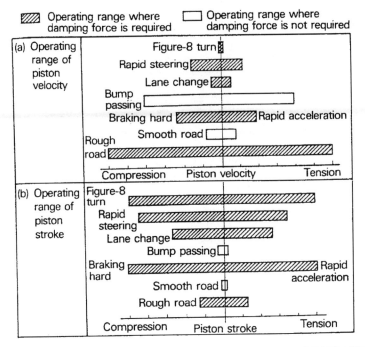

Figure 3.12.2 Damper operating range analysed by: (a) damper velocity; (b) damper stroke. Reproduced from Fukushima, Hidaka and Iwata (1983) Optimum characteristics of automotive shock absorbers under various driving conditions and road surfaces, *JSAE Review*, pp.62–69.

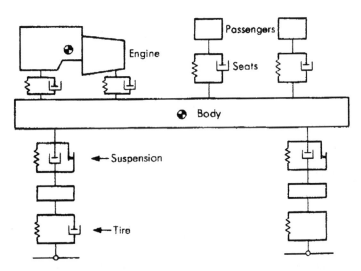

Figure 3.12.3 Simulation model used in velocity and stroke damper studies. Reproduced from Fukushima, Hidaka and Iwata (1983) Optimum characteristics of automotive shock absorbers under various driving conditions and road surfaces, *JSAE Review*, pp.62–69.

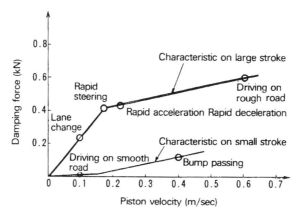

Figure 3.12.4 Optimum damper characteristics. Reproduced from Fukushima, Hidaka and Iwata (1983) Optimum characteristics of automotive shock absorbers under various driving conditions and road surfaces, *JSAE Review*, pp.62–69.

were made using the simulation model of Figure 3.12.3. As a result, the stroke sensitive vortex valve was introduced (Fukushima *et al.* 1984).

Figure 3.12.4 shows the optimum form of damper curve according to velocity and stroke as found by this study, where the short and long strokes are seen to require considerable differences of force, which is difficult to achieve.

The stroke-sensitive damper now has some currency in other guises, as position-dependent (long tapered rod in an orifice, or grooved pressure cylinder) or so-called frequency dependent. A fully controllable damper, such as magnetorheological ones, are of course capable of producing the required behaviour, but at a significant extra cost. Therefore passive hydraulic dampers with stroke sensitivity and similar refinements remain an attractive option for cost-conscious vehicle manufacturers.

3.13 Damper Asymmetry

The compression and extension forces exerted by a damper at a given speed in or out are, as an empirical fact of life, highly unequal. Typically the extension force is three to four times the compression force. In terms of the mean force F_m, the extension and compression forces F_E and F_C at a given speed magnitude are

$$F_E = (1 + e_D)F_m$$
$$F_C = (1 - e_D)F_m$$

where e_D is the damper force transfer factor for that particular speed. Given the actual forces F_E and F_C, the mean force F_m and the transfer factor e_D may be calculated from

$$F_m = \frac{F_E + F_C}{2}$$
$$e_D = \frac{F_E - F_C}{2F_m}$$

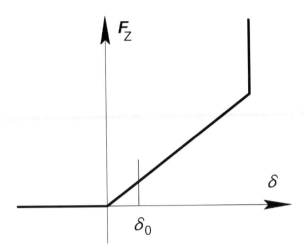

Figure 3.13.1 Tyre vertical force against tyre vertical deflection, for constant stiffness.

The transfer factor (asymmetry factor) varies considerably with particular dampers and with the operating speed, but is typically 0.5–0.6. In general, the value is positive (greater extension damping), but $e_D = -1$ would be pure compression damping. A value of $e_D = 0$ is symmetrical damping, and $e_D = +1$ would be pure extension damping. On passenger vehicles, the rear dampers in particular may have great asymmetry. There is some indication that competition dampers, particularly in racing, may have less asymmetry. One possible explanation is that the greater bump damping, although less comfortable, gives a better road feel to the driver. This has also been found on competition motorcycles.

The scientific research literature is devoid of a good explanation of the asymmetry of dampers. The usual theoretical investigations of ride quality are symmetrical, so they are inherently unable to provide an explanation. However, the actual asymmetry has a very long history, going back to the original mechanical snubbers, which acted purely in extension, presumably for simplicity. Advertisements for early 'shock absorbers' included claims that the vehicle would be prevented from leaping out of potholes, perhaps because the wheel was prevented from drooping fully, perhaps because the snubber actually restrained the body upward movement.

The author has questioned various vehicle dynamicists informally on this point, and received less-than-convincing replies. Generally, there is a belief that there is a simple explanation, but this was not actually forthcoming. A typical explanation is that the problem is asymmetrical because 'gravity acts downwards', but the exact implications of this were not offered, although it is true that the body cannot accelerate downwards at more than 1 g. Admittedly this is an asymmetry, but in the usual ride quality studies this does not appear, because all ride motions are about the equilibrium position, and of limited amplitude. Suspension asymmetries such as rising rate springs, bump stops or droop stops, and damper asymmetry, are design consequences to be explained, they are not the cause. When the road is so rough and the suspension travel is so great that the workspace is exhausted, it could be exhausted equally at both ends, and the bump and droop stops could remain symmetrical.

An actual asymmetry arises naturally when the tyre normal force becomes significantly nonlinear, in effect when the wheel leaves the ground or when the bump is so severe that the tyre is fully compressed and the wheel rim is impacted. In the former case, at this point it does indeed become a significant factor that gravity acts downwards. However, a rim impact event is unusually severe, and such events are not the basis of normal ride or handling optimisation. Figure 3.13.1 shows a representative function for tyre vertical force against tyre deflection. Evidently, tyre deflection amplitudes exceeding the mean deflection are required to introduce asymmetry.

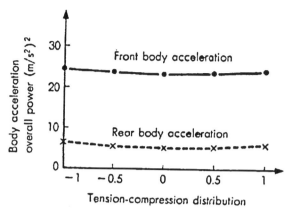

Figure 3.13.2 The effect of damper force transfer factor e_D on vehicle body vertical acceleration when driving on a rough road, simulation results. Reproduced from Fukushima, Hidaka and Iwata (1983) Optimum characteristics of automotive shock absorbers under various driving conditions and road surfaces, *JSAE Review*, pp.62–69.

The total available range is about 70 mm, roughly in the ratio 1 extension to 4 compression, i.e. 14 mm extension before the tyre leaves the ground, but 56 mm compression before the tyre is crushed against the rim. Even within the linear stiffness model, the extension range of 10–15 mm acts as an asymmetrical limit. Hence, continuous short wave sinusoids of amplitude 10 mm, or ramp-steps up or down of that size, are symmetrical, of 15 mm are slightly asymmetrical and of 20 mm or more are highly asymmetrical. Relative to the mean tyre vertical force, the greatest downward force is only about 1/4 of the greatest possible upward force. This seems to offer some justification for relieving the damper compression force at high velocities, but not at low ones. This also suggests that perhaps long damper compression strokes should be pressure relieved, contrary to present thinking about stroke sensitive damping.

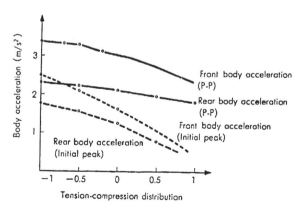

Figure 3.13.3 The effect of damper force transfer factor e_D on vehicle body vertical acceleration when driving over a solitary positive bump, simulation results. Reproduced from Fukushima, Hidaka and Iwata (1983) Optimum characteristics of automotive shock absorbers under various driving conditions and road surfaces, *JSAE Review*, pp.62–69.

Another possible factor is that road roughness itself is in fact not symmetrical, contrary to the usual road spectral analysis methods. Sometimes it is claimed that there are more bad holes than bad bumps. If this were indeed true, which is possible but unproven, then it may be a factor. The opposite explanation may also be offered–there are more bad bumps, so damper compression relief is necessary to ride them, but, supposedly, weak extension damping is not required in order to ride the bad holes. In any case, this would only justify asymmetry at high damper speed, not at normal speed.

The tyre enveloping characteristic is said to be a factor here. The tyre can ride over a short hole to some extent, without necessarily maintaining contact with the bottom of the hole, but a short bump must penetrate the tyre profile fully. However, the tyre need not make contact with the full profile of a bump, it may miss the internal corners. Also, the argument would only apply to short steep bumps and holes.

Hypothetically, a further possible factor is that passenger sensitivity is not symmetrical, and that sharp upward and downward accelerations are not equally uncomfortable.

The study by Fukushima *et al.* (1983), addressing desirable damper characteristics, dealt with this to some extent, Figures 3.13.2 and 3.13.3. It is notable that for driving on a rough road they found no favourable effect for asymmetry. For driving over an upward bump they found, understandably, that asymmetry was desirable. They gave no result for a trough (negative bump).

To the author, the explanations given above, are tentative at best. It may be felt that there must be a physical explanation, but perhaps, not. The explanation may be being sought in the wrong place. It may simply be that asymmetry is more convenient for the damper manufacturer, or, to put it bluntly, it is cheaper to provide extension damping than to provide compression damping. This is basically because of the constructional characteristics of the telescopic hydraulic damper, which has more difficulty in providing large compression forces without risking cavitation.

A definite disadvantage of damper asymmetry is the resulting damper jacking. This is analysed in Section 7.15.

It appears, then, that the usual ride quality studies on low-amplitude random roads can seek to explain and to optimise the mean damping coefficient, but that the asymmetry factor must be sought elsewhere, amongst:

(1) tyre range asymmetry on large deflections;
(2) road bump/trough asymmetry;
(3) passenger sensitivity asymmetry;
(4) damper internal construction asymmetry;
(5) tyre enveloping asymmetry;
(6) manufacturing costs.

Given a putative explanation (e.g. there are more holes than bumps), one must ask whether the removal of such a cause would result in future dampers being made symmetrical.

4
Installation

4.1 Introduction

The analysis of the vehicle presented in previous chapters, for example the heave-and-pitch motion, was based on the damping coefficient C_W effective at the wheel. However the damper itself is not installed at this point, but elsewhere on the suspension. Often, for a car, it operates on the bottom suspension arm. Sometimes it operates through a linkage, including various forms of rocker, especially on racing cars and motorcycles. This chapter considers how to relate the damping characteristic at the wheel to the characteristics of the damper itself. This involves:

(1) evaluation of the relative motion of the wheel and damper;
(2) consideration of the implications of this ratio.

The following sections therefore define the motion ratio, consider its consequences, and explain methods of evaluation for various kinds of suspension. An understanding of the analytical methods of analysis, such as the drawing of velocity diagrams, makes it possible to write a computer program to solve such problems, an instructing and interesting exercise in itself.

The damper free stroke is S_F. The installed stroke, possibly limited by bump or droop stops, is S_I. The damper stroke utilisation is

$$U_{DS} = \frac{S_I}{S_F}$$

4.2 Motion Ratio

At a given suspension bump position z_S from normal ride height, the damper compression is z_D. A small further suspension bump motion δz_S results in a corresponding further damper compression δz_D, Figure 4.2.1. The ratio of these is the displacement motion ratio for the damper at the position z_S. This is denoted $R_{D/S}$, (damper relative to suspension bump), abbreviated to R_D.

$$R_{D/S} = \frac{\delta z_D}{\delta z_S}$$

Note that this is the damper increment divided by the suspension bump increment, not *vice versa*. This value of R_D will be independent of z_S and δz_S only if the system is linear, i.e. if R_D is constant, which is generally not true. Therefore in general the above expression is usable only for small δz_S. With

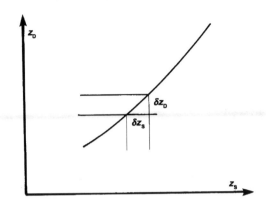

Figure 4.2.1 Damper displacement vs suspension displacement.

a suitable computer program to analyse the suspension geometry, the motion ratio may be evaluated in the above way, with δz_S being given a suitable small value, e.g. 1 mm or 0.1 mm, depending on the precision with which the suspension position can be calculated.

Mathematically, the motion ratio value is really the above ratio as δz_S tends to zero. The result is the derivative

$$R_D = \frac{dz_D}{dz_S}$$

This provides a possible means of evaluation if an explicit algebraic expression is available for z_D (damper compression) as a function of z_S (suspension bump displacement). For example, if it is known that, over the range of interest,

$$Z_D = A_1 Z_S + A_2 Z_S^2$$

then

$$R_D = \frac{dz_D}{dz_S} = A_1 + 2A_2 z_S$$

The derivative definition of R_D may be extended by inserting a time element:

$$R_D = \frac{dz_D}{dt} \times \frac{dt}{dz_S}$$

The two terms on the right-hand side are simply velocities, so R_D may be written as

$$R_D = \frac{V_D}{V_S}$$

The motion ratio is therefore identically equal to the quotient of damper compression velocity over suspension compression (bump) velocity.

The velocity ratio definition provides a means of evaluating R_D if the velocity diagram can be drawn. Frequently this is feasible, even easy, for a two-dimensional set-up, and may also be practical for a three-dimensional suspension layout if the parts can be treated sequentially as two-dimensional parts (e.g. a racing suspension with pushrod and rocker, in many cases). Within the computer, a three-dimensional velocity diagram can be 'drawn'.

Considering the displacement z_P of an intermediate pushrod, the overall motion ratio R_D is

$$R_D = \frac{\delta z_D}{\delta z_P} \times \frac{\delta z_P}{\delta z_S}$$

or, in the limit,

$$R_D = \frac{dz_D}{dz_P} \times \frac{dz_P}{dz_S}$$

$$= R_{D/P} \times R_{P/S}$$

so the overall motion ratio is the product of the sequential ratios. In terms of the velocities V_D, V_P and V_S,

$$R_D = \frac{V_D}{V_P} \times \frac{V_P}{V_S} = \frac{V_D}{V_S}$$

Hence a suspension mechanism can be solved by separate analysis of the sequential parts. This may be especially valuable where a complex suspension can be treated as separate sub-mechanisms, each analysable in two-dimensional motion, and each analysed in the most convenient way for that particular part.

In general the damper velocity ratio is a function of z_S, the bump position of the suspension:

$$R_D \equiv R_D(z_S)$$

For passenger cars the value may be approximately constant or increase slightly with bump, in the latter case usually also reducing with droop. For racing cars a rather rapidly increasing velocity ratio may be used. Where applied to a spring, this gives a rising stiffness with bump, and is called a rising-rate suspension. Applied to a damper, it is rising-rate damping. Obtaining rising rate by mechanism design is generally much easier and more controllable than doing so by manufacturing rising rate springs or position-dependent dampers. Often the spring and damper are fitted co-axially and have the same motion ratio. For rising rate, this conveniently gives the desirable increase of wheel damping coefficient with increasing wheel rate.

4.3 Displacement Method

One way to obtain the motion ratio value for one nominal bump position is by analysis of a pair of slightly different suspension positions. If the position analysis is undertaken by a drawing method it is prone to inaccuracy because of the relatively small difference of positions. Hence the drawing must be undertaken by an experienced draughtsman at a large scale. With less emphasis on accuracy, a wide spread of positions will give an average ratio over the movement which may be useful in some cases. Also a sequence of, say, eight to ten positions may be drawn throughout the bump motion, and the damper compression plotted as a graph against suspension position, with the curve smoothed through the points. This helps to reveal any errors. The motion ratio for any particular position is then the gradient dz_D/dz_S of the curve.

Drawing methods were the traditional technique in bygone days, but are regarded as somewhat archaic nowadays because of the availability of computers. However drawing methods may still be of some value, for example if the particular configuration cannot be handled by existing software. In that case, careful drawing or velocity diagram analysis may be more expedient than writing new software, unless the configuration is known to be one of lasting interest. Computer-aided drawing packages offer potentially enhanced accuracy, of course, and can even be used to solve simple installation ratio problems.

In some cases the position relationships may be found algebraically. In three dimensions the solutions are too unwieldy for hand solution in other than isolated cases, for example to check early results of a computer analysis. However, if the system can be treated in a two-dimensional sequence then an analytical expression for positions may be possible, with differentiation giving the velocity ratio.

On one occasion the author was asked to adjudicate between two commercial computer programs which gave substantially different results. A careful drawing solution clearly sided with one of the two programs, and also suggested the reason for the fault in the other program. Drawing methods may therefore still be of value, even in this computerised day and age.

4.4 Velocity Diagrams

A velocity diagram is a coordinate diagram on which any point (pair of coordinates x, y) corresponds to given velocity components V_X V_Y, drawn to a particular velocity scale, e.g. 10 mm s^{-1}: 1 mm. Hence any point on any component has a velocity which can be plotted on the diagram; in practice this is done for a few special points, essentially those at the joints between components, i.e. at the pivot points and at sliders. The distances between points in the velocity diagram represent velocity magnitudes. The velocity diagram is constructed point-by-point sequentially. It may drawn accurately to a given scale, or, and often more conveniently and more accurately, simply sketched and solved algebraically. Accurate drawing requires some numerical calculation anyway. A full description of the principle of velocity diagram construction appears in standard texts.

4.5 Computer Evaluation

Commercial and proprietary packages are available for analysing suspension geometry on computers. Most of these simply analyse a given configuration. At least one program, by the author, includes design facilities to specify, for example, a given bump steer for which the computer will choose suitable link dimensions.

When such numerical analysis packages work, they generally work well, and accurately. However not all packages can handle all possible configurations, especially for the more esoteric types used in racing. Also the 'assembly logic' of some packages is not always reliable, so that in some configurations there is a sudden total failure, usually manifested by a 'square-root of negative number' error.

A full three-dimensional analysis program is a considerable job to write well. Such a program is certainly very useful, but provides specific numbers rather than design insight, and is best considered as an adjunct to qualitative understanding and simple algebraic models rather than completely replacing them.

4.6 Mechanical Displacement

If the suspension to be analysed already exists, or it is viable to construct one, perhaps adapting some other design, then it may be useful to actually measure the displacement graph of spring position and damper position against bump position, using dial gauges. This graph may be compared with solutions achieved by drawing, analytical methods or computers. The results are apt to be

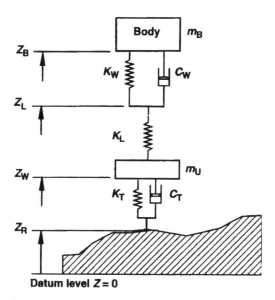

Figure 4.6.1 Heave-only suspension model including link compliance K_L.

disappointing in some cases. Fair agreement is usually obtained for passenger cars, but in the case of racing cars with long links and stiff springs the elasticity of suspension members, and even of the chassis at the mounting points, may give rise to substantial discrepancies. In that case, if it is accepted that the discrepancy is indeed due to compliance in the suspension, then a more elaborate suspension model may be required. Figure 4.6.1 shows such a model, in which the suspension linkage compliance K_L is assumed to have negligible damping, a fairly realistic approximation. The suspension bump deflection

$$z_{SB} = z_W - z_B$$

is no longer related directly to the damper or spring deflection by a simple geometric motion ratio. This figure assumes that the suspension spring and damper operate at the same point, as indeed is applicable for most racing cars, or else more than one suspension compliance must be included.

4.7 Effect of Motion Ratio

It is quite well known that the effect of motion ratio on effective spring stiffness as seen at the wheel is proportional to the spring motion ratio squared. It is also quite well known that the effect of motion ratio on effective damping coefficient is proportional to the damper motion ratio squared. The former statement is only true for a constant motion ratio. The latter statement is only true for a linear damper, and may be seriously in error for nonlinear cases as found on many real dampers.

Consider a nonlinear damper exerting a damper force F_D related to the damper velocity V_D by

$$F_D = C_1 V_D^n$$

where the exponent n has a value 1 for a linear damper (notionally pure viscosity), but may vary in practice from zero (effective Coulomb damping) to 2 (pure fluid dynamic damping).

At a damper motion ratio R_D and suspension bump speed V_S, the damper compression speed is

$$V_D = R_D V_S$$

The damper force F_D is

$$F_D = C_1 (R_D V_S)^n$$

and the force F_W at the wheel is

$$F_W = R_D F_D$$

Hence, the damping force at the wheel is

$$F_W = C_1 R_D^{1+n} V_S^n$$

Note that, in lever terminology, the mechanical advantage has been put equal to the velocity ratio, implying that linkage friction has been neglected. There may indeed be extra friction in rubber bushes, metal bushes or ball joints, but this can be dealt with separately.

From the last equation the shape of the influence of V_S on F_W is retained as V_S^n, but the actual coefficient is scaled by the damper coefficient ratio, which in this case is

$$R_{DC} = R_D^{1+n}$$

Some special cases to be considered are:

(1) Exponent $n = 0$, corresponding to dry Coulomb friction (old-fashioned snubbers) or to hydraulic dampers with a sudden-acting blow-off valve. In this case the damper velocity makes no difference, as long as it is moving, so the coefficient ratio and the force ratio are both equal to the velocity ratio:

$$R_{DC} = R_D \quad (n = 0)$$

(2) Exponent $n = 1$, corresponding to a linear damper:

$$R_{DC} = R_D^2 \quad (n = 1)$$

(3) Exponent $n = 2$, corresponding to a hydraulic damper with a valve that is fixed, (e.g. fully closed at low speed or fully open at high speed):

$$R_{DC} = R_D^3 \quad (n = 2)$$

(4) A representative intermediate case at damper speeds around the knee of the curve, with $n \approx 0.5$, giving:

$$R_{DC} = R_D^{1.5} \quad (n = 0.5)$$

It is common for the motion ratio to be around 0.7, sometimes even as low as 0.4, from which it will be apparent that the damper coefficient ratio will vary very widely for different damper force–speed relationships.

Installation

For a more general polynomial representation of damper behaviour,

$$F_D = C_0 + C_1 V_D + C_2 V_D^2 + \cdots + C_r V_D^r + \cdots$$
$$= C_0 + C_1 R_D V_S + C_2 R_D^2 V_S^2 + \cdots + C_r R_D^r V_S^r + \cdots$$
$$F_{DW} = R_D C_0 + C_1 R_D^2 V_S + C_2 R_D^3 V_S^2 + \cdots + C_r R_D^{r+1} V_S^r + \cdots$$

with each term scaling differently. This may seem to suggest that the shape of the damper force–velocity curve is altered as seen at the wheel. However this is not so, it merely suffers a scaling transformation.

Figure 4.7.1(a) shows an example damper force/speed characteristic. Considering that $V_D = R_D V_S$ and $F_D = F_{DW}/R_D$, this is also the shape of the curve seen at the wheel, but rescaled. Hence the curve shape is retained in Figure 4.7.1(b) as seen at the wheel, where, for $R_D < 1$, the curve is stretched by the ratio $1/R_D$ along the velocity axis and compressed by the ratio R_D on the force axis.

Considered in the opposite way, from a design aspect, the characteristic required at the wheel is the starting point, which is transformed by the motion ratio to the required damper characteristic. A low motion ratio therefore calls for a high damping coefficient for the damper itself.

Considering a linear damper, the various damper parameters, to provide a given characteristic at the wheel, scale as shown in Table 4.7.1. For example, if a motion ratio of 0.5 is used it means that the actual damper coefficient C_D required changes by a factor $0.5^{-2} = 4$ (twice the force must be produced at one-half of the speed). Because the piston area and stroke have compensating values of the scaling index, the total volume remains constant. Hence the volume and mass of fluid will be constant. The cylinder area varies with $R_D^{0.5}$ so the metal volume and mass tends to $R_D^{0.5}$ (but compromised by the ends). Hence for smaller R_D the mass of metal reduces a little. The thermal capacity is approximately constant, or slightly reduced. The surface area is reduced which makes the cooling worse. The use of a motion ratio less than 1.0 is therefore of no benefit as far as provision of the damper itself is concerned, but is used primarily for convenience of packaging the suspension. One possible advantage of a low motion ratio is that the

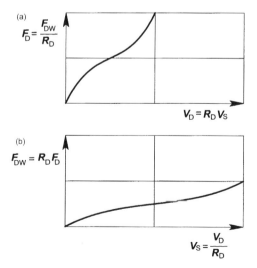

Figure 4.7.1 Damper forces with motion ratio: (a) at the damper; (b) at the wheel.

Table 4.7.1 Damper parameter scaling $(R_D)^p$

Parameter	Scale index p
Velocity	+1.0
Force	−1.0
Coefficient	−2.0
Pressure	+0.0
Piston area	−1.0
Piston diameter	−0.5
Stroke	+1.0
Fluid volume	+0.0
Cylinder area	+0.5
Metal volume	+0.4

range of motion of the rubber bushes is greater as seen at the wheels, so bushes of limited size can play a more significant part in isolating small-amplitude high-frequency road stimulus.

4.8 Evaluation of Motion Ratio

In view of the general imprecision of the damping performance of a worn damper in service conditions, and also of wide variations of opinion on the damping for optimum ride, the actual damping coefficient obtained at the wheel is not extremely critical. Hence, evaluation of the motion ratio is not too critical, especially in the early stages of design. Approximate methods, within a percent or two, are therefore of some value, and algebraic methods have the benefit of giving more insight into a design than do numerical methods. Hence, the following sections describe algebraic or drawing methods for common suspensions on passenger cars, racing cars and motorcycles. These methods may also be used for computer solution, in contrast to the usual finite increment of displacement method.

The main principle involved is the geometry of the lever, or rocker, which is therefore described first. Rigid arm suspensions, such as trailing arms or swing axles follow on the same principle, when viewed correctly. Double wishbones require a different kind of analysis, to relate the wheel motion to the motion of the end ball joint of the relevant arm, usually the lower one, which is carrying the spring or damper, giving the first factor of the total motion ratio. Once this is known, normal rocker analysis will give the other factor. For struts, unique analysis is required.

Motorcycle front forks are considered, having the same analysis as the slider suspension of old car designs. Some other once popular front motorcycle suspensions are included. Motorcycle rear suspensions nowadays often have interesting leverage mechanisms that can be dealt with by successive rocker methods.

4.9 The Rocker

The basic component in motion ratio evaluation is the rocker, or lever, which arises in principle, even if not explicitly as a rocker, in virtually all suspensions. In some cases, especially modern racing cars, a rocker is included as such, to link the basic suspension to the spring and damper. By virtue of including a pushrod, or pullrod, and rocker, the spring and damper can be better positioned, vertically inside the body, or, as is common nowadays, laid down horizontally on top of the body or along the upper sides of the gearbox. This improves the aerodynamics by removing the spring–damper unit from the high-speed

Installation

airflow, and the inclusion of a rocker in the system makes it very easy to change the motion ratio and rising rate simply by changing the rocker.

The function of a rocker, illustrated in Figure 4.9.1, can be specified by three aspects:

(1) the total rocker angular deviation between pushrods θ_{RD};
(2) the rocker motion ratio R_R;
(3) the rocker rising rate factor f_R.

Figure 4.9.1 shows a general rocker, of which there are numerous simpler special cases. The angle θ_{RD} is the rocker deflection angle, the total angular difference between input and output, which will actually vary a little over the range of motion, and is specified at the normal ride height. Figure 4.9.2 shows three examples of deflection angle θ_{RD} (0°, 90°, 180°). Intermediate deflection angles are equally possible.

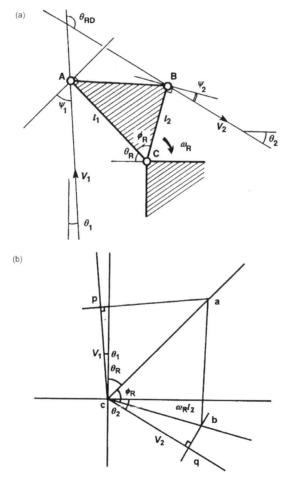

Figure 4.9.1 General rocker configuration; (a) geometry, (b): velocity diagram.

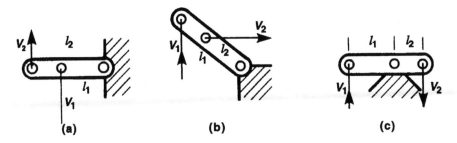

Figure 4.9.2 Simple rockers with various deflection angles.

The rocker motion ratio R_R, specified at the normal ride height, depends most obviously upon the arm lengths from the pivot axes to the input and output points, but also depends on the angular position of the input and output rods. Figures 4.9.3-4 illustrate some possible ways of achieving motion ratios V_2/V_1.

The rising rate is the proportional increase of motion ratio per unit of rocker rotation:

$$f_R = \frac{1}{R_R} \times \frac{dR_R}{d\theta_R}$$

To express this in terms of suspension motion at the wheel, i.e. as wheel rising rate, the geometry of the rest of the suspension must be known, so wheel rising rate is not necessarily a property of the rocker alone, but becomes so if the rest of the system is essentially linear.

The properties of the rocker may be deduced for special cases, as required, but in fact can be derived for the general rocker of Figure 4.9.1 quite easily, by virtue of an informed choice of representational parameters, arrived at by experience. It is assumed in the following analysis that the motion is planar. If the input or output rods are not parallel to the rocker plane, defined as a plane perpendicular to the axis

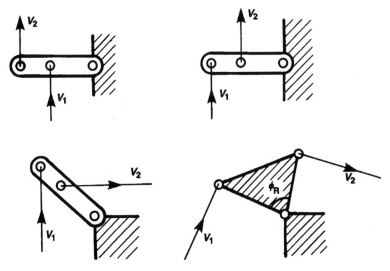

Figure 4.9.3 Simple rockers with various motion ratios.

Photographic Plates: The background is a 10 mm mesh in all cases.

Plate 1.01 Sectioned dampers from a small–medium family car (US compact). The front strut carries more load and takes side loads, so it is generally more robust, and in particular has a much larger rod diameter.

On the left the front strut, $D_P = 28.7$ mm, $D_R = 20.0$ mm, pressure tube thickness 1.2 mm, strut reservoir tube wall 2.5 mm. Piston length 15 mm, seal length 8 mm. Visible here is the large coil spring of the piston extension valve with its sleeve nut. The piston compression valve has a star shim pressing a sealing shim, not seen here. The castellated foot valve allows flow to and from the reservoir. The compression foot valve is a spool valve, of which the lower flange is seen below the foot valve body, and on top the retaining washer against the closure spring. The extension foot valve has a sealing shim held by a light coil spring, just visible immediately above the foot valve body.

The rear damper at the right is similar in general form. $D_P = 25.4$ mm, $D_R = 12.4$ mm, pressure tube wall thickness 0.9 mm, reserve tube wall thickness 1.25 mm. Also visible here are the piston compression valve shim and backing plate. The foot valve body is not castellated, requiring a shaped base for the outer tube, with three support points and three flow channels, not visible here.

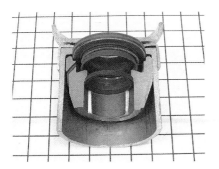

Plate 2.01 Header 1. Section of the top of a front strut. Visible here are the external body, on top, and the cap to hold the rubber bush. In the centre is the main casting or sintering which would connect to the pressure tube below. A part section of the bearing material insert may be seen. The primary seal and final seal are complete, the latter with spring tension radial load to ensure long-term sealing despite elastomeric seal wear and creep. Also visible at the left is one leakage return channel through the main body.

Plate 2.04 Header 4. This header unit is for a pressurised single-tube damper, requiring elaborate sealing for reasonable life. The various parts of the head bearing and seal are held in the main tube by circlips, visible in the tube at top left.

Plate 2.02 Header 2. Sectioned at left, complete unit at the right. The bearing material insert and the seal are visible. The protruding lower ring fits into the pressure tube.

Plate 2.05 Header 5. Cast aluminium header unit incorporating ball joint connector.

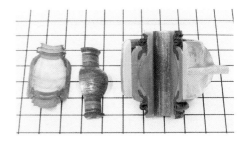

Plate 2.03 Header 3. Sectioned at the left, complete at the right. A separate seal is used in this case. The bearing material insert is visible. The section shows two leakage return channels, of a total of four visible at the right and one on the bottom left.

Plate 2.06 Header 6. Section of the ball joint connector of the previous item, showing metal ball-rod plus various elastomeric supports and seals.

Plate 3.01 Piston Valves 1. Four example pistons of traditional configuration., with simple star-shim controlled compression valves (not visible) and coil-spring controlled extension valves. The first and fourth exhibit cast iron piston rings. Piston and rod diameters are (1) 35.0/15.8, (2) 28.3/20.0., (3) 26.6/20.0, (4) 25.4/10.8 mm. The large rod/piston diameter ratios shown by the centre two reveal them as from struts. Wear patterns are visible on the first piston.

Plate 3.02 Piston Valves 2. The same pistons and rods as in the previous figure, seen from the rod (extension chamber) side.

Plate 3.03 Piston Valves 3. A strut piston and rod, somewhat unusual in using a diametral reduction above the piston. Piston diameter 25 mm, rod 17.0 and 12.2 mm. Conventional compression star-shim and extension coil spring.

Plate 3.05 Piston Valves 5. A different shim-pack piston seen from the rod (expansion chamber) side. Again, triangulation of the bottom shim has been used. This also facilitates three-wing bending. Six active shims in uniform diametral steps.

Plate 3.04 Piston Valves 4. A shim-pack piston seen from the compression chamber side. The sealing shim is triangulated to allow free flow through this pack during compression Five shims and a rigid washer.

Plate 3.06 Piston Valves 6. A third shim-pack piston showing the compression chamber side. Counter-flow is achieved in this case by the angled perimeter face. Three shims with spacer and rigid backing washer.

Plate 4.01 Foot Valves 1. Seven example foot valves seen from below (from the reservoir chamber side). Notable variations in the body form include the presence or lack of castellation, and the actual number of such (four and six visible here). Various foot compression valve types may be identified, e.g. at bottom right is a shim pack, at bottom left is a spool valve type. Pressure tube inner diameters are, from left to right, 25.4, 28.8, 25.4, 35.2, 27.0, 30.3 and 27.2 mm. Masses are 19, 37, 28, 54, 30, 37 and 30 g.

Plate 4.02 Foot Valves 2. The same seven foot valves seen from the top (compression chamber side). The active valves on this side are for extension, so all must be low pressure loss. The weak springs are apparent. At top left and bottom left, the spring is a single curved shim (planar curvature). At centre and top right may be seen concentric coil springs, the inner one controlling a spool valve, the outer one the expansion seal-shim.

Plate 5.01 Piston 1, compression valve side. Compression flow is up and out of the four holes in the outer annulus. This is sealed by a shim, with holes in its inner part to allow flow inwards over the inner rim during compression, and also down during extension to pass through the two holes in the inner annulus.

Plate 5.02 Piston 1, extension valve side. The sealing shim/disc reaches only the raised rim here, so flow into the outer holes, for compression, is free.

Plate 5.03 Piston 2, compression valve side. Similar operating principle to Piston 1, but with detail changes. Three holes for compression flow up into the outer annulus, six holes for extension flow down.

Plate 5.04 Piston 2, extension valve side. Again, the extension seal shim/disc reaches only the raised rim.

Plate 5.05 Piston 3, compression valve side. The standard operating principle, four holes in each direction.

Plate 5.06 Piston 3, extension valve side.

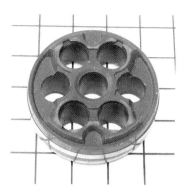

Plate 5.07 Piston 4, compression valve side. A 2/4 hole shim-pack type piston with four holes for compression flow, up into the raised area. The other two holes, from the lower area below shim level, on the left and right, are for downward extension flow.

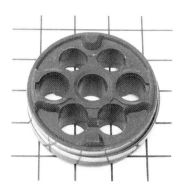

Plate 5.08 Piston 4, extension valve side. Two holes into the raised are subject to shim control on this side. Four holes down for compression flow to the other shim pack.

Plate 5.09 Piston 5, compression valve side. Basically a flat surface, with six holes for compression flow controlled by the shims on this side. The more complex holes are for extension flow down. The flat surface requires a triangulated bottom shim to clear the extension flow holes.

Plate 5.10 Piston 5, extension valve side. Three holes for extension flow controlled by the shims on this side. Again, triangulated shims are required to facilitate counter-flow.

Plate 5.11 Piston 6, compression valve side. A 3/6 hole shim-pack piston in machined bar-stock aluminium alloy. Six holes for compression flow. Three other holes with side notches allow extension flow without triangulating the shims.

Plate 5.12 Piston 6, extension valve side. Three holes for extension flow into the shim pack on this side. Six holes with side notches for downward compression flow.

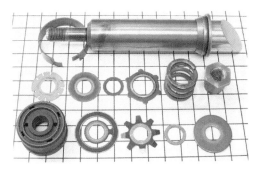

Plate 6.01 Piston Valve Set 1. Top, piston seal and strut rod. Centre row: the extension valve parts. The first, 0.10 mm, shim is notched to give a small 'leak' area. Then there is a 0.25 mm supporting shim and a small 0.20 mm spacer, and a 0.8 mm rigid support plate which is pressed by the coil spring, in turn located by the sleeve nut. These go below the piston as shown. Row 3 has the piston, compression valve side up, 0.20 mm compression seal shim, the 0.10 mm eight-leg star-shim dished to 0.8 mm, a small 0.40 mm spacer and a rigid 1.3 mm backing plate. The seal shim is extensively perforated to allow compression flow from the inner edge, and also to allow extension flow into the two holes down through the piston.

Plate 6.02 Piston Valve Set 2. Top, the strut rod. Middle row: the piston, compression valve side up, the 0.15 mm sealing shim, notched around the edge for leak area, and perforated for compression flow from the inner edge seal and for extension flow. Then a 0.20 mm supporting shim with holes to feed perforation of the sealing shim. Then a small 0.20 mm spacer, 0.15 mm three-leg star shim dished to 1.6 mm, 1.2 mm spacer, and rigid 1.2 mm backing plate with holes for flow in both directions. The bottom row has three 0.20 mm shims, possibly internally opening, full diameter 1.1 mm backing plate, coil spring and sleeve nut.

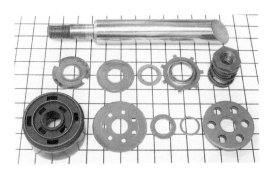

Plate 6.03 Piston Valve Set 3. Middle row: extension valve parts, notched 0.20 mm sealing shim, 0.20 mm support shim, a small 0.20 m spacer, 0.9 mm backing plate, coil spring and sleeve nut. Bottom row: piston, compression valve side up, 0.35 mm compression shim with holes for extension feed from the inner ring seal and feed to inner annulus during extension, two spacers of 0.9 mm and 0.20 mm, rigid 3.7 mm backing plate with feed holes.

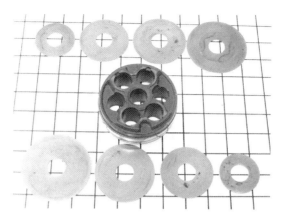

Plate 6.04 Piston Valve Set 4. A 2/4 shim-pack piston shown compression side up, with four compression holes and two extension holes, with, top and bottom rows. The compression shims are 30.6×0.40, 28.0×0.30, 25.0×0.30, 20.0×0.40. The extension shims are 31.4×0.40, 27.0×0.40, 24.0×0.4 and 18.5×0.60.

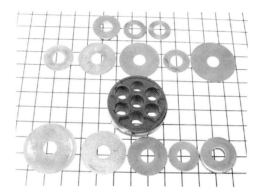

Plate 6.05 Piston valve Set 5. The same piston design as in Set 4, but with different valve shims. Top row: compression shims, four-stage design with thin spacer shim after the sealing shim at the right. $D \times t$ from top left 19.5×2.0 mm washer, 16×0.55, 16×0.55 again, 20×0.45, 25×0.30, 27×0.3, 18×0.10 spacer, 30.6×0.25 mm. Bottom row: conventional sequence of shims for extension, plus solid backing plate, from the left 31.4×0.35, 27.0×0.40, 24.0×0.4, 18.5×0.60 and 22×3.0 mm washer.

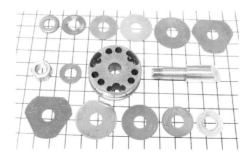

Plate 6.06 Piston Valve Set 6. Centre row: retaining nut, washer, piston compression valve side up, and rod. Top row: compression shims, triangulated sealing shim with conventional sequence of reducing diameter support shims and washer, 0.20, 0.25, 0.25, 0.30, 0.35 mm thickness largest to smallest. Bottom row: similar shims for extension, 0.30, 0.30, 0.35, 0.45 and 0.50 mm thickness largest to smallest.

Plate 7.01 Foot Valve Set 1. Top row: spool valve, to be inserted upwards into the cast or sintered foot valve body seen extension valve side up. Note the large extension flow area moulded into the body. Bottom row: extension valve sealing shim, conical extension valve spring, locating pressing, compression coil spring to go around and retain the stem of the spool valve. A small washer, absent, would be peened onto the top of the spool to retain the spring.

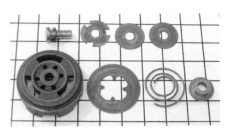

Plate 7.02 Foot Valve Set 2. Bottom row: Foot valve body seen extension valve side up, extension valve sealing shim with four-leaf clover centre hole to allow inner edge extension flow, and compression flow into the six body holes, with central shim location, flat spiral extension spring, sleeve washer (inverted) locates spring and shim centrally, retained by peening over of the small end of the pin, top left. This fixed pin also holds the three compression shims in place. The sealing shim is notched for leak area.

Plate 7.03 Foot Valve Set 3. Top row: pin with axial hole and also cross-drilling (hardly visible), and body, extension valve side up. Bottom row: extension sealing shim, curved shim spring, locating pressing, sleeve seal, spring for sleeve seal, retaining washer (peened on). The lower left three parts fit on top of the piston as shown, held by the pin head. On the compression valve side, the sleeve slides on the pin, opening against the coil spring.

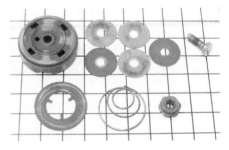

Plate 7.04 Foot Valve Set 4. Top row: foot valve body, shown extension valve side up, five extension valve shims to fit underneath and retaining screw. Bottom row: extension valve sealing shim with internal clearance for compression flow and locating points, conical spring, retaining sleeve nut, which locates the spring and the shim on its sleeve length, shown inverted.

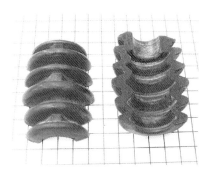

Plate 8.01 Sectioned rubber bump stop from a strut, in perfect condition after 10 years in service. 63 g.

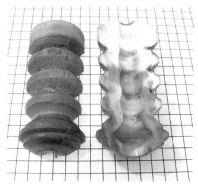

Plate 8.02 Sectioned polyurethane bump stop from a strut, after 10 years service showing significant deterioration, with mechanical degradation and oil soaking, but still fully functional. 73 g, lower density than natural rubber.

Plate 8.03 Rubber bush in a damper eye, featuring cavities at the top and bottom to give reduced rigidity in the line of action of the damper.

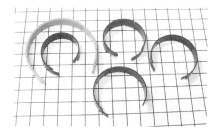

Plate 8.04 A selection of plastic piston rings, typically graphite-loaded PTFE. Cast iron, Tufnol and other materials have also been used.

Plate 8.05 Barrel from an adjustable damper. Rotation of the barrel brings one of six holes into registration with the body flow channel.

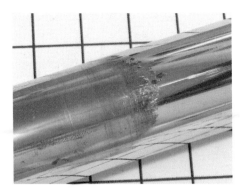

Plate 8.06 Severe wear on a strut rod after about 10 years in service. The wear has penetrated the hard chromium plating, and there is evidence of consequent corrosion. Strut replacement was required.

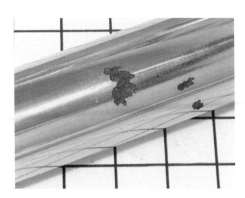

Plate 8.09 Bad pitting corrosion of a chromium-plated rod, probably caused by inadequate rod preparation before plating, with consequent poor adhesion of the chromium.

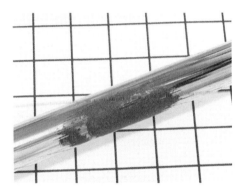

Plate 8.07 Severe wear on a damper rod, resulting in complete removal of the chromium plating, of inadequate thickness for the in-service side load.

Plate 8.10 Severe pitting and general corrosion at and near the exposed end of a damper rod.

Plate 8.08 Mild, and insignificant, discoloration and slight corrosion on the steel part of a damper rod, terminating at the chromium plating region.

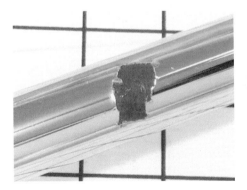

Plate 8.11 Flaking of chromium plating on a damper rod, probably caused by inadequate preparation of the base metal and poor adhesion of the plating.

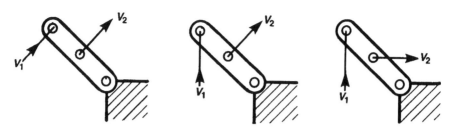

Figure 4.9.4 Simple rockers with various rising rates (zero in the first case).

of rotation of the rocker on the body, then the velocities in Figure 4.9.1 are the rocker plane velocities, related to the actual rod velocities by the cosine of the out-of-plane angle.

The parameters in Figure 4.9.1 are:

(1) rocker included angle ϕ_R, positive when the output leads the input as shown;
(2) the input and output rocker arm lengths l_1 and l_2;
(3) the input and output rod offset angles ψ_1 and ψ_2, between the rod and the tangent perpendicular to the corresponding arm radius;
(4) the rocker position θ_R, from some appropriate datum, usually the normal ride position;
(5) the input and output rod velocities, in the rod directions, V_1 in and V_2 out;
(6) the rocker angular velocity ω_R.

Of the above, lengths l_1 and l_2 are constant, as is ϕ_R, these three parameters being the essence of the rocker geometry. Other parameters will vary, although ω_R is normally deemed to be some constant value for the purpose of analysis.

The performance parameters of the rocker, θ_{RD}, R_D and f_R are derived as follows. The deviation angle is

$$\theta_{RD} = \psi_1 + \phi_R + \psi_2$$

In practice, for design purposes, this is required in the form

$$\phi_R = \theta_{RD} - \psi_1 - \psi_2$$

Figure 4.9.1 includes the velocity diagram for the rocker. At rocker angular speed ω_R, assumed for analysis, the tangential speed of point A is $\omega_R l_1$, perpendicular to line AC, so at angle θ_R from the vertical, giving point a representing V_A. Similarly, point b representing V_B is established, at $\theta_R + \phi_R$.

The velocity of A can be resolved into components parallel and perpendicular to the input rod at angle θ_1 to the vertical. To do this, construct in the velocity diagram the line at θ_1 from the vertical, parallel to the input rod, and drop the perpendicular from a, giving point p. The length cp represents the velocity along the rod, whilst pa represents the tangential velocity of one end relative to the other. Similarly for the output, so the actual damper compression speed on the output is represented by cq. The rocker motion ratio R_R is given by

$$R_R = \frac{V_Q}{V_P} = \frac{cq}{cp}$$

Hence, the velocity ratio of the rocker, defined by

$$R_R = \frac{V_2}{V_1}$$

is given by

$$R_R = \frac{\omega_R l_2 (\theta_R + \phi_R - \theta_2)}{\omega_R l_1 \cos(\theta_R + \theta_1)}$$

This may be expressed more concisely as

$$R_R = \frac{l_2 \cos \psi_2}{l_1 \cos \psi_1}$$

This result may be summarised very simply by a physical interpretation. The denominator is the perpendicular distance from the axis to the line of action of the input link, Figure 4.9.1. The numerator is the perpendicular distance from the axis to the line of action of the output link.

As an alternative interpretation, using the rocker arm length motion ratio

$$R_{RL} = \frac{l_2}{l_1}$$

and the rod angle motion ratio

$$R_{R\psi} = \frac{\cos \psi_2}{\cos \psi_1}$$

then the rocker motion ratio R_R then may be expressed as

$$R_R = R_{RL} R_{R\psi}$$

The first of these new parameters, R_{RL}, is constant. The second, $R_{R\psi}$, varies as the rocker moves, and provides the basis of design for rising rate.

The input and output have tangential velocities, $V_{tan,in}$ and $V_{tan,out}$, one end relative to the other, of

$$V_{tan,in} = \omega_R l_1 \sin \psi_1$$
$$V_{tan,out} = \omega_R l_2 \sin \psi_2$$

These rotate the rods, thereby also altering the ψ angles, in a way which depends on the length of the input and output rods, or of the damper. This can be dealt with separately, and is, in any case, generally a much smaller effect than the basic rocker rising rate effect. Considering the input and output rods to remain parallel to their starting angle, and considering ψ to be positive, as shown in Figure 4.9.1, at rocker angle position θ clockwise from normal ride height,

$$\psi_1 = \psi_{Z1} + \theta$$
$$\psi_2 = \psi_{Z2} - \theta$$

where ψ_{Z1} and ψ_{Z2} are the values of ψ at normal ride height (zero deflection).

The rocker angle motion ratio factor is then

$$R_{R\psi} = \frac{\cos(\psi_{Z2} - \theta)}{\cos(\psi_{Z1} + \theta)}$$

Depending on the application, it may be convenient simply to think in terms of the motion ratios at two positions, say zero bump and some expected bump position. However, the mathematical rising rate factor, at zero bump defined earlier, was

$$f_R = \frac{1}{R_R}\frac{dR_R}{d\theta_R}$$

with

$$R_{R\psi} = R_{R\psi 0}(1 + f_R\theta)$$

The rocker angle motion ratio at θ is

$$R_{R\psi} = \left[\frac{\cos(\psi_{Z2} - \theta)}{\cos(\psi_{Z1} + \theta)}\right]$$

The rising rate factor f_R is given by

$$f_R = \frac{1}{\theta}\left(\frac{R_{R\psi}}{R_{R\psi,0}} - 1\right)$$

which, by considering infinitesimal θ, with substitution and condensation, becomes

$$f_R = \tan\psi_1 + \tan\psi_2$$

Hence, the rising rate is governed by the angles ψ_1 and ψ_2. Note that these are positive as defined in Figure 4.9.1 and the signs must be respected; also they may well change sign within the range of motion of the rocker. Physically, the reason for the rising rate is that the input moment arm is increasing, and the output one is decreasing.

At rocker deflection θ the motion ratio R_R is therefore estimated to be

$$R_R = R_{RL}R_{R\psi}(1 + f_R\theta)$$

This method of rising rate analysis is useful for a preliminary appraisal, giving a first estimate of the required values for ψ_1 and ψ_2 in combination, which then, in conjunction with the required rocker angular deflection θ_{RD}, gives the rocker included angle

$$\phi_R = \theta_{RD} - \psi_1 - \psi_2$$

Practically, because of the packaging problems of large rockers, fairly large rocker angular movements are used in practice, so once a first estimate has been made it is more practical to work with two or more actual rocker positions, e.g. normal ride height and a bump position, and to obtain two corresponding values of motion ratio.

The usual situation is that a given increase of ratio is required for a given rocker angular displacement, estimated from a given wheel bump motion, intermediate ratio and rocker input arm length. The required increase of rate may be shared between input and output. The required increase of rate on the input, say, is then known. Let the ratio be $r = R_2/R_1$. From the angle motion ratio

$$R_{R\psi} = \frac{\cos(\psi_{Z2} - \theta)}{\cos(\psi_{Z1} + \theta)}$$

may be obtained

$$r = \frac{\cos \psi_{Z1}}{\cos(\psi_{Z1} + \theta)}$$

This may be solved for ψ_{Z1} as

$$\tan \psi_{Z1} = \frac{1}{\sin \theta} \left(\cos \theta - \frac{1}{r} \right)$$

giving the correct initial angular position for the arm. In practice it may be difficult to obtain a low enough ratio with a practical rocker size, and ψ_{Z1} may be negative. This gives a rate which will fall slightly before then rising to the required value.

For design purposes, rocker design is a matter of juggling the lengths and angles according to the given input pushrod motion and the required rising rate. Two points to bear in mind are that for a linear damper it is the square of the motion ratio that controls the damping coefficient at the wheel, and that a motion ratio rising from normal ride height will also generally fall when below the normal ride height. Hence large motion ratio changes must be designed with caution.

4.10 The Rigid Arm

Rigid arm suspensions, such as a trailing arm, have a single arm from a pivot axis, the wheel camber angle being rigidly fixed relative to the arm. It is possible, but unusual, to have steering with such a system. Rigid arms may be classified in various ways. From the geometric point of view, the important distinction is the angle, ψ_A, between the arm pivot and the vehicle centre line in plan view, Figure 4.10.1(a). Sometimes there is a nonzero angle of the arm pivot, ϕ_A, in rear view Figure 4.10.1(b).

The basic classifications by ψ_A are:

(1) trailing arm (90°);
(2) semi-trailing arm (e.g. 70°);
(3) leading arm (90°);
(4) swing axle (0°);
(5) Transverse arm (0°);
(6) Semi-trailing swing axle (e.g. 45°).

The spring and damper usually act directly on the arm. In any case, it is necessary to obtain the relationship between the suspension bump velocity, i.e. of the vertical wheel velocity at the contact patch, relative to the body, and the angular velocity of the arm. The radius of action of the wheel is l_{WP} in plan view, Figure 4.10.1(a). For an arm angular speed ω, the tangential speed of the wheel is $\omega \, l_{WP}$, but this is not vertical in rear view, so the actual suspension wheel bump velocity V_S is

$$V_S = \omega_A l_{WP} \cos \phi_A$$

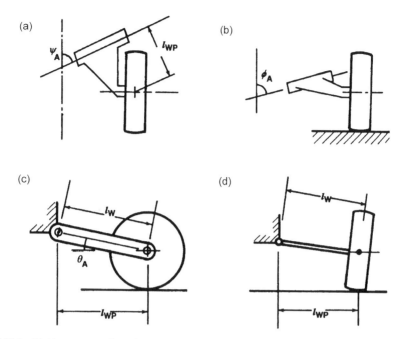

Figure 4.10.1 Rigid arm suspensions: (a) plan view of semi-trailing arm; (b) rear view of semi-trailing arm; (c) side view of trailing arm; (d) rear view of swing axle.

The velocity ratio of arm to wheel $R_{A/W}$ is therefore

$$R_{A/W} = \frac{\omega_A}{V_S} = \frac{1}{l_{WP}\cos\phi_A}\ [\text{rads}^{-1}/\text{ms}^{-1}]$$

Note that the pivot axis plan angle θ_A does not appear in this expression. Neither does any influence of the angle of the arm in side view, because this has been incorporated by using the plan length l_{WP}, which may in fact vary somewhat through bump movement, because

$$l_{WP} = l_W \cos\theta_A$$

For any given bump position, θ_A follows, whence l_{WP} and $R_{A/W}$:

$$R_{A/W} = \frac{1}{l_W \cos\theta_A \cos\phi_A}$$

With some consideration, the above can be applied to any of the rigid arm suspensions listed above.

The second part of the damper motion ratio then follows from the position of the damper; Figure 4.10.2 shows the rigid arm in elevation viewed along the pivot axis. The damper may not be in the plane of the elevation, but will be close to it, with out-of-plane angle α_D.

The rigid arm analysis is now easily completed as for analysis of the output of the rocker:

$$V_2 = \omega_A l_2 \cos\psi_2$$

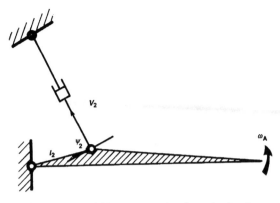

Figure 4.10.2 Rigid arm suspension shown in elevation.

Allowing for the out-of-plane angle α_D,

$$V_D \cos \alpha_D = V_2$$

so

$$V_D = l_2 \frac{\cos \psi_2}{\cos \alpha_D} \omega_A = l_2 \frac{\cos \psi_2}{\cos \alpha_D} \frac{V_B}{l_W \cos \theta_A \cos \phi_A}$$

Hence, the damper motion ratio R_D is

$$R_D = \frac{V_D}{V_S} = \frac{l_2}{l_W} \left(\frac{\cos \psi_2}{\cos \theta_A \cos \phi_A \cos \alpha_D} \right)$$

This is very similar to the expression for a simple rocker, but includes effects from ϕ_A and α_D.

As in the case of the rocker, judicious choice of ψ_2 at zero bump will give a rising rate, or not, as desired, rising rate occurring for ψ_2 as shown, the angle reducing with bump action, increasing the effective damper moment arm. If anything, this is easier to design than an extra rocker because the angular motion of the suspension arm will generally be less than for a rocker, and the smaller angle makes a more linear progression possible. For example, a bump deflection z_s of, say, 100 mm on an effective arm length l, which may be as much as 1.3 m for a transverse arm or swing axle, gives an angular bump motion of the arm of about 4–5°. For a trailing arm of length 0.4 m, the angle is 15°.

4.11 Double Wishbones

The double wishbone or double A-arm suspension is a little more difficult to solve than the simple rigid arm. As before, it is necessary to establish a motion ratio between the suspension bump velocity and the angular velocity of the arm which operates the spring or damper, or operates the pushrod to the rocker for a racing car. Figure 4.11.1(a) shows the basic configuration.

If the bump camber coefficient of the suspension is already known, then a particularly simple method is possible. The bump camber coefficient ε_{BC} is the rate of change of wheel camber angle γ with

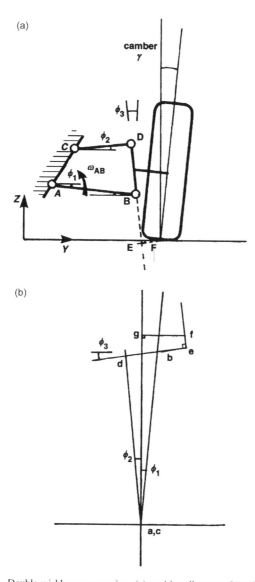

Figure 4.11.1 Double wishbone suspension: (a) position diagram; (b) velocity diagram.

suspension bump, arising from suspension geometry:

$$\varepsilon_{BC} = \frac{d\gamma}{dz_S} \quad (\text{rad/m})$$

For a suspension bump velocity V_S, then for a reasonably constant ε_{BC}, usually a good approximation for the present purpose, the wheel camber angular velocity is

$$\frac{d\gamma}{dt} = \varepsilon_{BC} \frac{dz_S}{dt} = \varepsilon_{BC} V_S$$

From Figure 4.11.1, the vertical velocity of B differs from the vertical velocity of F by the camber angular velocity multiplied by the lateral difference of position e, where

$$e = X_F - X_B$$

The vertical velocity of B is therefore

$$V_B = V_S - e\varepsilon_{BC}V_S = V_S(1 - e\varepsilon_{BC})$$

Hence, the motion ratio $R_{B/S}$ of ball joint B to suspension bump is

$$R_{B/S} = 1 - e\varepsilon_{BC}$$

Realistic values are $e = 0.1$ m and $\varepsilon_{BC} = 1$ rad/m, which will give $R_{B/S}$ a value of 0.9, a substantial deviation from 1.0 which should certainly be included in the analysis.

In the absence of prior information on the bump camber coefficient, a velocity diagram may be considered, as in Figure 4.11.1(b). This is more easily constructed by initially assuming an angular velocity ω_{AB} for the lower link, rather than a bump velocity of the wheel. A and C are fixed points, therefore appearing at the origin of the velocity diagram. The tangential velocity of B relative to A is $\omega_{AB}l_{AB}$, and the line ab in the velocity diagram is perpendicular to link AB, the length of ab being the tangential velocity at the diagram scale. This establishes point B. Line cd is perpendicular to CD, and bd is perpendicular to BD; the intersection gives point d.

To obtain the velocity of F, at the bottom of the notionally rigid wheel, in the position diagram project line DB and drop perpendicular from F, giving E. In the velocity diagram, the rigid wheel with the wheel upright is solved by scaling. Hence

$$\frac{be}{db} = \frac{BE}{DB}$$

giving point e. Draw the perpendicular from e. DBEF is a left turn, so dbef is also to the left. Use

$$\frac{ef}{de} = \frac{EF}{DE}$$

to give point f. Finally, drop the perpendicular from f to the vertical axis, giving point g.

This completes the velocity diagram to some convenient scale for some angular velocity ω_{AB} of the lower arm. This methodology may, of course, form the basis of a computer program where repeated analysis is desired.

The velocities of interest may now be read from the diagram.

(1) The vertical velocity of the point F, i.e. the suspension bump velocity, V_S, is represented by ag ($V_S = V_{G/A}$).
(2) The wheel scrub (lateral) velocity V_{WS} is represented by fg.
(3) The tangential velocity of D relative to B, $V_{D/B}$, is represented by bd.
(4) The tangential velocity of D relative to C, $V_{D/C}$, is represented by cd.

Hence, the following may be deduced:

(1) The motion ratio of the lower arm to suspension bump, in rad s^{-1}/m s^{-1}

$$R = \frac{\omega_{AB}}{V_S} = \frac{V_{B/A}}{l_{AB} V_S}$$

(2) The camber angular velocity

$$\frac{d\gamma}{dt} = \frac{V_{D/B}}{l_{DB}}$$

(3) The bump camber coefficient

$$\varepsilon_{BC} = \frac{1}{V_S}\frac{d\gamma}{dt} = \frac{V_{D/B}}{l_{DB} V_S} = -\frac{l_{db}}{l_{DB} l_{ag}}$$

(4) The basic roll centre height (unrolled)

$$h_{RC} = \tfrac{1}{2} T \frac{l_{fg}}{l_{ag}}$$

In the present context, it is the velocity ratio that is of interest. The lower arm may then be analysed as a rocker output for the damper drive, as was the rigid arm, to give the overall damper motion ratio.

4.12 Struts

The strut suspension is a the usual choice nowadays for the front of passenger cars. The use of a strut at the rear is a little unusual, but has featured in several cases.

The usual strut incorporates the damper into the body of the strut, and has a surrounding spring. An alternative design, the damper strut, has only the damper in the strut body, with the spring acting separately on one of the arms. Geometrical considerations are the same, although, of course, in the latter case it is the arm which must be analysed for the spring motion ratio.

Overall, the method of analysis is similar to that of the double wishbone suspension. Figure 4.12.1(a) shows a strut suspension. This is in fact the simpler version where the strut axis passes through the ball joint at B.

If the bump camber coefficient is already available, then the vertical velocity V_{ZB} of B is given by

$$V_{ZB} = V_S(1 - e\varepsilon_{BC})$$

where

$$e = X_F - X_B$$

The tangential velocity of B is then

$$V_{B/A} = \frac{V_{ZB}}{\cos\phi_1}$$

and the damper compression velocity V_D is

$$V_D = V_{B/A} \cos(\phi_1 + \theta_2)$$

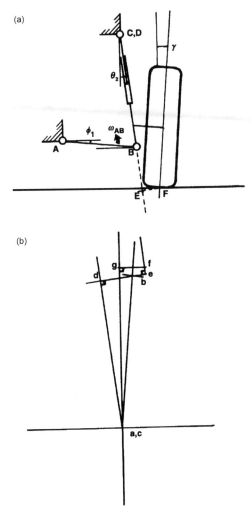

Figure 4.12.1 Strut suspension: (a) position diagram; (b) velocity diagram.

Hence

$$R_D = \frac{V_D}{V_S} = (1 - e\varepsilon_{BC})\frac{\cos(\phi_1 + \theta_2)}{\cos \phi_1}$$

Realistic values may give a motion ratio below 0.9, in contrast to the naive expectation of a value close to 1.0.

The velocity diagram is shown in Figure 4.12.1(b). Construction begins by assuming an angular velocity ω_{AB} for the bottom link:

$$V_{B/A} = \omega_{AB} l_1$$

perpendicular to AB, giving point b representing V_B in the velocity diagram. Velocity $V_{B/C}$ is the vector sum of longitudinal and tangential components, so construct a line through b perpendicular to CB and through c parallel to CB to intersect, giving point d. Point d represents the velocity of point D, which is a point notionally fixed to the lower part of the strut, and instantaneously coincident with the upper fixture point C. The damper compression velocity is represented by cd.

To obtain the suspension bump velocity, extend DB and drop the perpendicular from F to E. Use

$$\frac{be}{bd} = \frac{BE}{BD} \quad ; \quad \frac{ef}{de} = \frac{EF}{DE}$$

to give e and f, and drop the perpendicular from f onto the vertical axis to give g. Point f represents the motion of the base of the wheel, and g represents its vertical component.

From the velocity diagram can be obtained:

(1) the suspension bump velocity represented by ag;
(2) the wheel scrub velocity (fg);
(3) the tangential velocity of B relative to D (db).

Hence, the following may be calculated:

(1) the motion ratio R of the lower arm to suspension bump

$$R = \frac{\omega_{AB}}{V_S} = \frac{V_{A/B}}{l_{AB} V_{A/G}}$$

(2) the camber angular velocity

$$\frac{d\gamma}{dt} = \frac{V_{B/D}}{l_{BD}}$$

(3) the bump camber coefficient

$$\varepsilon_{BC} = \frac{d\gamma/dt}{V_S} = \frac{V_{B/D}}{l_{BD} V_S} = -\frac{l_{bd}}{l_{BD} l_{ag}} \quad [\text{rads}^{-1}/\text{ms}^{-1}]$$

(4) the basic roll centre height (unrolled)

$$h_{RC} = \tfrac{1}{2} T \frac{l_{fg}}{l_{ag}}$$

On front suspensions in particular, the damper axis is frequently aligned such that it does not pass through the ball joint at B, but rather inside or outside it. In that case the preceding analysis is still applicable, with the following provisos:

(1) The angle θ_2 used is that of the damper axis, not that of the steering axis CB.
(2) To obtain e and f still use the steering axis line CB extended.

4.13 Pushrods and Pullrods

In formula racing cars in particular, it is normal practice nowadays to use double wishbones with a pushrod driving a rocker which operates the spring and damper. These can be solved readily by the processes already described, treating the motion ratio as the product of the sequence of ratios arising from the particular system. This generally involves:

(1) the ratio from wheel displacement to bottom arm angle;
(2) the ratio from bottom arm angle to pushrod displacement;
(3) the rocker ratio.

For a practical racing suspension, the wishbones are long and the pushrod angle may be quite low, giving a low velocity ratio. This gives proportionally larger pushrod forces, but the reduced stroke allows a compact rocker. Pullrods have been used in the past, but have now largely given way to pushrods. The pullrod analysis is very similar to that of the pushrod.

The rocker axis may be rotated to lay the damper along the vehicle. With the pushrod system this allows the front dampers to lie above the driver's legs, permitting the best aerodynamic shape for the front part of the vehicle.

In these more complex systems, when analysed as a series of two-dimensional sub-mechanisms, it may be necessary to incorporate some cosine factors to correct for out-of-plane motions, but these are usually quite small.

4.14 Motorcycle Front Suspensions

On nearly all modern motorcycles, the front suspension uses hydraulic forks, a slider system, whereas cars rarely use sliders nowadays. Historically there have been many other motorcycle front suspension systems, with possible advantages, including leading link, trailing link and girder forks. In general machine design, a slider is rarely preferred over a pivot, because of the greater friction of the former, so the prevalence of the slider fork is possibly due mainly to its clean styling. A disadvantage of the slider fork is that it suffers from severe dive on braking, a problem that can be overcome by other systems.

Figure 4.14.1(a) shows the slider fork geometry. The angle of the slider θ_D is in principle separate from that of the steering axis, although in practice these angles are normally the same, with a value in

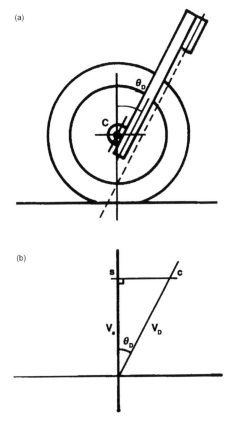

Figure 4.14.1 Motorcycle sliding forks: (a) position diagram; (b) velocity diagram.

the range 25–30°. For a vertical wheel displacement z, the damper must move a distance $z/\cos\theta_D$. The velocity diagram is as shown, with

$$R_D = \frac{1}{\cos\theta_D}$$

and is likely to be around 1.10–1.15. Hence the motion ratio exceeds 1.0, unlike most other installations.

Figure 4.14.2 shows the line position diagram (a) and velocity diagram (b) for a trailing link front suspension. Of course, the steering axis, not shown, will still intersect the ground somewhat forward of the wheel centre, but still within the tyre contact patch. The entire component ABD steers. Suspension action is by rotation of the link DC about D. The (spring-) damper unit is connected to the link DC at E and to the main link at B. In the velocity diagram, motion is relative to the frame ABD, so the corresponding velocity points a, b and d are at the origin, with zero velocity. Consider a known vertical velocity of the wheel, usually taken as unity, 1 m/s, drawn as a horizontal line in Figure 4.14.2(b) at the

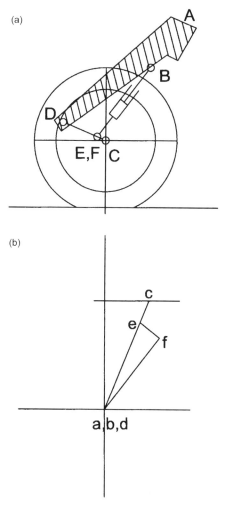

Figure 4.14.2 Motorcycle front trailing link: (a) position diagram; (b) velocity diagram.

appropriate scale and height above the origin. The motion of point C relative to D is perpendicular to CD, so line dc may be drawn, intersecting the horizontal line, giving point c, the velocity of C. The connection point E is on DC. The physical link DEC has an image dec in the velocity diagram, the same shape as DEC, but rotated 90°, enabling e to be determined. If E is on the line DC then simple proportion may be applied to find e, as in this example. Consider now the imaginary ('coincident') point F, which is at the same position as E, but is treated as being fixed to the upper part of the damper. As seen in the velocity diagram, the line ef is the tangential component of the velocity of E relative to B, and bf is the centripetal velocity, i.e. the desired damper compression velocity. The lines ef and bf are perpendicular. The velocity bf must be along the damper. Further, ef, the tangential velocity, must be perpendicular to EB, being tangential. This suffices to construct lines from e and b intersecting at f. The motion ratio is thereby determined.

Figure 4.14.3 shows a leading link system. The steering pivot axis remains unchanged. The method for solution of the velocity diagram is similar to that for the trailing link. For a given vertical velocity of the wheel, c is easily found, dc being perpendicular to DC. Point e is then found from proportion or as

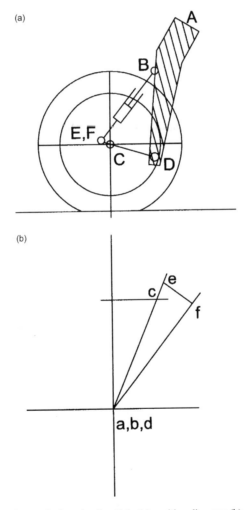

Figure 4.14.3 Motorcycle front leading link: (a) position diagram; (b) velocity diagram.

an image of the link DCE. In this example, the connection point E is on DC extrapolated, so e is beyond c in the velocity diagram. Point f is found as before, giving the compression velocity bf.

The once popular girder fork system is shown in Figure 4.14.4. The girder is link CDE, supported on links AE and BD. This exhibits some similarities to the front view of a car double wishbone suspension, with considerable scope for altering the characteristics by changing the link lengths and angles. The spring-damper unit can often be conveniently contained in the upper part, possibly acting between D and A. In general this has a more tricky solution. Often, however, the links are parallel and equal. With parallel links, at the point of analysis, the motion of the girder is pure translation, with zero rotation, so the velocity image of CDE is one point, points c, d and e being identical. In this case, the girder fork acts similarly to a slider fork sliding perpendicular to the links AE and BD, although with reduced static friction.

With nonparallel links, there are two good methods of solution of the velocities. One is to project AE and BD to an intersection point H, which is the instantaneous centre (centro) of rotation of CDE. The velocity of C is then perpendicular to HC, and the solution may proceed as for the other systems, i.e.

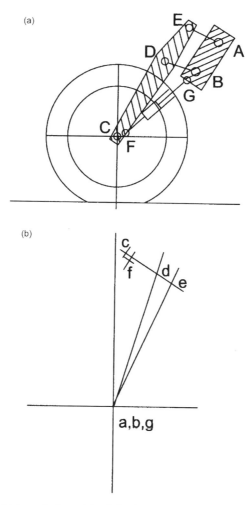

Figure 4.14.4 Motorcycle front girder forks: (a) position diagram; (b) velocity diagram.

rather like a very long leading or trailing link. The theoretical point H may be at infinity, when the links are parallel, in which case the velocity of C is simply perpendicular to AE and BD as stated. Also, H may be below the road level, with no particular implications. An alternative general solution is to begin instead with an angular velocity of one of the links, say AE. Then from the length AE the velocity of E, point e, is immediately determined, ae being perpendicular to AE. Then, line bd is perpendicular to BD and ed is perpendicular to ED, giving intersection at d. The girder EDCF then has a velocity image edcf giving c and f. Thus the vertical velocity of C and the damper compression velocity fg are known, and the motion ratio is known.

4.15 Motorcycle Rear Suspensions

The basic form of motorcycle rear suspension is a simple rocker as shown in Figure 4.15.1, with one spring-damper unit on each side. By forming a pyramid ACD, as in Figure 4.15.2, a single spring-

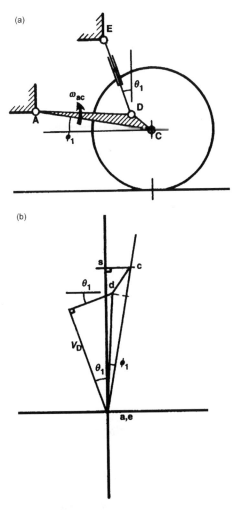

Figure 4.15.1 Motorcycle rear suspension, Type 1.

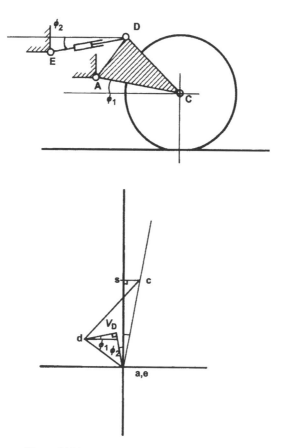

Figure 4.15.2 Motorcycle rear suspension, Type 2.

damper unit can be used, placed under the seat. This also solves some problems associated with pairs of dampers being badly matched side-to-side, which causes wheel camber and steer compliance angles due to imperfect rigidity of the arms and frame. The velocity diagrams for the above are straightforward, on the principles previously explained, and are of undoubted accuracy because the mechanisms are definitely two-dimensional in action.

On more expensive passenger and sports motorcycles, and especially on racing motorcycles, more elaborate mechanisms may be used. In some cases these do have advantages regarding overall packaging of the suspension.

There are three main pullrod concepts; these place the spring-damper unit either vertically in front of the wheel, as in Figures 4.15.3 and 4.15.4, or horizontally beneath the frame, as in Figure 4.15.5. For the vertical spring, the two options are rather similar in principle, but are implemented slightly differently. All these systems use a rocker, and can be given the required motion rates and rising rate by analysis along the lines previously described.

Known production variations of the type of Figure 4.15.3 include actuating the damper from part-way along EB, or even from an extension of EB above B.

Essentially there are two pushrod systems, as in Figures 4.15.6 and 4.15.7 (p. 164–5). In the former, the pushrod operates the damper through rocker DEF giving control of ratio and rising rate.

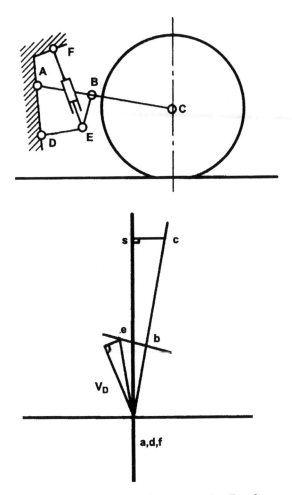

Figure 4.15.3 Motorcycle rear suspension, Type 3.

Alternative 4.15.7 is interesting in that the spring and damper forces act on the suspension from both ends. This does not have any magical properties, but does create some extra complexity in evaluating the motion ratio. This configuration may have arisen because, with the vertical spring-damper in front of the wheel it is actually easier to mount the lower end on the main trailing link than to provide a fixture for it on the frame. In other words, it may be done simply for packaging reasons. This mechanism can be analysed for the motion ratio to each end of the damper, giving R_{D1} and R_{D2}.

The total compression velocity of the damper is then $(R_{D1}+R_{D2})V_S$, resulting in a damper force F_D. Considering the mechanical advantage, this force acting on one end at motion ratio R_D, gives suspension force

$$F_{S1} = R_{D1}F_D$$

and the other end gives

$$F_{S2} = R_{D2}F_D$$

Installation

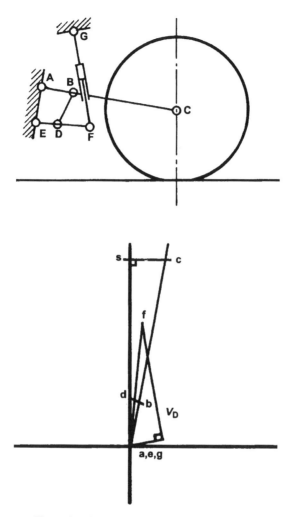

Figure 4.15.4 Motorcycle rear suspension, Type 4.

Hence, the total effective suspension force is

$$F_S = F_{S1} + F_{S2} = (R_{D1} + R_{D2})F_D$$

The result is therefore the same as any mechanism giving a total motion ratio

$$R_D = R_{D1} + R_{D2}$$

In other words, for the multiple-acting system the motion ratios simply add together.

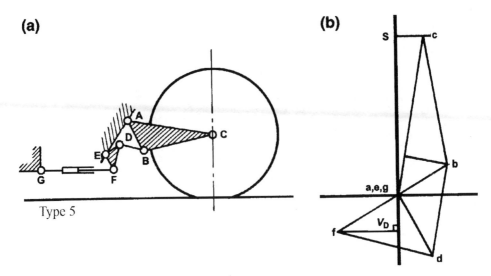

Figure 4.15.5 Motorcycle rear suspension, Type 5.

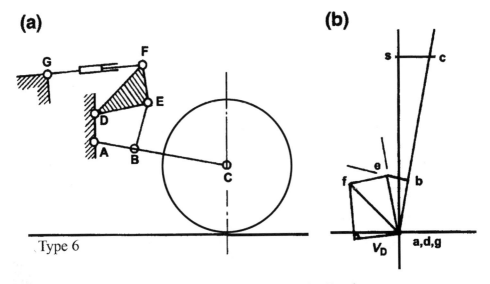

Figure 4.15.6 Motorcycle rear suspension, Type 6.

Installation

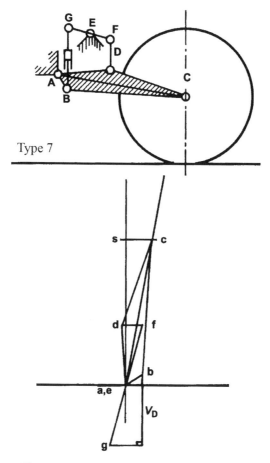

Figure 4.15.7 Motorcycle rear suspension, Type 7.

4.16 Solid Axles

Solid car axles, driven or undriven, may be located either by leaf springs—the Hotchkiss axle if driven ('live')—or, more often nowadays, by links. The springs and dampers may act on the axle itself, or on the links, as in Figure 4.16.1 (on p. 166).

For bump analysis, if the damper acts directly on the axle, as in Figure 4.16.2 (on p. 167), above the wheel centreline, and the dampers are vertical, then the motion ratio is very close to 1.0. However the dampers are sometimes angled, inwards at the top by angle θ_D, in which case the bump motion ratio is $\cos\theta_D$.

One purpose of this angling is to provide some damping of the axle in lateral (sway) vibrations. The velocity ratio for sway motion is $\sin\theta_D$, so even quite small angles are significant.

If the damper connection point to the axle is forward of the wheel centreline by a distance e, (or aft, negative e) then the axle needs to be checked for pitch in heave. Similarly to the case of wheel camber affecting motion ratio, an axle pitch/heave coefficient ε_{APH} (rad/m) will cause a motion ratio factor

$$R_{APH} = 1 + e\varepsilon_{APH}$$

This is usually small.

If the dampers act on the locating links, then the velocity diagram needs to be drawn, as in Figure 4.16.1. With this configuration, some rising rate can be achieved by angling the dampers in side view. The vertical wheel velocity is found from the wheel centre, and its relationship to points B and C. The axle angular speed ω_A is

$$\omega_A = \frac{V_{B/C}}{l_{BC}}$$

and the pitch/heave velocity ratio or coefficient ε_{APH} is

$$\varepsilon_{APH} = \frac{\omega_A}{V_S} \, [\text{rad/m}]$$

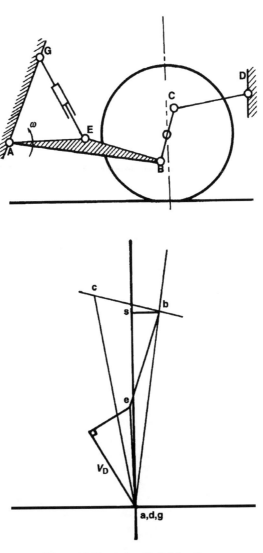

Figure 4.16.1 Axle with link location.

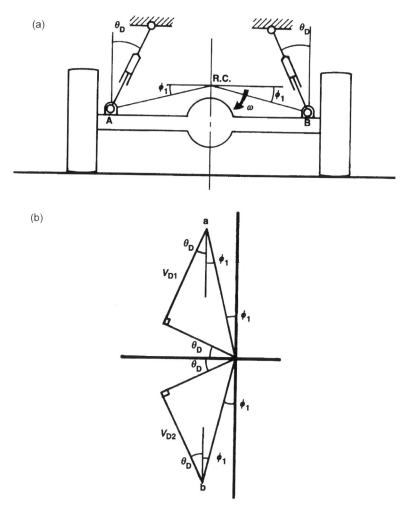

Figure 4.16.2 Inclined dampers on an axle, for axle sway vibration damping: (a) rear view; (b) velocity diagram.

To analyse the roll damping of the vehicle, for a simple axle, consider the vehicle to roll about the roll centre for the axle, which is the point of lateral location in the centre plane. The position of the roll centre is a complex issue, for which reference will need to be made to vehicle handling texts. However, in the case of the use of a Panhard rod for lateral location, it is the point at which the rod pierces the longitudinal central vertical plane. The velocity diagram in Figure 4.16.2. shows that the damper velocity V_D is

$$V_D = \omega l_1 \cos(\phi_1 + \theta_D)$$

so the damper velocity ratio in roll is

$$R_{D\phi} = l_1 \cos(\phi_1 + \theta_D) \quad [\text{ms}^{-1}/\text{rads}^{-1} = \text{m/rad}]$$

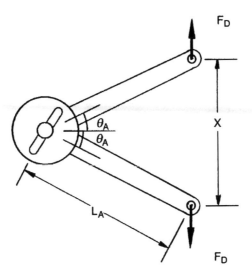

Figure 4.17.1 Geometry of the scissor-type snubber.

This will generally be substantially less than would be achieved with the dampers acting directly at the wheels. This is because l_1 is typically only about 0.7 of half the track, to allow the dampers to clear the wheels. This seriously reduces the roll damping.

For dampers acting on the links instead of the axles, the above process gives the velocity of the end of the link (without θ_D). The link is then analysed to obtain the actual damper velocity.

4.17 Dry Scissor Dampers

The dry scissor damper is not used nowadays, of course, although an analysis remains of interest. The disc pack creates a friction moment M_F which results in a damping force F_D at the end of the arms, given by

$$F_D = \frac{M_F}{L_A \cos\theta_A}$$

where L_A is the arm length and θ_A is the arm angle.

The friction moment depends upon the number of frictional surfaces N_F, their effective coefficient of limiting friction μ_F, the pack diameter D, the pack compression force F_{PC} and the radial distribution of the consequent contact pressure P. To give the best wear characteristic, a uniform pressure is preferred. In that case

$$P = \frac{4F_{PC}}{\pi D^2}$$

$$M_F = \frac{\pi}{12} N_F \mu_F P D^3 = \tfrac{1}{3} N_F \mu_F F_{PC} D$$

Hence, the scissor disc damper acts symmetrically with a Coulomb friction characteristic. This is one of the main objections to it, plus the fact that it is poor at dissipating the heat, hence running hot, which leads to a loss of friction when friction is most needed. When hydraulic dampers were first introduced, dry scissor dampers were sometimes added to overcome the problem that the early hydraulics had too little damping at low speed.

5
Fluid Mechanics

5.1 Introduction

Damper oil is usually a selected light mineral oil, sometimes instead a synthetic oil which is more expensive but which may have reduced variation of viscosity with temperature. The usual damper mineral oil contains sulphur compounds, giving it a lingering noxious smell. The relevant fluid properties include those in Table 5.1.1. The primary values are the density, typically varying from 850 to 860 kg/m^3, and the viscosity varying from 5 to 100 mPa s. Both of these are significantly temperature dependent. In a damper, the viscosity helps with lubrication, but is otherwise a nuisance. Lower viscosity oils would perform well, but have a higher vapour pressure and would be prone to cavitation. The old vane type dampers used very high viscosity oil to minimise leakage. Parallel-piston lever arm dampers reduced the viscosity requirement, and the telescopic type has reduced it further. Nevertheless, there is significant variation in the oils chosen.

The influence of liquid properties on the related problem of metering in carburettor jets has been studied by Bolt *et al.* (1971), and actually for dampers by Dalibert (1977).

The compressibility of the pure oil is small, less than 0.05 %/MPa, because it depends on distortion of the molecules, but in service conditions with absorbed gas, minute gas bubbles, etc. the practical liquid compressibility may be much greater, and important. A gas, such as air, has high compressibility, so its density varies easily with pressure variations. Nevertheless, the methods of 'incompressible flow' analysis may be applied to either, provided that the density varies little. Modelling a gas as incompressible really means that the pressure variations are so small that the density is nearly constant, so the phrase 'incompressible flow' really means 'approximately constant density under conditions applying'.

Bernoulli's equation makes this modelling assumption of effectively constant density fluid. Despite the small variations of density that always occur, incompressible flow methods form the basic approach to analysis, with compressibility to be dealt with as a deviation from this ideal model of the damper oil.

Even an ordinary passenger car damper must be able to withstand hot conditions (air temperature 30°C) coupled with severe actuation, raising fluid temperatures to over 100°C, giving a considerable operating temperature range. Manufacturers may specify that the damper should operate satisfactorily from −40 to 130°C. Reduced damper forces at high fluid temperatures have often been observed in damper tests (e.g. −0.1%/C to −0.3%/C), and it is often stated that this is due to the viscosity reduction giving a higher discharge coefficient by an increased velocity coefficient. In fact density variation due to thermal expansion is also important. The volumetric ('cubical') expansion coefficient of oil is about 0.1%/°C. Thus a 50°C temperature rise, easily created in vigorous testing, gives a 5% density reduction, which for a given volumetric flow rate would reduce the force by 5%, giving a variation

---

The Shock Absorber Handbook/Second Edition   John C. Dixon
© 2007 John Wiley & Sons, Ltd

Table 5.1.1 Representative damper oil properties (basic mineral oil)

1.	Density at 15°C	ρ	≈ 860	kg/m³
2.	Viscosity at 15°C	μ	≈ 40	mPa.s
3.	Temperature range	T	≈ -40 to $+130$	°C
3.	Pressure range	P	≈ 0 to 20	MPa
4.	Compressibility	$(d\rho/dP)/\rho$	≈ 0.05	%/MPa
5.	Thermal conductivity	k	≈ 0.14	W/m.K
6.	Thermal capacity	c_P	≈ 2.5	kJ/kgK
7.	Thermal expansion	$-(d\rho/dT)/\rho$	≈ 0.1	%/°C
8.	Viscosity–temperature sensitivity	$(d\mu/dT)/\mu$	≈ -2	%/°C
9.	Viscosity–pressure sensitivity	$(d\mu/dP)/\mu$	$\approx +3$	%/MPa
10.	Surface tension	σ_S	≈ 25	mN/m
11.	Air absorption coefficient	k_A	≈ 1.0	kg/m³MPa

coefficient of $-0.1\%/°C$. Thus, although temperature rise has a much more drastic effect on the viscosity than it has on the density, the consequences of density change may be just as important. High damper oil temperature is hardly a problem for a normal passenger car, but it may arise from hard working of the damper at speed on rough roads. Obviously this can be a problem for competition rally cars. Racing cars may also have problems, particularly at the rear, because engine radiator cooling air may pass over them and they may be in close proximity to the exhaust system. Desert racing 'buggies' in high air temperatures are a classic extreme case, and may require water cooling of the dampers by a dedicated radiator. At the other extreme, a damper may be required to operate after having been left overnight in cold weather conditions. The basic density and viscosity of the fluid are therefore specified at some known reference temperature, e.g. 15°C, with the important variation with temperature accounted for by appropriate equations and coefficients.

The fluid flows through passages of various shapes and through deliberately restrictive valves. The analysis of pressure losses and flow rates may involve all three basic principles of fluid mechanics:

(1) the Principle of Continuity — volume or mass;
(2) energy analysis — Bernoulli's equation covers variations of pressure, speed and height for constant density flow; and may be extended to include friction losses;
(3) momentum analysis — to investigate forces on valve parts.

The pressure losses are obviously of prime importance in a damper, but, despite the wealth of information available on the such topics, in practice it may be difficult to calculate the damper behaviour accurately from first principles. This is because the flow passages are of complex shape, and the losses are somewhat unpredictable. Nevertheless, theory does provide useful grounding for understanding the behaviour of fluid flow in dampers.

Fluid mechanics is a complex subject. The summary material here is offered only as an *aide memoire* and relevant handy reference, for those with existing experience of this field of study.

5.2 Properties of Fluids

The term 'fluids' means liquids or gases. Automotive dampers are based on liquid, although in some cases including some emulsified gas. Separated gas is frequently present in the damper. Some nonautomotive dampers do actually use gas alone, and the use of air has some obvious advantages, especially where only small forces are needed, as for damping camera mechanisms, eliminating the problem of leakage which occurs with liquids. Air dampers have also been used on pedal bicycles with suspensions.

The properties of the damper liquid, normally a mineral oil with some additives, may be classified under several headings, basically chemical, mechanical, thermal and others. In more detail these are:

(1) chemical structure and additives;
(2) density;
(3) thermal expansion;
(4) compressibility;
(5) viscosity;
(6) thermal capacity;
(7) thermal conductivity;
(8) vapour pressure;
(9) gas density;
(10) gas viscosity;
(11) gas compressibility;
(12) gas absorbability;
(13) emulsification.

5.3 Chemical Properties

Chemical properties involve the basic specification of the liquid in terms of its molecules. Hence it may be specified as a mineral oil (i.e. a hydrocarbon) refined from crude oil, with a given mean relative molecular mass (molecular weight), or perhaps as some synthetic oil such as a silicon-based oil, or polybutene etc. Quantities of additives, by mass or volume, and their individual specification, will be included, e.g. anti-wear additives, anti-foaming agents, and so on. This detailed specification is primarily the province of the producer of the damper oil; the damper engineer himself will usually just specify one or more allowable products by the oil manufacturer's name and specification number. Included here may be safety-related data such as combustibility, ignition temperatures, toxicity, etc., or other hazards. Damper oils are combustible, but this is rarely a significant fire hazard. However, the oxidisability of the oil can cause long term deterioration, which is why anti-oxidants may be included.

For a typical damper oil the relative molecular mass ('molecular weight') is about 350 kg/kmol, the empirical chemical formula being approximately $C_{25}H_{52}$. Avogadro's number is 6.0225×10^{26} molecules/kmol, so one molecule of such oil has a mass of 0.580×10^{-24} kg. From the oil density, it may be calculated that the average volume occupied by such a molecule is 0.676 nm^3.

5.4 Density

The density, represented by ρ (rho) is, of course, the mass per unit volume:

$$\rho = \frac{m}{V}$$

The liquid density has an important effect on damper performance. It is likely to be around 860 kg/m^3 (0.860 g/cm^3). Larger hydrocarbon molecules have a higher ratio of carbon to hydrogen, making the density greater.

The density must be distinguished from the relative density d (previously known by the obsolete term specific gravity) which is the density normalised and nondimensionalised by a reference density. This reference density used is conveniently 1000 kg/m^3 (which is the density of pure water at its greatest, at 4 °C). Hence

$$d = \frac{\rho}{\rho_{ref}} = \frac{\rho}{1000 \, \text{kg/m}^3} \quad \text{[nondimensional]}$$

Hence, the relative density d is numerically equal to the density when expressed in grams per cubic centimetre (g/cm³), but has no dimensional units, so is typically 0.860.

5.5 Thermal Expansion

For most practical damper liquids, over the temperature range of interest the density can be considered to be a reducing linear function of temperature. This may be written as

$$\rho = \rho_1 \{1 - \alpha(T - T_1)\}$$

or as

$$\rho = \frac{\rho_1}{1 + \alpha(T - T_1)}$$

where the reference specification is ρ_1 at T_1 (e.g. 15°C). For realistic small expansions these two equations amount to much the same thing. The former expression is perhaps clearer in meaning, but the second form may be slightly more accurate over a wide temperature range, and preferred, particularly when the compressibility is also considered. Parameter α (alpha) here represents the coefficient of volumetric thermal expansion or so-called 'cubical' thermal expansion with a value for oils around

$$\alpha \approx 0.001 \, \text{K}^{-1} (\text{or} \, ^\circ \text{C}^{-1}) \approx 1000 \, \text{ppm/K}$$

For a solid isotropic material, the volumetric expansion is three times the linear expansion, but a liquid cannot really be said to have a linear expansion in the same sense as a solid. The specified complete operating temperature range of the liquid, which may be 170 K or more (−40 to 130°C) then gives an approximately 17% change of volume and density. Although this density change contributes to damper fade, it may also be turned to positive effect by using it to adjust the valve to compensate for viscosity change.

5.6 Compressibility

The density of a liquid is affected by the pressure, in a fairly linear way for practical damper pressures (a few megapascal):

$$\rho = \rho_1 [1 + \beta(P - P_1)]$$

Here, β is the compressibility, the reciprocal of the bulk modulus K. A mineral damper oil has long chain hydrocarbon molecules which do not pack efficiently together. This allows a higher compressibility than a liquid such as water because the long molecular chains can distort.

The normal reference pressure P_1 used is usually one standard atmosphere, 101 325 Pa, around 0.10 MPa. For a pure oil the bulk modulus is around 1.5 GPa, so the compressibility has a value of around

$$\beta \approx 670 \times 10^{-12}/\text{Pa} \approx 670 \times 10^{-6}/\text{MPa} \approx 0.07\%/\text{MPa}$$

for conventional damper mineral oils in clean new condition. Hence a practical working pressure of 5 MPa will increase the density of a pure oil by only 0.35%. Compliance of the steel pressure cylinder may have more effect on the volume than that (see Section 7.14). In service conditions, the compressibility is greatly increased by small quantities of emulsified gases, considered later, so the value for pure clean oil is not of great practical significance for automotive dampers. To show up as a visible effect in the measured F(V) loop, the compressibility must be 0.5%/MPa or more. Oils under service conditions, generally with entrained air bubbles, are sufficiently compressible for it to be significant factor in the damper forces obtained in high frequency operation.

Aircraft undercarriage oleo legs are designed with oil compressibility as an important factor, but the pressures used are very high. The compressibilities assumed for design purposes are typically 0.07%/MPa. Values may be as high as 0.10%/MPa for a specially compressible oil, i.e. 7 and 10% at 100 MPa, and these rates reduce at higher pressures.

5.7 Viscosity

The viscosity of a fluid is the resistance to shearing motion. It is a particularly important parameter for a damper fluid. It is sometimes described as the 'stickiness' of a liquid, but this is scientifically incorrect. This confusion arises because common liquids such as syrup are viscous and also adhesive to the skin. Considering two plates, the upper one with a tangential motion at velocity V, with liquid between, for a simple constant viscosity liquid the velocity distribution between the plates is linear, with a velocity gradient

$$\frac{dV}{dy} = \frac{V_1}{h}$$

The plates of area A then have a shear force F opposing the motion, with

$$F = \mu A \frac{dV}{dy}$$

where μ is the coefficient of dynamic viscosity, with the SI unit N s/m^2 or Pa s (Pascal seconds). This happens to be numerically equal in value to the dynamic viscosity in the older unit of centipoise. Implicit in the above is that the viscosity is independent of the speed, and hence that the resistant shear force is proportional to speed. This is not exact; it is only a model of fluid behaviour. It is a good model for many liquids in practical conditions, and is suitable for damper oils. Some liquids have viscosity values which vary considerably with velocity, and even with the recent history of their motion.

The coefficient of dynamic viscosity μ is also known loosely as just the dynamic viscosity, and even just as the viscosity. In fluid dynamic problems the quotient μ/ρ occurs frequently, and this has been given the name kinematic viscosity, represented by the Greek letter nu:

$$\nu = \frac{\mu}{\rho}$$

The reason for the name is that the SI unit of kinematic viscosity is m^2/s, purely kinematic units, contrasting with N s/m^2 = Pa s for dynamic viscosity.

The viscosity of mineral and other oils depends very much upon their molecular structure, and is also sensitive to temperature, which can cause problems. The basic mechanisms of viscosity are:

(i) Thermal agitation of the fluid results in molecular exchange between adjacent layers at different velocities, giving momentum exchange and energy dissipation.
(ii) Weak secondary bonds are continuously formed between adjacent molecules and then broken, giving energy dissipation.
(iii) Long molecules become tangled, and are then pulled apart, giving energy dissipation.

Hence, it is understandable that the size and nature of the molecular structure is highly influential in viscosity. In fact, for oils generally, larger molecules generally mean higher viscosity, and also higher density, so viscosity is found to be broadly related to the oil density.

Because of the considerable effect of high temperature, which reduces viscosity, the viscosity of an oil should be quoted for a specific temperature, although 'room temperature', 15°C, is often taken as a default value. For a representative damper oil, which would usually be a light mineral oil, the dynamic viscosity is typically around 40 mPa s at 15°C. For comparison, the viscosity of water is about 1 mPa s. Higher room-temperature viscosities are problematic at very low temperatures, but low room-temperature viscosity is troublesome at high temperatures due to high vapour pressure and lack of lubrication.

Over a narrow temperature range, the effect of temperature on dynamic viscosity may be represented approximately by a linear reduction

$$\mu = \mu_1[1 + k_{\mu T}(T - T_1)]$$

with $k_{\mu T}$ having value around $-0.02/°C$ or $-2\%/°C$. Hence a 10°C temperature increase may give a 20% viscosity reduction.

Because of the large variation of viscosity, this simple linear expression is not accurate over the full range of service temperatures. The viscosity may then better be represented by the de Guzmann–Carrancio equation

$$\mu = \mu_0 e^{E/(RT)}$$

where E is a characteristic energy value (in practice 1/3 to 1/4 of the latent heat of vaporisation), R is the universal gas constant, and T is the absolute temperature. This may be re-expressed as

$$\mu = \mu_1 e^{C\left(\frac{1}{T} - \frac{1}{T_1}\right)}$$

where C will be a positive coefficient, expected to be approximately constant at a value around E/R, and μ_1 is the viscosity at absolute temperature T_1. Then

$$\log \mu = \log \mu_1 + \frac{C}{T} - \frac{C}{T_1}$$

$$\log \mu = A + \frac{C}{T}$$

Hence, this model predicts that the logarithm of viscosity plotted against the reciprocal of absolute temperature will be a straight line, which proves to be a fairly good approximation for most real liquids.

For unimproved mineral oils the temperature sensitivity coefficient increases as the molecular weight increases and the viscosity at room temperature increases. A reasonable estimate may be made from

$$C = 5693 - 304 \log_{10}(\mu_{15}) - 646 \log_{10}^2(\mu_{15})$$

where μ_{15} is the viscosity at 15°C in Pa s, applicable for $0.003 < \mu_{15} < 0.300$ Pa s. Additives may reduce the sensitivity. Silicon oils may have a sensitivity around half that of mineral oils.

A representative viscosity for a light oil is $\mu = 10$ mPa s at 288 K, with $\log_{10}(\mu_{15}) = -2$, predicting a temperature sensitivity $C = 3717$ K (for use with logarithms to base e in the de Guzmann–Carrancio equation). This is a worst case unimproved oil. Figure 5.7.1 illustrates how the viscosity varies dramatically with temperature over the required operating temperature range.

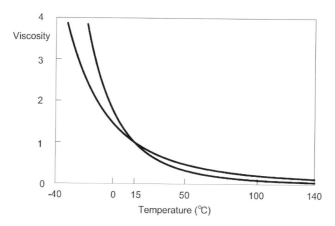

Figure 5.7.1 Mineral oil viscosity relative to its room-temperature viscosity (viscosity sensitivity $C = 2000$ and 3000 K).

In the USA, the empirical Walther viscosity equation is considered good for mineral oils. This equation is

$$\log\{\log(v + 0.6)\} = m \log T + b$$

(using a logarithm of a logarithm) where v is the kinematic viscosity in centistokes, T is the temperature in degrees Rankine, and m and b are constants for a given oil.

The effect of pressure on viscosity is typically about $+3\%$/MPa. However, in a damper, at the point of high velocity the static pressure will not be very high (by Bernoulli's equation) so the viscosity–pressure sensitivity is unlikely to have much effect.

For the viscosity of a mixture of liquids, an estimate may be made by Kendall's equation:

$$\log \mu = X_1 \log \mu_1 + X_2 \log \mu_2$$

where the X values are mass fractions, volume fractions or mole fractions, whichever is convenient, indicating the approximate nature of the equation.

For the viscosity of a dilute suspension of solid spheres in a liquid, Einstein's equation may be used:

$$\frac{\mu}{\mu_0} = 1 + 2.5\varphi$$

where φ is the volume fraction of the spheres. However, this makes no allowance for the effect of the particle size and shape, which appears to be significant in practice.

5.8 Thermal Capacity

The specific thermal capacity c_P of light mineral oil is around

$$c_P \approx 2500 \text{ J/kg K}$$

The thermal capacity of a given mass m of oil is then

$$C_P = m c_P \quad [\text{J/K}]$$

The c_P value may be significantly different for the more esoteric oils occasionally used, e.g. silicone oils as used in racing. However, for damper engineering purposes the specific thermal capacity can be considered constant over the temperature range of interest. A mass of 160 g of light oil may then be considered to have a constant thermal capacity of about 400 J/K.

5.9 Thermal Conductivity

The thermal conductivity of light mineral oil is typically around

$$k = 0.15 \text{ W/m.K}$$

5.10 Vapour Pressure

The vapour pressure of a liquid increases very rapidly with temperature, until, in free air, it equals the external pressure, at which point boiling occurs, the liquid converting freely to gas at a rate controlled by the supply of energy and the specific evaporation energy (latent heat) of the oil. The boiling temperature therefore varies with the external pressure; unless otherwise indicated, a boiling temperature applies to a pressure of one standard atmosphere (101.325 kPa).

The vapour pressure is usually represented by the Antoine equation:

$$\log_{10} P_V = B - 0.2185 \frac{A}{T}$$

where A and B are constants (approximately) and T is the absolute temperature. In SI units, $B \approx 8.73$ and A is the molar heat of vaporisation, about 3700 kJ/kg.

If a liquid boils at low temperature because the applied pressure is reduced to the vapour pressure, this is called cavitation. Any sort of vaporisation is problematic, because an increase of pressure can almost instantly reconvert the vapour to liquid. When this occurs for small vapour bubbles at or very near to a solid surface, the result is a small but very powerful jet of liquid directed at the surface, which causes erosion. This has been known to cause rapid damage to damper pistons. A vapour pocket or vapour bubble formed during one stroke of a damper is liable to collapse suddenly near to the end of the stroke, or at reversal, giving extreme harshness and noise, high structural stresses and accelerated deterioration of the oil.

Hence, even if the boiling point of the oil is above the damper operating temperature, low pressures within the damper, i.e. pressures less than atmospheric, must be avoided.

5.11 Gas Density

Gas is included in dampers, separately from the oil, to provide compressibility to allow for the rod displacement volume. The gas may be air, although dampers may be charged with nitrogen to reduce corrosion and oil oxidation. The absolute pressure, density and temperature are related by the perfect gas equation:

$$PV = mRT_K$$

where T_K is the absolute (kelvin) temperature:

$$T_K = T_C + 273.15 \text{ K}$$

and R is the specific gas constant. For air this is

$$R_A = 287.05 \text{ J/kg K}$$

and for nitrogen

$$R_N = 296.80 \text{ J/kg K}$$

The density of the gas is then

$$\rho_G = \frac{m}{V} = \frac{P}{RT_K}$$

5.12 Gas Viscosity

Gas viscosity is sensitive to temperature, varying approximately as the absolute temperature to the power 1.5. Typically it is represented by Sutherland's equation

$$\mu = \mu_1 \left(\frac{T}{T_1}\right)^{1.5} \frac{C + 273.15}{C + T}$$

where T is the absolute temperature. The factor C provides a correcting refinement. In automotive dampers the viscosity of the gas is of little significance.

5.13 Gas Compressibility

The compressibility of a gas or liquid is defined as

$$\beta = -\frac{1}{V}\frac{dV}{dP} = \frac{1}{\rho}\frac{d\rho}{dP}$$

which, using the perfect gas equation, becomes

$$\beta = \frac{1}{P}$$

Hence, for an ideal gas, including air and nitrogen, the compressibility equals the reciprocal of the absolute pressure. The bulk modulus of any material is defined as the reciprocal of the compressibility:

$$K = \frac{1}{\beta}$$

so for a gas the bulk modulus equals the pressure. The bulk modulus and compressibility of a gas are therefore far from constant. For accurate modelling, the above equations should be used, although simpler constant values may be of use for modelling the damper as a component in the context of the whole vehicle.

5.14 Gas Absorbability

The quantity of gas that can dissolve in a liquid depends on the particular gas and liquid. If there is any chemical affinity between the two, the amount may be considerable. The extreme case is when the gas is the vapour of the liquid in which case the absorbability may be infinite. For nonreacting materials, the maximum absorbable mass of gas m_{GA} is usually modelled by the linear Henry's equation, often known as 'Henry's Law' (William Henry, 1774–1836):

$$m_{GA} = C_{GA} V_L P_G$$

where m_{GA} is the mass of gas absorbed of the gas in the liquid, C_{GA} is the gas absorption coefficient, V_L is the volume of liquid, and P_G is the partial pressure of the particular gas above the liquid. In other words, the absorption is proportional to the pressure of the gas above. For example, at 15 °C for oxygen in water the absorption coefficient is 0.38 kg/m³MPa, for nitrogen 0.18 kg/m³MPa. The absorption coefficient is often expressed by its reciprocal, called Henry's coefficient. The solubility reduces quite rapidly with temperature.

For air in mineral oil, the absorption coefficient is about

$$C_{GA} = 1 \text{ kg/m}^3\text{MPa}$$

This means that at 1 MPa the oil will absorb roughly its own volume of air as seen at normal pressure. This is only a problem if the oil is first kept at high pressure in the presence of air allowing the absorption to take place, and the pressure is then reduced allowing the gas to emerge into bubbles throughout the liquid. Two well known examples of this are (1) CO_2 from lemonade when the bottle top is removed, (2) diver's 'bends' when nitrogen desorbs from the blood.

This absorbed mass of gas increases the mass of liquid:

$$m_L = m_{L0} + m_{GA}$$

It also increases the volume of liquid somewhat. Allowing simultaneously for the compression through the bulk modulus, the resulting liquid volume is

$$V_L = V_{L0} - \frac{V_{L0}}{K_L}P + C_{GLV}\frac{m_{GA}}{m_{L0}}V_{L0}$$

where C_{GLV} is the gas absorption volume coefficient (not to be confused with C_{GA}). The value of C_{GLV} depends on how well the different gas and liquid molecules pack together, but it can be expected to be of order 1.0. Substituting for m_{GA} gives

$$V_L = V_{L0} - \frac{V_{L0}}{K_L}P + \frac{C_{GLV}C_{GA}V_{L0}^2}{m_{L0}}P$$

where, evidently, the increase of volume due to absorption offsets the reduction of volume due to compressibility of the base liquid. Substituting reasonable values indicates that these effects may roughly compensate. In fact they are equal when

$$C_{GLV} = \frac{\rho_{L0}}{K_L C_{GA}} \approx 1$$

Absorption of the gas may take some time, depending on the area of liquid surface exposed to the gas, e.g. many small bubbles with a large total surface area. If the liquid near to the gas becomes saturated, then agitation of the liquid will encourage further absorption by bringing unsaturated liquid into proximity with the gas.

Absorption of gas into the liquid reduces the mass of free gas and has a significant effect on the density and compressibility of an emulsion.

At a pressure of 10 MPa, the maximum absorption is such that the numerical ratio of air to oil molecules is about 0.14. A sphere of diameter 10 nm will contain about 800 oil molecules and 110 air molecules.

When the pressure is reduced, the maximum mass of soluble gas may fall below the actual dissolved gas mass. The initial situation is then the existence of a supersaturated solution. The surplus gas can come out of solution, causing extensive cavitation-type problems. However, this process may be delayed, because there may be no energetically favourable route for the desorption to occur. If a few molecules of air try to combine into a bubble, it will be an extremely small one. There would then be a very high bubble internal pressure due to surface tension, which would be sufficient to cause reabsorption of the gas.

The surface tension of oil against air is small, $\sigma_S \approx 25$ mN/m, and its effects on a large scale may be modest, but on a small scale the effect may be enormous. Consider a bubble of diameter D, sectioned across a great circle. The total surface tension force around the periphery, pulling the two halves together, applied to the cross-sectional area, gives the additional pressure inside the bubble. The force is

$$F = \pi D \sigma_S = \frac{\pi}{4} D^2 P_{ST}$$

so the surface tension bubble pressure P_{ST} is

$$P_{ST} = \frac{4\sigma_S}{D}$$

This pressure acts in addition to the static pressure of the oil. For a normal bubble, say $D = 0.1$ mm, the surface tension pressure is only about 1 kPa. However, for a very small bubble of a few molecules trying to desorb, the surface tension pressure may be enormous. To generate a pressure of 10 MPa, the bubble diameter is 10 nm, with a volume of 524×10^{-27} m^3 = 524 nm^3. At 10 MPa pressure and 310 K, the density of air is 112.4 kg/m^3. The mass of one air molecule is 48.1×10^{-27} kg (Appendix B) so the number of molecules in such a 10 nm diameter bubble is

$$N = \frac{\rho V}{m_1} = 1224$$

Smaller bubbles than this cannot easily desorb, because the surface tension pressure is too high.

Desorption therefore depends on the existence of nucleation points to provide an energetically favourable route to adequately sized bubbles. This may be seen in a lemonade bottle when a stream of bubbles forms from one point, with relative stillness of the bulk of the liquid. This depends on the existence of sharp points or edges of dirt or wear debris in the oil. Therefore the time for the desorption process is not readily predictable.

The size of a bubble for N air molecules is

$$D = \sqrt{\frac{3 m_1 R_G T_{KG} N}{2\pi \sigma_S}}$$

where $m_1 = 48.1 \times 10^{-27}$ kg is the mass of one air molecule, e.g. $N = 100$, $D = 2.86$ nm, $P_{ST} = 35$ MPa.

5.15 Emulsification

Commonly, an emulsion is a combination of two essentially immiscible liquids, in which one of the liquids is divided into very fine droplets suspended fairly uniformly in the other liquid. The liquids are immiscible in the sense that they do not blend uniformly at the molecular level. The term emulsion is

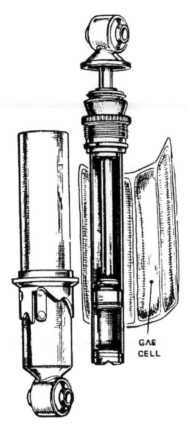

Figure 5.15.1 The Girling Arcton damper, which used a nylon sleeve containing the gas to prevent emulsification of the oil.

also applied to a finely dispersed gas in a liquid, with the gas in very small bubbles. If the volume of gas much exceeds that of the liquid, an emulsion is considered to be a foam (also known as a froth). Characteristics of bubbly oil are considered by Hayward (1966). If emulsification occurs in a damper, there can be a dramatic reduction in damper force, especially for short strokes, because the emulsion compressibility inhibits the generation of chamber pressures. The importance of this problem is illustrated by Figure 5.15.1, in which the gas is separated from the liquid, an effective but expensive solution. The problem is also solved by the single-tube damper, in which a separator piston is used.

To qualify as an emulsion, a further factor is that the gas content must not be easily absorbed by the liquid. Hence the presence of fine bubbles of oil vapour does not constitute an emulsion; their behaviour is totally different because, however great the volume of oil vapour or however finely divided, the vapour is easily reconverted to liquid oil by a modest increase in pressure.

In contrast, a true gas–liquid emulsion is highly stable. The volume of gas is reduced by increased pressure, of course, perhaps considerably so, but the gas can be absorbed into the liquid only to a well-defined limited extent. Absorption can only be fast if the liquid has the residual absorption capacity and the gas is in the form of many small bubbles with a large total surface area. Once a fine gas–liquid emulsion is formed, it is difficult to separate out the fine gas bubbles. Bubbles will gradually rise in a liquid because of their buoyancy, forming a surface foam and slowly clearing the liquid, but with small bubbles the drift speed is very low, and damper action agitates the liquid, discouraging separation. A standard double-tube damper may suffer inadvertent emulsification and consequent reduction of damping effect. A moderate degree of emulsification may be beneficial to ride quality, having a similar effect to rubber bushes, giving some stroke sensitivity. Some dampers are designed to operate in the emulsified state, with the valves set appropriately for the condition of the emulsion, with its lower density. Even in that case, the mass fraction of the gas remains small.

Gas bubbles (an emulsion) or vapour bubbles (not an emulsion) can appear in a liquid in several ways:

(1) boiling — vapour bubbles occur due to energy input and rising temperature at constant pressure, when the vapour pressure exceeds the static pressure;
(2) cavitation — vapour bubbles occur due to reduced static pressure at constant temperature, when the vapour pressure exceeds the static pressure;
(3) desorption of previously absorbed gas, due to a reduction of static pressure;
(4) mechanical agitation, entraining gas at the liquid/gas surface and mixing it throughout the bulk of the liquid.

Emulsion Density

Consider a mass m_G of gas (probably air) emulsified with mass m_L of liquid oil, of standard density ρ_L. At an absolute pressure P and absolute temperature T_K (kelvin), using the perfect gas equation the gas density ρ_G is

$$\rho_G = \frac{P}{R_G T_K}$$

where R_G is the specific gas constant for the gas ($8314.5/28.965 = 287.05$ J/kg K for air). The volume of gas V_G is

$$V_G = \frac{m_G}{\rho_G} = \frac{m_G R_G T_K}{P}$$

At standard temperature and pressure (15°C = 288.15 K, 101.325 kPa) the density of air is 1.2256 kg/m³. At a pressure of 10 MPa the density becomes 121 kg/m³ at the same temperature, still much less than that of the oil. For normal bubbles, the pressure can be considered the same as that of the liquid. This is not true for very small bubbles, as discussed later.

The gas volume is very nonlinear with pressure because of the pressure range, so for computer calculations it is best to do independent calculations for the gas and liquid volumes. However, simple algebraic analysis may give some useful insight.

Without any absorption of gas into the liquid, the masses of liquid and gas are constant, and the volumes of the liquid and gas are

$$V_L = \frac{m_L}{\rho_L}$$

$$V_G = \frac{m_L}{\rho_G}$$

The total emulsion values are then

$$m_E = m_L + m_G$$
$$V_E = V_L + V_G$$

and the emulsion density ρ_E is

$$\rho_E = \frac{m_E}{V_E} = \frac{m_L + m_G}{V_L + V_G}$$

The mass fraction of gas f_{Gm} is

$$f_{Gm} = \frac{m_G}{m_L} = \frac{\rho_G}{\rho_L}\frac{V_G}{V_L}$$

The volume fraction of gas f_{GV} is

$$f_{GV} = \frac{V_G}{V_L} = \frac{\rho_L}{\rho_G} f_{Gm}$$

For small bubbles, heat transfer between the gas and liquid is good, and it is appropriate to use a gas temperature equal to that of the liquid. At a varied pressure P instead of reference P_0, and constant temperature, the emulsion density becomes

$$\rho_E = \frac{m_E}{V_E} = \frac{m_E}{V_{L0}[1 - \beta_L(P - P_0)] + V_{G0}P_0/P}$$

Neglecting the small effect of the pure liquid compressibility, and the small gas mass contribution,

$$\rho_E \approx \frac{m_L}{V_{L0} + V_{G0}P_0/P} \approx \frac{\rho_L}{1 + \left(\frac{V_{G0}}{V_{L0}}\frac{P_0}{P}\right)}$$

Figure 5.15.2(a) shows the density of oil/air emulsion without absorption, over a practical range of damper pressures. Each curve is for one value of the mass ratio m_G/m_{L0}.

The above analysis is oversimplified, in that the emulsion volume is in fact affected significantly by absorption of the gas into the liquid. The maximum gas mass absorption (Section 5.14) is

$$m_{GAmax} = C_{GA}V_{L0}P$$

If the total mass of gas is less than this, then there can be complete absorption and no free gas at all. Then

$$m_G \leq m_{GAmax}$$

$$m_{GA} = m_G$$

$$m_{GF} = 0$$

$$m_L = m_{L0} + m_{GA}$$

Allowing for the liquid compressibility and the absorption, the emulsion volume is then

$$V_E = V_L = V_{L0} - V_{L0}\frac{P}{K_L} + C_{GLV}\frac{m_{GA}}{m_{L0}}V_{L0}$$

where C_{GLV} is the gas absorption volume coefficient (Section 5.14, not to be confused with C_{GA}). The 'emulsion' density is then just

$$\rho_E = \rho_L = \frac{m_L}{V_L} = \frac{m_{L0} + m_G}{V_E}$$

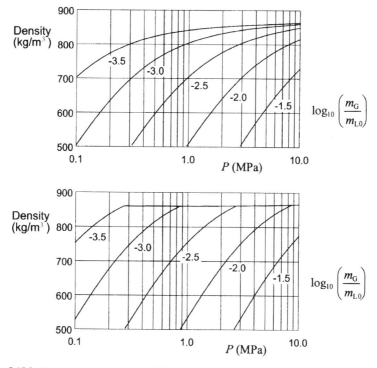

Figure 5.15.2 The mean density of an oil/air emulsion: (a) without absorption; (b) with absorption.

In the more complex case, with a greater mass of gas, or at lower pressure, some of the gas remains unabsorbed:

$$m_G \geq m_{GAmax}$$

$$m_{GA} = m_{GAmax}$$

$$m_{GF} = m_G - m_{GAmax}$$

$$m_L = m_{L0} + m_{GA}$$

Allowing for the liquid compressibility and absorption, the liquid volume is

$$V_L = V_{L0} - V_{L0}\frac{P}{K_L} + C_{GLV}\frac{m_{GA}}{m_{L0}}V_{L0}$$

The free gas volume is

$$V_{GF} = \frac{m_{GF}R_G T_{KG}}{P}$$

where T_{KG} is the absolute gas temperature. If the gas is in small bubbles then the compression may be close to isothermal, but large gas pockets may be adiabatic. The total volume is

$$V_E = V_L + V_{GF}$$

and the emulsion mean density is

$$\rho_E = \frac{m_E}{V_E} = \frac{m_{L0} + m_G}{V_L + V_{GF}}$$

Figure 5.15.2(b) shows the density of oil/air emulsion with absorption. The discontinuity in these curves is the point of complete absorption, beyond which, at higher pressure, the emulsion is simply a liquid.

Emulsion Compressibility/Bulk Modulus

In the context of dampers, the concept of compressibility, β_E, of an emulsion is of limited quantitative value, as it is either too small to matter or too nonlinear an effect to be modelled by a bulk modulus. However, it has some use as an indicator of whether compressibility will be significant or not.

For small pressures, neglecting absorption, the mass of free gas is constant, so the compressibility of the emulsion is

$$\beta_E \approx \beta_L + f_{GV}\beta_G$$

where the gas volume fraction f_{GV} is evaluated at the pressure of interest, and β_E applies locally at that pressure. For practical cases, with significant gas content, this reduces simply to

$$\beta_E \approx f_{GV}\beta_G$$

Because of the large pressures in dampers, the effective compressibility will vary considerably.

The ratio of the compressibility of the emulsion to that of the pure liquid is

$$\frac{\beta_E}{\beta_L} \approx 1 + f_{GV}\frac{\beta_G}{\beta_L}$$

This shows particularly clearly that, because the compressibility of a gas, $\beta_G = 1/P$, is, even at high pressure, much greater than that of the liquid, even a small mass fraction of gas will have a drastic effect on the compressibility of the emulsion.

The pure oil compressibility is only 0.05/GPa. It requires a compressibility of about 5/GPa to show up in the $F(X)$ loop. This occurs for a gas volume content of about 0.05%.

Figure 5.15.3(a) shows how the emulsion bulk modulus varies with pressure for an oil/air emulsion over a practical damper pressure range, without absorption.

Taking gas absorption into account, if the total mass of gas is less than the absorbable mass,

$$m_G \leq m_{GAmax} = C_{GA}V_{L0}P$$

then there is no free gas and the bulk modulus of the emulsion is approximately

$$K_E \approx K_L$$

where the small but uncertain effect of the absorbed molecules on the liquid compressibility has been neglected.

If the total mass of gas exceeds the absorbable quantity, as shown earlier, neglecting the effect of absorption on the liquid volume V_L, it is

$$V_L = V_{L0} - V_{L0}\frac{P}{K_L}$$

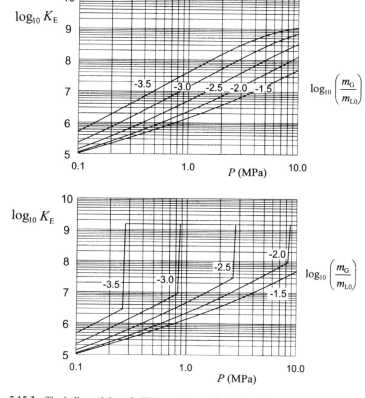

Figure 5.15.3 The bulk modulus of oil/air emulsion: (a) without absorption; (b) with absorption.

and the free gas volume is

$$V_{GF} = \frac{(m_G - m_{GA})R_G T_{KG}}{P}$$

with

$$m_{GA} = C_{GA} V_{L0} P$$

giving

$$V_{GF} = \frac{m_G R_G T_{KG}}{P} - C_{GA} V_{L0} R_G T_{KG}$$

so the compressibility of the gas fraction is

$$\frac{dV_{GF}}{dP} = -\frac{m_G R_G T_{KG}}{P^2} = -\frac{V_{GT}}{P}$$

where V_{GT} is the volume that the total amount of gas would occupy at the prevailing pressure. Surprisingly, the compressibility of the gas fraction is independent of the absorption coefficient C_{GA}.

Although the mass of free gas is less at high pressure, reducing the compressibility, this is just compensated by the direct effect of pressure increase on further absorption.

The emulsion volume is

$$V_E = V_L + V_{GF}$$

Isothermally, then,

$$\frac{dV_E}{dP} = -\frac{V_{L0}}{K_L} - \frac{V_{GT}}{P}$$

The emulsion bulk modulus is, by definition,

$$K_E = -V_E \frac{dP}{dV_E}$$

which is

$$K_E = \frac{V_E P K_L}{P V_L + K_L V_{GT}}$$

If the effect of absorption on the liquid volume is included, the bulk modulus result is

$$K_E = \frac{V_E}{\frac{V_{L0}}{K_L} + \frac{m_G R_G T_{KG}}{P^2} - \frac{C_{GLV} C_{GA} V_{L0}^2}{m_{L0}}}$$

Figure 5.15.3(b) shows the bulk modulus for an oil/air emulsion including absorption, to be contrasted with (a) without absorption. In particular, with absorption there is a large discontinuity at the pressure for full absorption. The process of absorption itself contributes to the effective compressibility.

At the point of full absorption, the step change in K_E may be analysed as follows:

$$m_G = C_{GA} V_L P$$

$$V_{GT} = \frac{m_G R_G T_{KG}}{P} = C_{GA} V_L R_G T_{KG}$$

$$\frac{K_E}{K_L} = \frac{P}{P + K_L C_{GA} R_G T_{KG}} \quad \text{(at full absorption point)}$$

For example, with $C_{GA} = 1$ kg/m^3 MPa, $R_G = 287$ J/kg K, $T_{KG} = 310$ K, $K_L = 1.5$ GPa, $P = 1$ MPa, then $K_E/K_L = 1/134.5$, a step change in bulk modulus of more than two orders of magnitude.

In summary, the general behaviour is that for a pressure less than the fully absorbing pressure, the bulk modulus is as though the gas were not absorbed at all. At the full absorption pressure, the bulk modulus is sharply discontinuous, changing to that of the liquid.

Emulsion Viscosity

The viscosity of gas/liquid emulsions, as found in dampers, does not seem to have been the subject of published investigations, although individual manufacturers may have proprietary information. In view of the negligible viscosity and relatively small mass fraction of the gas, it seems likely that the emulsion viscosity μ_E could be expressed by

$$\mu_E = \mu_L (1 + k_\mu f_{GV})$$

where k_μ is an empirical constant, resulting from actual flow patterns in the presence of gas bubbles. A linear relationship between μ_E and f_{GV} seems highly likely for small gas volume fractions, but this remains speculative.

Some evidence points to an increase rather than decrease of viscosity with gas content, with k_μ having a value of 1.5 for f_{GV} up to 0.30. This may be because the surface tension maintains the bubbles close to spherical, and despite the lack of a no-slip condition at the bubble surface the liquid must follow a convoluted path around the bubbles with greater shearing action. Beyond this value of f_{GV}, the emulsion becomes a froth (foam) with quite different properties.

De-emulsification

De-emulsification is a slow process, even when the damper is stationary, because of the slow speed at which small bubbles rise in the liquid. The shape of a rising bubble depends on the ratio of gravity forces to surface tension forces. The gravity force creates a pressure difference $\rho g D$ from top to bottom, which distorts the bubble, trying to make it lenticular. A large bubble has a high rising velocity and assumes a somewhat hemispherical shape, as may be observed in films of aqualung divers.

In contrast, the surface tension tries to make the bubble circular. The ratio of forces is

$$\frac{F_G}{F_{ST}} = \frac{(\rho_L - \rho_G)gD\frac{\pi}{4}D^2}{\sigma_S \pi D} \approx \frac{\rho_L g D^2}{4\sigma_S}$$

This is commonly represented by the Bond number:

$$N_{Bo} = \frac{(\rho_G - \rho_L)gD^2}{\sigma_S}$$

which is defined to be negative for a gas bubble in a liquid. A large bubble has a large ratio of gravity force to surface tension force, and is badly distorted. A small bubble is dominated by surface tension forces and remains nearly spherical. The distinguishing diameter is

$$D = \sqrt{\frac{4\sigma_S}{\rho g}}$$

For an air bubble in oil of density 860 kg/m³ and surface tension 25 mPa/m, the distinguishing diameter is 3.4 mm. Simply shaking a transparent container of oil and air shows that bubbles of 3 mm and less do indeed remain close to spherical, so a spherical model is appropriate.

A rising bubble is subject to a reducing static pressure, so it increases progressively in size. The importance of this effect is represented by the expansion number

$$N_{EX} = \left(\frac{gD}{U_B^2}\right)\left(\frac{\rho_L - \rho_G}{\rho_L}\right)$$

This does not seem to be a significant factor in dampers.

It is the bubbles of less than 1 mm that are slow to rise, and therefore the ones of most interest. These may have speeds of 10 mm/s or less relative to the liquid, so having a Reynolds number, based on the liquid properties, of less than 1. Tentatively applying Stokes' equation (due to G. G. Stokes, 1901), which would be applicable to a solid sphere of density equal to that of the gas, the drag coefficient would be given by 24/Re, and the actual viscous drag force by

$$F_D = 3\pi \mu D U_B$$

The bubble buoyancy force is simple the density difference times the volume times g, so the bubble rising speed could then be expressed as

$$U_B = C_V \frac{\rho_L g D^2}{18\mu}$$

The velocity coefficient is included to allow for the dubious applicability of Stokes' equation in this case. The bubble is not a solid sphere. The liquid will not form a boundary layer in the same way as it would around a solid sphere, much of the shearing could occur in the gas bubble. On the other hand, the liquid must shear to change shape as the bubble passes. According to Hadamard (1911, discussed in Batchelor, 1967), a liquid sphere of internal viscosity μ_S will have a velocity coefficient

$$C_V = \frac{3\mu + 3\mu_S}{2\mu + 3\mu_S}$$

In the case of a gas, the viscosity within the sphere is insignificant, so then $C_V = 3/2$.

The quadratic dependence of velocity on diameter shows how smaller bubbles will be very slow to clear the liquid. Applied to a bubble of diameter 0.1 mm in an oil of viscosity 10 mPa s, the above equations predict a rising velocity of less than 1 mm/s.

5.16 Continuity

The Principle of Continuity is really a statement of the conservation of mass for a fluid. For a control volume, in steady state the mass flow rate into the control volume equals the mass flow rate out. In unsteady state, the rate of increase of mass inside the control volume is equal to the mass inflow rate minus the mass outflow rate.

From a practical point of view, this is the basis of many assumptions in fluid flow analysis, including the idea that, for example, the mass flow rate along the inside of a pipe is the same at all transverse sections of the pipe.

A common assumption in many fluid flow problems, often justified, is that the density is constant. With this assumption, the Principle of Continuity can easily be expressed in terms of volume flow rate rather than mass flow rate; the volume of fluid is then conserved, and the volumetric flow rate at each section of the pipe will be the same. The degree to which this is applicable to a damper depends on the degree of emulsification.

From the practical point of view, dampers under test frequently exhibit characteristics which can be explained by fluid compressibility, especially at high stroking frequencies. The compressibility obviously manifests itself as volume changes of the bulk fluid in the chambers, but may also affect the flow in the valves as the gas expands at regions of low static pressure.

5.17 Bernoulli's Equation

Bernoulli's equation may be considered to be an expression of the conservation of energy for a flowing fluid at constant density. However, it is usually more convenient to think of it as an equation about pressures. In a steady state condition it may be applied to a series of points along a streamline or a series of sections of a streamtube, provided that the losses are negligible. First, then, the streamline or streamtube, and the relevant two points or sections must be specified. For uniformity of static pressure over a flow cross-section, the streamlines passing through the section should be straight and parallel. Bernoulli's equation may then be expressed as

$$P_1 + \tfrac{1}{2}\rho u_1^2 + \rho g h_1 = P_2 + \tfrac{1}{2}\rho u_2^2 + \rho g h_2$$

For damper analysis, the potential pressure term $\rho g h$ does not change significantly, so this term may be omitted. Hence Bernoulli's equation simplifies to

$$P_1 + \tfrac{1}{2}\rho u_1^2 = P_2 + \tfrac{1}{2}\rho u_2^2$$

The pressure terminology is:

(1) P: static pressure;
(2) q: dynamic pressure $q = \tfrac{1}{2}\rho u^2$ at fluid speed u;
(3) P_{St}: stagnation pressure $P_{St} = P + q = P + \tfrac{1}{2}\rho u^2$.

The static pressure is the pressure actually experienced by the liquid, including, for example, a small particle being swept along with the liquid. Bernoulli's equation therefore gives a relationship between the static pressure and the flow velocity. Considering flow from a reservoir (i.e. a region of zero or negligible velocity) to a point where there is a velocity u_2, then, neglecting losses, Bernoulli's equation gives

$$P_1 = P_2 + \tfrac{1}{2}\rho u_2^2$$

$$u_2 = \sqrt{\frac{2(P_1 - P_2)}{\rho}}$$

Considering a pressure difference $P_1 - P_2$ equal to 5 MPa, as may be required in a damper, at a density of 860 kg/m^3, the resulting velocity is $u_2 = 108$ m/s. This will be required for a damper speed of about 1 m/s, indicating that the valve exit flow area will be around 1% of the piston area, or about 5 mm^2.

Bernoulli's equation may be extended to allow for energy losses between the two sections being analysed:

$$P_1 + \tfrac{1}{2}\rho u_1^2 = P_2 + \tfrac{1}{2}\rho u_2^2 + \Delta P_{St}$$

where ΔP_{St} is the loss of stagnation pressure arising from friction at the walls, energy dissipation due to extra turbulence at bends, etc. between sections 1 and 2.

A further improvement to Bernoulli's equation is to allow for the effect of the velocity profile, as described in Section 5.7. Because the velocity is not uniform, the mean kinetic energy per unit volume entering a stream tube or passing through a pipe section is not exactly ½ρu^2, ideally an energy correction factor should be applied.

The pressure loss ΔP_{St} is often related to some particular dynamic pressure by the loss coefficient K:

$$\Delta P_{St} = K\,q = K\,\tfrac{1}{2}\rho u^2$$

Obviously the loss coefficient in any particular case depends not just on the pressure loss but also on the particular reference dynamic pressure, which must be specified. In the case of a pipe or other channel of uniform cross-sectional area, this is based on the mean speed at that cross-sectional area. Sometimes the viscous losses are absorbed into an approximate Reynolds-number-dependent K value, sometimes they are accounted for separately.

5.18 Fluid Momentum

Fluid momentum analysis arises in the design of valves, and in particular in consideration of the forces on a valve. A mass of moving fluid, just like a solid object, has momentum, a vector, the product of mass and velocity, with units kg m/s, which are also N s, the units of impulse. Analysis of fluid momentum is generally in terms of the rate of passage of momentum through a section, the momentum flux, which is the mass flow rate times the velocity, with units kg m/s², which are also the unit N (newton).

A jet of fluid at velocity u through area A has a mass flow rate $\rho u A$. The momentum flux, for uniform velocity, is

$$\dot{M} = \dot{m}u = \rho u^2 A$$

In contrast to dynamic pressure, note the absence of a factor of 0.5 here. The velocity across the section is not really uniform. To allow for the effect of the velocity profile, as described in Section 5.7, ideally a momentum correction factor should be applied.

To produce this momentum flux, a force is required on the fluid. For example, in the case of a nonsymmetrical spool valve, there will be a significant side force that may cause friction problems and inconsistent valve area versus pressure.

Figure 5.18.1 shows an example spool valve. To analyse the position of the valve given the reservoir pressures P_1 and P_2, the basic approach is to select the metal of the spool as a free body, and to calculate all the forces on the spool, including all the normal and tangential fluid forces, both inside and outside. Often, it is much more convenient to choose a larger volume, including some of the fluid, particularly that inside the spool, often also some outside, simplifying the analysis. The result is no longer a free-body analysis, but a fluid control volume analysis, with flow over some of the control surfaces. The fluid momentum must then be included.

The basic forces are the structural (solid mechanical) ones and the forces due to static pressure in the fluid at the boundaries of the control volume, F_{FS}. In addition, there are the momentum transfers wherever fluid passes over the boundary. For steady-state analysis, the fluid momentum flux, in kg m/s², can be expressed simply as effective forces in newtons. In steady state there is equilibrium, and there is no change of the momentum of the contents of the control volume, meaning that the applied forces including momentum fluxes must be in equilibrium. This allows calculation of the solid applied forces, e.g. the spring force required to maintain a particular valve position.

Consider the simplified circular-section spool valve in Figure 5.18.1(a). A possible control volume is shown in (b). This control volume includes the fluid inside the spool, which is opened by the total fluid

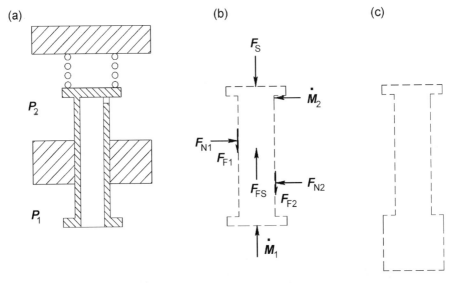

Figure 5.18.1 Force-momentum analysis example of a simple spool valve: (a) physical configuration; (b) control volume; (c) alternative control volume profile.

static force F_{FS} which is resisted vertically by a position-dependent spring force F_S. It is positioned laterally by normal reaction forces F_{N1} and F_{N2} spaced nominally at the thickness of the base material holding the valve, with Coulomb friction forces F_{F1} and F_{F2}. These friction forces may change direction, of course, depending on the attempted direction of motion. The fluid static pressures are P_1 well away from the inlet and P_2 well away from the outlet. The dominant force opening the valve is the fluid static pressure force F_{FS} which is the integral of the static pressure forces over the control volume surface, but modified by the momentum fluxes.

By analysis of the resistances in the flow passage, an estimate of the volumetric flow rate may be made, using Bernoulli's equation with losses. This also establishes the momentum flux values M_1 and M_2 at inlet and exit. The momentum flux force vector is always into the control volume, regardless of the direction, in or out, of the mass flow. Mathematically, this is because at the outlet the mass flow rate into the CV is negative. The static pressure at the outlet will be close to P_2. The upper part of the spool is subject to a force from the static pressure P_2. The lower part of the CV is subject to the static pressure of the liquid over the surface of the CV, but this is not simply P_1, there may be quite high velocities around the entrance, with associated reduced static pressures.

Figure 5.18.1(c) shows an alternative control volume, extended out into the first reservoir. If this is made large enough, the velocities through its surface will be small. The static pressure over this surface is then just P_1, and also the inlet momentum flux is negligible. This illustrates that in this case the momentum flux and reduction of static pressure in (b) just compensate at the inlet.

It will be appreciated that to carry through a successful analysis of this type requires a good knowledge of fluid flow and momentum analysis. A particular point to be emphasised here is that the asymmetrical outlet momentum may be a significant factor leading to large reaction forces and friction forces. For example, *in extremis*, at a flowrate of 1 kg/s and an exit velocity of 100 m/s the momentum flux is 100 kg m/s = 100 N. The values for the supporting reactions depend on their spacing and the position of the exit flow, but evidently the total of the reactions would be at least equal to the exit momentum, and possibly substantially more. The resulting friction forces could cause bad hysteresis in the damper $F(V)$ curve. Obviously, therefore, any such asymmetry of the valves requires careful consideration. If valve flutter and vibration is a problem, the deliberate introduction of a small amount of friction may be beneficial.

5.19 Pipe Flow

The properties of fluid flow in simple circular pipes have been extensively studied, and this provides some guidance on flow behaviour in more complex flow passages.

As first investigated by Osborne Reynolds, these are found to be two types of flow, laminar and turbulent. The criterion for this is the Reynolds number

$$Re = \frac{\rho u D}{\mu}$$

with, for a circular or approximately circular section,

$$Re < 2000 \quad \text{Laminar flow}$$
$$Re > 4000 \quad \text{Turbulent flow}$$

For $2000 < Re < 4000$ the flow may be either laminar or turbulent, and is not easily or reliably determined, depending on the pre-existing turbulence of the flow at the entrance to the pipe, and any bends or other abnormal shaping of the pipe. Actually, laminar flow has been obtained up to a Reynolds number of 10,000 in the laboratory, but for practical engineering conditions 4000 is the upper limit.

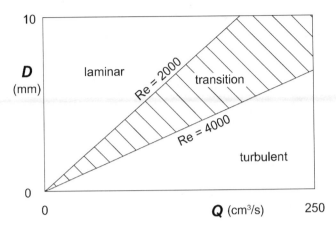

Figure 5.19.1 Laminar / transition / turbulent flow regimes for various combinations of circular pipe diameter and volumetric flow rate. Density 860 kg/m³, viscosity 10 mPa s.

The volumetric flow rate through a circular section is

$$Q = \frac{\pi}{4} D^2 u$$

so the Reynolds number may also be expressed as

$$Re = \frac{4\rho Q}{\pi \mu D}$$

Figure 5.19.1 shows the regions of laminar and turbulent flow to be expected for various combinations of diameter and volumetric flow rate. This diagram covers the practical range of parameters for ordinary damper operation, and it may be seen that both types of flow will occur. The uncertain transition area is potentially a problem, and able to cause inconsistent results.

For turbulent flow, the pressure drop along a pipe is approximately proportional to the square of the mean velocity, so the following parameters are useful.

The mean speed is

$$u = \frac{Q}{A}$$

The dynamic pressure is

$$q = \tfrac{1}{2}\rho u^2 = \frac{\rho Q^2}{2A^2}$$

The stagnation pressure loss ΔP_{St} is

$$\Delta P_{St} = Kq$$

where K is the pressure loss coefficient. The pressure loss is a loss of stagnation pressure, but for a pipe of constant cross-sectional area the dynamic pressure is constant, so ΔP_{St} is seen as the change of static pressure.

The pipe friction factor f is related to the pressure loss coefficient K by

$$K = f \frac{L}{D}$$

The pressure drop is therefore

$$\Delta P_{St} = f\left(\frac{L}{D}\right)\tfrac{1}{2}\rho u^2$$

which is known as the Darcy–Weisbach equation (Henry Darcy, 1803–1848, French civil engineer, Julius Weisbach, 1806–1871, German hydraulics researcher). It is necessarily correct, since it is in effect a definition of the friction factor f. Its accuracy in practical application depends on how well the friction factor can be predicted. The numerical value of the friction factor f is the number of dynamic heads lost along a length equal to one diameter, and $1/f$ is the L/D value to lose one dynamic head of pressure, with a value around 20, this varying somewhat with Reynolds number.

For laminar flow, the pressure drop along a circular pipe is given by

$$\Delta P_{St} = \frac{32\mu u L}{D^2}$$

This is known as the Hagen–Poiseuille equation, and may be demonstrated by straightforward analysis of viscous flow in a pipe of circular section. The velocity u is the mean speed over the cross section, so the volumetric flow rate is

$$Q = \frac{\pi}{4}D^2 u$$

The Hagen–Poiseuille equation can therefore also be expressed as

$$\Delta P_{St} = \frac{128\mu L Q}{\pi D^4}$$

The pressure drop in laminar flow is therefore proportional to the mean speed or to the volumetric or mass flow rate (rather than the approximate proportionality to u^2 or Q^2 of turbulent flow).

The loss coefficient K_L of laminar flow is given by

$$K_L = \frac{64\mu L}{\rho D^2 u}$$

In terms of the friction factor f, for laminar flow

$$f_L = \frac{64}{Re}$$

with

$$K = f\frac{L}{D}$$

and

$$\Delta P_{St} = K\tfrac{1}{2}\rho u^2$$

as usual. However, this equation may be misleading in this context. The pressure drop in laminar flow is not proportional to the square of the velocity. The loss coefficient K decreases with the velocity,

Table 5.19.1 Typical surface roughness

Material	e(m)
Drawn metal tubing	1.5×10^{-6}
Commercial steel	40×10^{-6}
Wrought iron	50×10^{-6}

being inversely proportional to it in the laminar flow regime. This form of equation is used because it suits the common turbulent pipe-flow case so well.

Surface roughness, unless extreme, has little or no effect on the pressure loss in laminar flow, so normally no attempt is made to account for this factor. In turbulent flow, the situation is more complex, with the loss coefficient being sensitive to both Reynolds number and surface roughness. The latter factor is expressed by an effective roughness for the surface, and by the relative roughness e/D for a particular pipe. Examples of surface roughness values are given in Table 5.19.1. Values vary considerably between samples. For good machining, the effective value should be better than 5×10^{-6} m (0.005 mm).

The dependence of friction factor f on Re and e/D may be read from a Moody diagram or Colebrook diagram, Figure 5.19.2. This shows the friction factor for laminar flow to the left, with multiple lines for turbulent flow at higher Reynolds numbers according to the effective surface roughness. The transition region between Reynolds numbers of 2000 and 4000 is uncertain, but an approximate transition line is shown as an indication. This is given by

$$\log_{10} f = -4.8 + \log_{10} Re$$

As an alternative to Moody's diagram, several equations are available to evaluate f for turbulent flow, all simply empirical, unlike the laminar flow analysis. In common use is Moody's equation, adequate for most purposes, giving the friction factor as

$$f = 0.0055 \left[1 + \left(20,000 \frac{e}{D} + \frac{10^6}{Re} \right)^{1/3} \right]$$

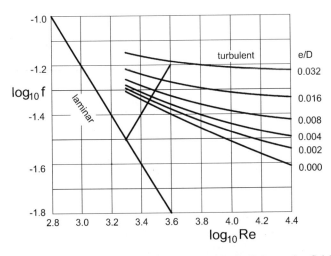

Figure 5.19.2 The Colebrook Diagram for pipe friction factor f (Moody diagram using Colebrook's equation).

A somewhat more accurate expression is Colebrook's original equation, usually expressed as

$$\frac{1}{\sqrt{f}} - 2\log_{10}\left(\frac{D}{e}\right) = 1.14 - 2\log_{10}\left(1 + \frac{9.28}{Re\frac{e}{D}\sqrt{f}}\right)$$

This is not as easy to use because the friction factor appears on both sides of the equation, in nonlinear terms, so f cannot be made an explicit subject of the equation. However, the equation can be rearranged as

$$f = \left[1.14 - 2\log_{10}\left(\frac{e}{D} + \frac{9.28}{Re\sqrt{f}}\right)\right]^{-2}$$

This provides the basis for a good iterative solution, converging to excellent engineering accuracy in three to ten attempts, depending on the Reynolds number and relative roughness. The initial attempt may be made by Moody's equation above, or just using 0.05. Moody diagrams are usually prepared using Colebrook's equation because of its superior accuracy to Moody's equation.

Many other empirical equations have been proposed for the turbulent friction factor, based on Colebrook's data, as may be seen in various fluid dynamic texts. They are argued to be more accurate than Moody's equation but easier to use than Colebrook's equation. In view of the uncertainties such as surface roughness, Moody's equation is probably good enough anyway. If more accuracy is desired, modern computer evaluation makes the use of Colebrook's equation easy.

Noncircular Sections

For noncircular sections with laminar or turbulent flow, the concept of the hydraulic diameter may be used. This is defined as

$$D_H = \frac{4A}{C}$$

where A is the liquid cross-sectional area and C is the wetted circumference. For a circular pipe running full, the hydraulic diameter equals the physical internal diameter. (Some textbooks use the hydraulic radius, which is defined as A/C, and is only one quarter of the hydraulic diameter, a fruitful source of confusion.) The hydraulic diameter is calculated and then the resistance of the system is analysed as for a circular pipe of that diameter.

The Reynolds number may then be based on D_H:

$$Re = \frac{\rho u D_H}{\mu}$$

In the case of laminar flow, the pressure loss is proportional to the first power of the flow rate, so the friction factor reduces with Re. Many such cases can be solved analytically, with some effort, and may be found in reference tables, e.g. Blevins (1984). For example, for a fully wetted semicircle (*not* a pipe running half-full):

$$D_H = \frac{2\pi R}{\pi + 2} \quad f = \frac{63.1}{Re}$$

For a square section of side a:

$$D_H = a \quad ; \quad f = \frac{56.9}{Re}$$

For wide parallel plates close-spaced at spacing a:

$$D_H = 2a \qquad f = \frac{96}{Re}$$

The pressure loss coefficient is then $K = fL/D_H$.

The use of a loss coefficient that declines inversely proportional to flow rate is inherently unsatisfying, and is used only because the dynamic pressure loss coefficient is a good approach for turbulent flow. With laminar flow, it may be more useful to obtain the constant in $P = CQ$. This may be found by direct analysis, or by substitution in the above reference data. For example, to take the case of flow between parallel plates, using the above analysis, by obtaining algebraic expressions for the Reynolds number, friction factor, loss coefficient and pressure loss in turn,

$$\frac{P}{Q} = \frac{12\mu L}{wa^3}$$

This result is also obtained by direct analysis of the laminar flow with integration. For a thin annulus of mean radius R and thickness a, this gives

$$\frac{P}{Q} = \frac{6\mu L}{\pi R a^3}$$

For noncircular sections with turbulent flow, the friction factor is found as for a circular pipe of the hydraulic diameter, using e.g. Colebrook's equation. For example, for a rectangle a by b

$$D_H = \frac{2ab}{(a+b)}$$

which, with an equation or graph for the turbulent friction factor is sufficient. For other values, see fluid dynamics handbooks such as Blevins (1984).

5.20 Velocity Profiles

Because of the effect of viscosity and turbulence, the velocity distribution across a section of a flow is not uniform — for a straight flow there is a V_{max} in the centre and zero velocity at the walls (the no-slip condition). This means that the expressions for momentum flux and kinetic energy flux given in earlier sections, assuming a uniform velocity, or based on a mean velocity, are not quite correct. For turbulent flow the corrections are relatively small, with multiplying factors close to 1.0, and are often neglected. For laminar flow, the corrections may be large. The correction factors may be deduced from the flow geometry and the type of flow, i.e. from the velocity profile.

For laminar flow of a Newtonian fluid, which has constant viscosity, the velocity profile is given by the expression

$$V = V_{max}\left(1 - \frac{y^2}{s^2}\right)$$

where s is the radius for a circular pipe, and for flow between parallel plates s is one half of the spacing, Figure 5.20.1. This may be demonstrated by the application of basic physical principles given the constant viscosity, i.e. it is mathematically correct.

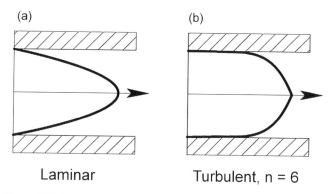

Figure 5.20.1 Velocity profiles for: (a) laminar; (b) turbulent flow ($n = 6$).

For turbulent flow, an equation proposed by Prandtl is used. This is not deduced directly from physical principles (although influenced by Blasius' boundary layer theory), but is empirical:

$$V = V_{max}\left(1 - \frac{y}{s}\right)^{1/n}$$

where n, the profile reciprocal index, is often taken to be 7 in general fluid dynamics. Experimental investigation of flow in smooth pipes (J. Nikuradse) shows that n varies with Reynolds number. For smooth or non-smooth pipes n may be related to the friction factor f. For turbulent flow there is a theoretical logarithmic velocity profile, due to von Karman for smooth pipes, Prandtl and von Karman for fully rough pipes, and to Colebrook for the combined case. This is not very conveniently analytic for obtaining the momentum and energy correction factors, but by matching the ratio of V_{mean}/V_{max} for the logarithmic and power index profile equations a quadratic equation relating f and n is obtained:

$$\{2.86\sqrt{f}\}n^2 - 3n - 1 = 0$$

This is easily solved for n using the standard solution with the positive square root. If f is required it can be made the explicit subject. To a good engineering approximation, the relationship may be expressed more simply as

$$n = \frac{1.11}{\sqrt{f}}$$

For extremely rough pipes at Re = 2000, the friction factor could approach an extreme value of 0.08 with a consequent $n = 4$. For damper passages, $f = 0.035$ (according to Re and e/D) is a reasonable value with $n = 6$.

Given an algebraic representation of the velocity profile, the relationship between the mean and maximum velocities, and the momentum and energy correction factors, may be derived mathematically by integration. The mean velocity is given by

$$\frac{V_{mean}}{V_{max}} = \frac{1}{A}\int \frac{V}{V_{max}} dA$$

The momentum correction factor is found by integrating the actual momentum flux across the section and comparing it with the value according to the simple calculation using the mean velocity, giving

$$\beta = \frac{1}{A}\int \left(\frac{V}{V_{mean}}\right)^2 dA$$

Table 5.20.1 Velocity profiles and correction factors

	Laminar	Turbulent
Velocity profile	$V = V_{max}\left(1 - \dfrac{y^2}{s^2}\right)$	$V = V_{max}\left(1 - \dfrac{y}{s}\right)^{1/n}$
	$s = t/2$ (parallel plates), $s = r$ (pipe)	
	$n = 1.11/\sqrt{f}$ (pipe)	
$V_{mean}/V_{max} \quad \dfrac{V_{mean}}{V_{max}} = \dfrac{1}{A}\int \dfrac{V}{V_{max}} dA$		
Parallel plates	$\dfrac{2}{3}$	$\dfrac{n}{1+n}$
Circular pipe	$\dfrac{1}{2}$	$\dfrac{2n^2}{(1+n)(1+2n)}$
Momentum coefficient $\beta = \dfrac{1}{A}\int\left(\dfrac{V}{V_{mean}}\right)^2 dA$		
Parallel plates	$\dfrac{6}{5}$	$\dfrac{(1+n)^2}{n(2+n)}$
Circular pipe	$\dfrac{4}{3}$	$\dfrac{(1+n)(1+2n)^2}{4n^2(2+n)}$
Energy coefficient $\alpha = \dfrac{1}{A}\int\left(\dfrac{V}{V_{mean}}\right)^3 dA$		
Parallel plates	$\dfrac{54}{35} \approx 1.543$	$\dfrac{(1+n)^3}{n^2(3+n)}$
Circular pipe	2	$\dfrac{(1+n)^3(1+2n)^3}{4n^4(3+n)(3+2n)}$

The kinetic energy correction factor is similarly obtained by integrating the kinetic energy transfer, giving

$$\alpha = \frac{1}{A}\int\left(\frac{V}{V_{mean}}\right)^3 dA$$

In the case of laminar flow, simple numerical values are obtained. In the case of turbulent flow, expressions may be obtained in terms of the profile index n. Table 5.20.1 summarises the results. The

Table 5.20.2 Turbulent flow correction factor example values

	$n = 6$	$n = 7$
V_{mean}/V_{max}		
Parallel plates	0.8571	0.8750
Circular pipe	0.7912	0.8167
Momentum correction factor		
Parallel plates	1.0208	1.0159
Circular pipe	1.0269	1.0204
Energy correction factor		
Parallel plates	1.0586	1.0449
Circular pipe	1.0768	1.0584

mean velocity divided by the maximum is less than or equal to 1.0, of course. The correction factors, Table 5.20.2, are always greater than or equal to 1.0. In the special case of a uniform velocity, the factors are all exactly 1.0. The unlikely fraction 54/35 for the kinetic energy correction of laminar flow between parallel plates is the exact result of an integral.

5.21 Other Losses

Apart from the basic friction loss at the walls of a conduit, described earlier, there are additional losses arising from:

(1) entry;
(2) bends;
(3) change of section;
(4) the exit.

These are dealt with by obtaining a loss coefficient (K value) for each feature, and summing the K values for the complete conduit, including basic pipe friction. Information on these extra losses may be found in handbooks, frequently under the title of 'minor losses'. In the context of a long pipeline, they may indeed be minor losses, but they are important for the passages in dampers. They will be dealt with briefly, in turn, here.

Entry Loss

An additional entry loss arises from two possible causes. The stable velocity profile has to develop from an initial uniform velocity, with high surface shear stress. Also there may be flow separation due to entry shape. There is therefore a minimum entry K of about 0.1 for a bellmouthed entry with edge radius r not less than $D/7$. Other shapes are worse, depending on the details, Figure 5.21.1. A conical entry is easy to machine consistently, and usually quite effective where a reasonably low loss is desired.

Bends

The pressure loss of a bend is greatly reduced by a smooth radius, Figure 5.21.2. In contrast, a sharp-mitred bend has $K = 1.1$, Figure 5.21.3(a), ameliorated by a faired bend with vanes, 5.21.3(b).

Change of Section

A sharp-edged contraction from A_1 to A_2, Figure 5.21.4, gives

$$K = 0.50\left(1 - \frac{A_2}{A_1}\right)$$

$$\Delta P_{St} = Kq_2$$

with K within about 0.03 of most test data. The above is a purely empirical equation.
A sharp-edged sudden expansion, Figure 5.21.5, gives

$$K = \left(1 - \frac{A_1}{A_2}\right)^2$$

$$\Delta P_{St} = Kq_1$$

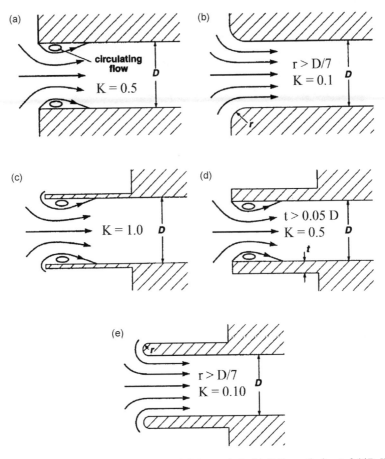

Figure 5.21.1 Entry losses: (a) Sharp-edged, $K = 0.5$ (separation), (b) Bell-mouthed, $r \geq 0.14D, K = 0.10$ (no separation), (c) Sharp re-entrant, $K = 1.0$ (separation), (d) Sharp thick-edged re-entrant, $t \geq 0.05D, K = 0.5$ (separation as in (a)), (e) Rounded re-entrant, $r \geq 0.14D, K = 0.10$ (no separation).

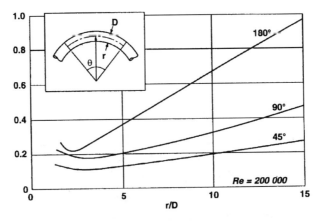

Figure 5.21.2 Bend losses.

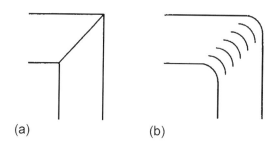

Figure 5.21.3 (a) Sharp-corner losses; (b) ameliorated by a faired bend with vanes.

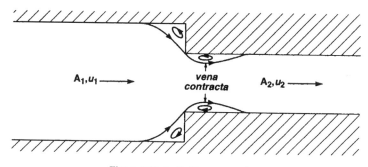

Figure 5.21.4 Sudden contraction.

This result is obtained from a theoretical analysis including continuity, momentum and energy, and is quite accurate. Although there is a stagnation pressure loss, the static pressure increases, and is easily found by the extended Bernoulli equation, giving

$$P_2 - P_1 = \eta_R q_1$$

where η_R is the static pressure recovery coefficient, defined by

$$\eta_R = \frac{P_2 - P_1}{q_1}$$

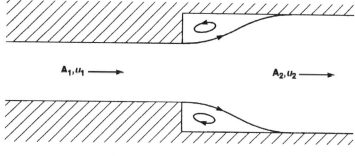

Figure 5.21.5 Sudden expansion.

with a value

$$\eta_R = 2\left(\frac{A_1}{A_2}\right) - 2\left(\frac{A_1}{A_2}\right)^2$$

or

$$\eta_R = 1 - \left(\frac{A_1}{A_2}\right)^2 - \left(1 - \frac{A_1}{A_2}\right)^2$$

Considered as a static-pressure-recovering stepped diffuser, the diffuser recovery coefficient η_D, to be distinguished from η_R, is defined as

$$\eta_D = \frac{P_2 - P_1}{q_2 - q_1}$$

which may be expressed as

$$\eta_D = \frac{\eta_R q_1}{q_1 - q_2} = \frac{\eta_R}{\eta_R + K}$$

Although a damper is unlikely to have a passage explicitly designed to give pressure recovery, tapered and stepped diffusers may occur inadvertently.

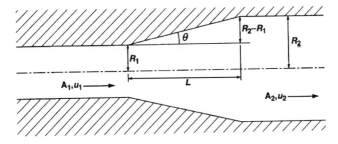

Figure 5.21.6 Diffuser.

A conical, or other smooth shape, is used to give a more efficient pressure recovery, Figure 5.21.6. The stagnation pressure loss is

$$\Delta P_{St} = K q_1 = \left(1 - \frac{A_1}{A_2}\right)^2 \varepsilon q_1$$

where ε (epsilon) is called the diffuser loss factor. It has a value of 1.0 for a sharp step, but much less for a good diffuser design. Hence

$$K = \left(1 - \frac{A_1}{A_2}\right)^2 \varepsilon$$

For a reasonably good diffuser, the conical half-angle θ is in the range 2–10°, in which case

$$\varepsilon = 0.140 + 0.0066(\theta - 3.5)^2$$

Actually the optimum half-angle θ (3.5°) and minimum ε (0.140) are somewhat dependent upon Reynolds number and surface finish.

Fluid Mechanics

The recovery coefficient, from Bernoulli's equation, becomes

$$\eta_R = 1 - \left(\frac{A_1}{A_2}\right)^2 - K$$

$$\eta_R = 1 - \left(\frac{A_1}{A_2}\right)^2 - \varepsilon\left(1 - \frac{A_1}{A_2}\right)^2$$

Exit Loss

At the exit plane of a pipe, entering a large area 'reservoir', the fluid has the specific flow kinetic energy $u^2/2$ (J/kg) (at a streamline, possibly improved by an energy correction factor for a stream tube). This energy is then dissipated by turbulence of the jet into the bulk of fluid in the reservoir, with the fluid eventually reaching negligible speed and kinetic energy. The exit mechanical energy per unit volume $\frac{1}{2}\rho u^2$ is therefore lost, becoming thermal energy by turbulent dissipation. However, whether or not it appears as a loss in the application of the extended Bernoulli equation depends upon the choice made for the end of the streamline or streamtube. If section 2 is at the exit plane, then the energy has not yet been lost, it is merely the $\frac{1}{2}\rho u_2^2$ term of the final dynamic pressure, and should not be included in the losses. On the other hand, if the streamline is chosen to pass into the bulk of fluid in the reservoir, then it appears as lost energy, at a point where there is no longer any kinetic energy. The final speed is then $u_2 = 0$, with $\frac{1}{2}\rho u_2^2 = 0$, and the dissipated energy must be included in the losses. In that case, the loss coefficient based on exit speed is $K = 1.0$. To avoid either omitting this term, or double accounting for it, the position of the end of the streamline, section 2, must be carefully defined.

5.22 The Orifice

Figure 5.22.1 shows a basic flow through a sharp-edged orifice, from reservoir 1 to reservoir 2. Features to note are:

(1) contraction of the flow as it approaches the hole;
(2) the *vena contracta*, the neck in the flow where the cross-sectional area of the flow is a minimum and the velocity is a maximum, and the static pressure equals that of the exit reservoir;
(3) subsequent turbulent dissipation of the kinetic energy in a submerged jet.

The volumetric flow rate Q of liquid, including damper oil in practical conditions, through such an orifice, is generally described by the equation

$$Q = C_d A u_T$$

where C_d is the discharge coefficient, A is a reference area, generally the passage minimum area, and u is the exit speed. A theoretical speed u_T is found from Bernoulli's equation, which applied between the reservoirs gives

$$P_1 + \tfrac{1}{2}\rho u_1^2 = P_2 + \tfrac{1}{2}\rho u_2^2$$

Taking the streamline to begin in the first reservoir at zero velocity, and to end at the exit *vena contracta*, where the static pressure equals that of the second reservoir, then

$$\tfrac{1}{2}\rho u_T^2 = P_1 - P_2$$

$$u_T = \sqrt{\frac{2(P_1 - P_2)}{\rho}}$$

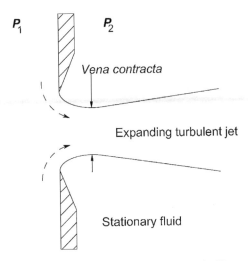

Figure 5.22.1 Flow through a sharp-edged orifice.

where $(P_1 - P_2)$ is the pressure difference across the orifice or tube (between reservoirs) and ρ is the density, treated as a constant. This speed is only an ideal theoretical one. The real speed is less because of friction due to turbulence and viscosity, giving a velocity coefficient C_V defined as the ratio of the mean speed at the *vena contracta* over the ideal speed, so:

$$u = C_V u_T$$

Also, the effective exit area, A_E, at the *vena contracta*, is less than the orifice area A, the ratio of areas being the area coefficient C_A, so:

$$A_E = C_A A$$

For example, at high Reynolds number the area coefficient for a circular sharp-edged orifice is theoretically (potential flow theory, zero viscosity) $C_A = \pi/(\pi + 2) = 0.611$, a value in good agreement with suitable experiments. At practical normal values of Reynolds number for dampers, an area coefficient around 0.8 is observed. This is because the viscosity slows the fluid approaching the orifice along the walls, which reduces the radial inwards momentum of the fluid, so reducing the contraction.

The discharge coefficient is the product of the velocity and area coefficients:

$$C_d = C_V C_A$$

Under practical damper operating conditions and scale the discharge coefficient has a value typically in the range 0.6–0.8.

This analysis is appropriate for orifices, or for any short tubes with turbulent flow or laminar flow, and in practice suitable for real damper passages with a length/diameter ratio L/D up to at least 10. For most practical damper conditions, the Reynolds number is sufficiently high for C_d to be only a moderate function of Re, so the discharge volume for a given pressure depends primarily on the density, the viscosity having some influence, this being treated through its effect on C_d (for example Lichtarowicz *et al.* (1965).

Data for some useful cases may be found in fluid dynamics handbooks, e.g. Blevins (1984) or Idelchik (1986), but the range of possible geometries is very great and complete design data is not available.

Some useful early papers, include that by Zucrow (1928) who studied the effect of Reynolds number and chamfering on small jets for carburettors, the one by Iversen (1956) who reviews low Reynolds number data for short orifices, that by Stone (1960) who studied conical valves, and that by Kastner and McVeigh (1965) who considered various geometries of orifice at low Reynolds number.

For an orifice of cross-sectional area A and pressure difference P flowing an incompressible zero-viscosity fluid of density ρ at a volumetric flow rate Q, using Bernoulli's equation with a discharge coefficient the pressure drop required is given by

$$P = \tfrac{1}{2}\rho \left(\frac{Q}{C_d A}\right)^2$$

Alternatively, from a pipe-flow type of analysis with a dynamic pressure loss coefficient K, the required pressure is given by

$$P = K \tfrac{1}{2}\rho \left(\frac{Q}{A}\right)^2$$

The pressure loss coefficient K and the discharge coefficient C_d are therefore related by

$$K = \frac{1}{C_d^2} \qquad C_d = \frac{1}{\sqrt{K}}$$

Published investigations of orifice flow may therefore be expressed in terms of K or C_d. It is stated by Segel and Lang (1981) that a C_d of 0.7 can be used for dampers, which is a useful first approximation.

From the actual orifice area and a discharge coefficient may then be deduced the effective area of the orifice, A_E. For a pressure drop P across the orifice the following simple equations then summarise the flow in terms of the ideal (theoretical) Bernoulli speed U_T:

$$U_T = \sqrt{\frac{2P}{\rho}}$$

$$A_E = C_d A$$

$$Q = A_E U_T$$

$$Q = A_E \sqrt{\frac{2P}{\rho}}$$

$$\tfrac{1}{2}\rho Q^2 = A_E^2 P$$

The last is a particularly useful formulation when solving damper valve flows.

Damper flow passages are often holes with length equal to several diameters, so a very useful investigation is that by Lichtarowicz et al. (1965), of the discharge coefficient for circular cylindrical sharp-edged orifices with L/D ratios of 0.5 to 10 and Reynolds numbers from 10 to 10^5. One notable result is that the discharge coefficient increases from 0.61 to 0.81 (at $Re > 2 \times 10^4$) as L/D increases from zero to 2; i.e. in this range the resistance falls considerably as the length increases, because there is a diffuser effect with reattachment and pressure recovery.

For $Re > 2 \times 10^4$, C_d is given, within about 2%, by

$$C_d = 0.61 + 0.16(L/D)^2 \qquad \text{for} \quad 0 < L/D <= 1$$
$$C_d = 0.730 + 0.040(L/D) \qquad \text{for} \quad 1 < L/D < 2$$
$$C_d = 0.827 - 0.0085(L/D) \qquad \text{for} \quad L/D >= 2$$

The discharge coefficient C_d tends to be unstable for L/D around 0.5, so it is better to avoid this region. The above C_d values reduce substantially as the Reynolds number decreases.

Dodge (1966), in Yeaple (1966), gives a review of fluid throttling methods. Dickerson and Rice (1969) investigated L/D ratios of 1 to 4 at low Reynolds number. According to Shabazov and Ashikhmin (1973) the discharge coefficient from a nozzle with a drowned efflux is less than that for an undrowned efflux, and may be adjusted by

$$C_{du} = \frac{C_{dd}}{\sqrt{1 + 0.03 C_{dd}^2}}$$

Akers (1984) considers the effect of cavitation parameter C_C on discharge coefficients for L/D of 4 and 10, and found a reduction of C_d of as much as 5% for $C_C = 30$, where

$$C_C = \frac{P_1 - P_2}{P_2 - P_{vap}}$$

where P_1 and P_2 are the pressures at the two sides of the orifice, and may be for example 4 MPa and 1.6 MPA for a pressurised damper giving $C_C = 2.5$, or 2.0 MPa and 0.1 MPa for an unpressurised damper giving $C_C = 20$.

The discharge coefficient C_d can be substantially increased by small changes of geometry at the entry, e.g. by rounding or chamfering, the effects of which are documented for high Reynolds number by McGreehan and Schotsch (1988) and Idelchik (1986). For example, with an entry radius r equal to or exceeding about $D/6$ there is negligible separation of flow and no *vena contracta*, giving a higher C_d, and presumably eliminating the increase of C_d with length which is observed for a sharp-edged entry, and the instability at $L = D/2$.

McGreehan and Schotsch (1988) give the following equations for the effect of inlet radius with turbulent flow for a short orifice:

$$C_{dr} = 1 - f(1 - C_{d0})$$
$$f = 0.008 + 0.992 e^{-5.5(r/D) - 3.5(r/D)^2}$$

The effect of entry edge radius becomes more complex for long orifices because of pressure recovery after the *vena contracta*. The effect at low Reynolds number for all geometries is not well documented. Presumably the fact that the *vena contracta* is generally larger at low Reynolds number (larger area coefficient) implies that separation could be prevented by a smaller corner radius than would be required at higher Reynolds number.

Because damper operation covers a wide range of speeds, it also covers a wide range of Reynolds numbers in the flow orifices. At damper piston speed V, the volumetric flow rate in the piston valves for incompressible flow is

$$Q = V(A_P - A_R)$$

Flowing this through N orifices of diameter d, the total orifice area is

$$A_O = \frac{\pi N d^2}{4}$$

and the mean flow speed is

$$u = \frac{V(A_P - A_R)}{A_O}$$

giving

$$Re = \frac{\rho u D}{\mu} = \frac{4 \rho V (A_P - A_R)}{\pi \mu N d}$$

The ratio R_V of fluid velocity to damper velocity is

$$R_V = \frac{u}{V} = \frac{(A_P - A_R)}{A_O}$$

and hence is equal to the effective area ratio.

For example, piston area 500 mm², rod area 100 mm², and three passages of diameter 2 mm, gives an R_v value of 42.4. A piston speed of 2 m/s, gives an orifice flow speed of 84.8 m/s. A density of 860 kg/m³ and a dynamic viscosity of 10 mPas then gives a Reynolds number of almost 15 000. Thus the flows in a damper are those in the difficult region where Re most significantly influences C_d, and even where variable transition behaviour may be problematic.

Because of the sensitivity of C_d to the geometry of the entry and to Re, it is difficult to achieve a really accurate representation of C_d (and hence the simple constant value of 0.7 adopted by Segel and Lang, 1981).

Also, the sensitivity of C_d to production tolerances such as entry radius or burrs and hole surface finish mean that it is difficult to achieve consistent performance. Possible 'solutions' include

(1) accept performance variations for low cost;
(2) exert tight tolerances;
(3) calibrate individual jets by flow measurement;
(4) design to reduce sensitivity.

Design for reduced sensitivity might, for example, mean including a chamfer at the entry to a small hole. This does not increase the C_d as much as a radius, but is a more repeatable and cheaper machining operation, and reduces the problem of varying burrs.

5.23 Combined Orifices

For two orifices in parallel, Figure 5.23.1, the flow rates are simply added.

$$Q_{A+B} = Q_A + Q_B$$

The pressure differences are equal so

$$C_{d,A}A_A + C_{d,B}A_B = C_{d,A+B}A_{A+B}$$

If the discharge coefficients are equal ($C_{dA} = C_{dB}$) then the flow areas can simply be added:

$$A_{A+B} = A_A + A_B$$

For two orifices in series in steady state flow, Figure 5.23.2, by continuity the volumetric flow rates are equal (for constant density) and the sequential pressure drops simply add, so

$$P_{1-3} = P_{1-2} + P_{2-3}$$

$$\tfrac{1}{2}\rho\left(\frac{Q}{A}\right)^2 = \tfrac{1}{2}\rho\left(\frac{Q}{C_{d,A}A_A}\right)^2 + \tfrac{1}{2}\rho\left(\frac{Q}{C_{d,B}A_B}\right)^2$$

where A is the equivalent effective flow area (with discharge coefficient 1.0). Therefore,

$$\frac{1}{A^2} = \frac{1}{(C_{d,A}A_A)^2} + \frac{1}{(C_{d,B}A_B)^2}$$

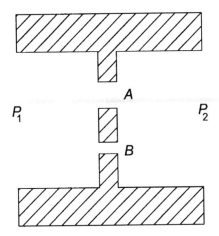

Figure 5.23.1 Parallel orifices.

This permits calculation of an equivalent area A for two orifices in series, and is immediately extendable to three or more in series.

For a known total pressure drop, the equivalent area A permits a solution of the flow rate. The intermediate pressure may then be solved if required by analysis of either single orifice, or sequentially from either end for more than one.

The triple orifice is a series pair A_S and A_L with one in parallel, A_P, Figure 5.23.3. The basic equations, using Bernoulli and Continuity, are

$$\tfrac{1}{2}\rho Q_P^2 = A_P^2 P_1 \qquad (1)$$

$$\tfrac{1}{2}\rho Q_S^2 = A_S^2 (P_1 - P_2) \qquad (2)$$

$$\tfrac{1}{2}\rho Q_L^2 = A_L^2 P_2 \qquad (3)$$

$$Q_S = Q_L \qquad (4)$$

$$Q = Q_P + Q_L \qquad (5)$$

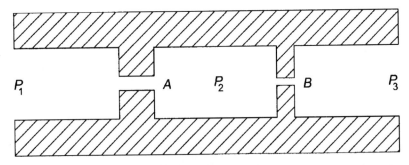

Figure 5.23.2 Series orifices.

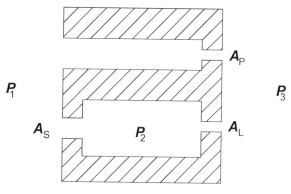

Figure 5.23.3 The triple orifice.

Equations (2), (3) and (4) give

$$A_S^2(P_1 - P_2) = A_L^2 P_2$$

so the pressure ratio is

$$\frac{P_2}{P_1} = \frac{A_S^2}{A_S^2 + A_L^2} \tag{6}$$

If P_1 is given then the total flow rate is just the sum of the two streams,

$$Q = A_P \sqrt{\frac{2P_1}{\rho}} + A_L \sqrt{\frac{2P_2}{\rho}} \tag{7}$$

If Q is given, then substitute Equation (6) into (7) to eliminate P_2, giving an explicit expression for P_1:

$$P_1 = \frac{\frac{1}{2}\rho Q^2}{\left(A_P + \dfrac{A_S A_L}{\sqrt{A_S^2 + A_L^2}}\right)^2}$$

5.24 Vortices

A vortex is a fluid flow pattern in which, ideally, the flow is circular about a central axis, as seen in Figure 5.24.1. Vortices are of great importance in aerodynamics, being intimately associated with wing lift. More prosaically, the flow pattern around a bath plug hole is a vortex, although in that case also with some small inward radial velocity and therefore also with some axial velocity to remove water from the central region. Once the speed builds up, the pressure distribution causes the free surface to distort, lowering in the centre. Although this is a useful illustration of the pressure, the bath vortex before the free surface deflects noticeably is nearer to the ideal vortex.

A vortex is often said to 'induce' the flow around it. This is a useful term, but possibly misleading. The vortex does not mysteriously influence the fluid—the fluid moves obeying the usual laws of mechanics, and the vortex is a representation of the motion that occurs.

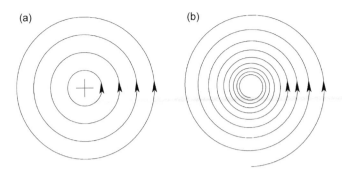

Figure 5.24.1 The circular flow pattern of a vortex (a) simple, (b) spiralling in.

Two main types of vortex are recognised:

(1) the forced vortex;
(2) the free vortex.

The free vortex, the one in common flow, has a tangential velocity inversely proportional to the radius. Obviously, this is only a model of the flow, and the velocity cannot really become infinite at the centre. This is the basic type of vortex at a plug hole or a tornado. In a continuous mass of fluid, near to the centre the simple model of inverse proportionality fails due to viscosity and the presence of the other velocity components. Naturally, viscosity is important near to the centre, where the velocity gradient would be highest, Figure 5.24.2.

Considering a free vortex with a control volume from radius R_1 to R_2, having a small radial velocity component inwards and negligible viscosity, in steady state the angular momentum inside the control volume must be constant, so the angular momentum entering the control volume at the outside edge must equal that leaving at the inside edge. The mass flow is the same at the two sections, so the tangential velocity giving the angular momentum must be inversely proportional to the radius:

$$\frac{V_{T2}}{V_{T1}} = \frac{R_1}{R_2}$$

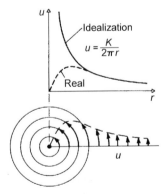

Figure 5.24.2 Ideal and real vortices.

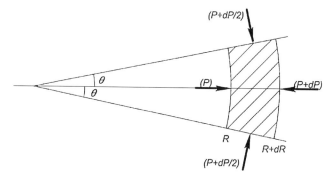

Figure 5.24.3 An element of a free vortex, a sector of an annulus.

or, the vortex strength K_V is a constant:

$$K_V = 2\pi V_T R$$

with units m²/s. The extra factor of 2π appears because the vortex strength is defined as the integral of the tangential velocity component around the loop. The inward mass flow must go somewhere, of course: down the plug hole, or upwards at the centre into the higher atmosphere the case of a tornado.

For a vortex of given axial length, with some inward flow, neglecting the axial velocities the radial mass flow rate is constant. At constant density, the inward volumetric flow rate must be constant, so the radial velocity component is inversely proportional to the radius, as is the tangential component. Therefore, the fluid spirals in at a constant angle, typically at 1–2° to the perpendicular to the radius.

The radial pressure gradient in a vortex may be determined as follows. Consider part of an annulus of the fluid, as shown in Figure 5.24.3, subtending a small angle 2θ. The net radial force must create the centrifugal acceleration, so considering radial force components for a length L (along the axis), with small θ, including the forces on the radial sections at angle θ from the normal to the central line of action,

$$(P + dP)(2\theta)(R + dR)L - P(2\theta\, R)L - 2(P + \tfrac{1}{2} dP)\theta\, dR\, L = (\rho 2\theta\, R\, dR\, L)\frac{V_T^2}{R}$$

Dividing by $2\theta L$, neglecting secondary quantities (products of infinitesimals), and simplifying, gives

$$\frac{dP}{dR} = \frac{\rho V_T^2}{R}$$

This may be integrated to give the pressures, according to the tangential velocity distribution of the particular vortex. For a free vortex, use

$$V_T = \frac{K_V}{2\pi R}$$

giving

$$P_2 - P_1 = \frac{\rho K_V^2}{8\pi^2}\left(\frac{1}{R_1^2} - \frac{1}{R_2^2}\right)$$

Relative to a remote point where $V_1 = 0$ and $P_1 = 0$,

$$P = -\frac{K_V^2}{8\pi^2 R^2}$$

For a real vortex with some radial flow, the flow spirals in at a small angle. Bernoulli's equation along the spiral indicates that the falling static pressure will be associated with an increasing velocity, in agreement with the $V = K_V/2\pi R$ result from the angular momentum analysis.

In the context of dampers, vortices may occur at various flow separation points, but are of particular interest for the vortex valve. In that case, the vortex occurs within a cylindrical chamber, filling it, also with radial and axial velocity components. In the case of the vortex valve, viscosity is also a significant factor, and the ideal vortex is too simple a model for accurate performance prediction, but does give a useful qualitative understanding.

5.25 Bingham Flow

A Bingham plastic or liquid is a material with a yield stress that subsequent to yield behaves as a liquid with a viscosity. This is of interest to damper analysis because electrorheological and magnetorheological fluids do this. Their yield stress depends on the electrostatic field or magnetic field respectively.

Considering the flow of a Bingham material through a circular pipe, for a small pressure there will be no flow. The pressure must be sufficient to cause yielding at some point. Considering a circular rod element along the pipe, the driving force is proportional to the pressure and the circular end area, whereas the shear resistance is proportional to the yield stress and the circumference. Therefore, yielding will first occur at the largest radius. At higher driving pressures, there will be an unyielded central core flowing as a solid slug, with an annular shearing region around it.

Some analysis of Bingham flow between plates and in a circular pipe is given in Appendix G.

5.26 Liquid–Solid Suspensions

A finely divided solid powder within a carrier liquid is called a suspension. These arise in ER (electrorheological) and MR (magnetorheological) liquids for controllable dampers. Typically they use aluminium silicate and soft iron particles respectively, in a low-viscosity oil. Invariably, of course, the suspension has the general property

$$\rho = \frac{m}{V}$$

Using subscripts for S, L and M for solid, liquid and mixture:

$$\frac{m_S}{m_L} = \frac{\rho_S}{\rho_L}\frac{V_S}{V_L}$$
$$V_M = V_S + V_L$$
$$m_M = m_S + m_L$$
$$\rho_M = \frac{m_S + m_L}{V_M} = \frac{\rho_S V_S + \rho_L V_L}{V_M}$$

As a matter of interest, the volume summation is not always very accurate when two liquids are mixed because of packing of the two shapes of molecules, and the small discrepancy may be of practical significance in some cases, particularly when attempting to deduce the mixture ratio from *a posteriori* measurements of the mixture density. However, it is sufficiently accurate in the present context. Note that the solid phase density to be used in the above equations is the material density, not the bulk powder density which is significantly lower because of the imperfect packing of the individual fragments in bulk.

It is not quite so straightforward to determine the volume or mass ratio of constituents when the volume and density of the mixture are known, given in addition the density of the solid and liquid phases alone of course. Then the two convenient and relevant equations are conservation of volume and mass through the mixing process:

$$V_M = V_S + V_L$$
$$\rho_M V_M = \rho_S V_S + \rho_L V_L$$

These must be solved simultaneously, giving

$$\frac{V_S}{V_M} = \frac{\rho_M - \rho_L}{\rho_S - \rho_L}$$
$$\frac{V_L}{V_M} = \frac{\rho_S - \rho_M}{\rho_S - \rho_L}$$

This calculation is not very well conditioned when the densities are similar. It is useful to premix known ratios of the constituents and to measure and plot the resulting density.

For a mean particle volume V_P and mean spacing between particle centres X, the cubical cell volume is X^3. The mixture density is then

$$\rho_M = \rho_L + (\rho_S - \rho_L)\frac{V_P}{X^3}$$

For a spherical particle diameter, or a mean effective diameter, D, then the mixture density is

$$\rho_M = \rho_L + \frac{\pi}{6}(\rho_S - \rho_L)\left(\frac{D}{X}\right)^3$$

The mean particle spacing X may therefore be estimated from

$$\frac{X}{D} = \left\{\frac{\pi}{6}\frac{(\rho_S - \rho_L)}{(\rho_M - \rho_L)}\right\}^{1/3}$$

The mixture compressibility β_M combines the component properties by volume (to a good approximation), so

$$\beta_M = \frac{V_S \beta_S + V_L \beta_L}{V_M}$$

The solid compressibilities are insignificant in relation to that of the oil, so in effect

$$\beta_M = \beta_L V_L / V_M$$

The volumetric thermal expansion properties of the mixture are also proportional to the volumetric constitution, so using the volumetric expansion properties of the solid and liquid:

$$\alpha_M = \frac{V_S \alpha_S + V_L \alpha_L}{V_M}$$

The specific thermal capacity (specific heat) properties of the mixture are obtained in the mass ratio of the constituents, so

$$c_{PM} = \frac{m_S c_{PS} + m_L c_{PL}}{m_M}$$

The thermal conductivity of a suspension is not easy to analyse, and would be obtained experimentally. In practice, a coarse estimate may be made according to the conductivities of the solid and liquid, but the material in non-homogeneous. In the case of ER and MR liquids, with the field on the properties are non-isotropic.

In a liquid suspension, the solid particles and base liquid are usually of significantly different density, from that of the base liquid, so separation will occur when the liquid is stationary. For fine spherical particles, Stokes' equation may be applied. The drag force is

$$F_D = 3\pi \mu D V$$

This corresponds to a drag coefficient of 24/Re. For a nonspherical particle, the drag may be somewhat higher, so an extra shape coefficient C_S should really be applied. The downward driving force is the weight minus the buoyancy, so at terminal speed

$$\frac{\pi}{6} D^3 (\rho_S - \rho_L) g = C_S 3 \pi \mu D V$$

This gives the settlement velocity

$$V = \frac{D^2 (\rho_S - \rho_L) g}{18 C_S \mu}$$

The time for settlement may then be estimated.

5.27 ER and MR Fluids

Practical ER and MR liquids have a high proportion of dense solids in small particles (a few micrometres) carried in a lower-density liquid, giving a very high density mixture, as high as 4 g/cm³ for MR. Example values are shown in Table 5.27.1. The component properties are given in Table 5.27.2

The solid phase, being denser, is subject to some settlement and separation over time, particularly when there are larger particles with a large density difference. This is more of a problem for MR liquids, but with appropriate additives the problem is not severe, and remixing occurs very rapidly.

For practical numbers in an MR liquid, at a solid mass content of 80%, the solid volume content is 40.64 %. The mixture density is 2844 kg/m³, so the spacing ratio X/D is 1.22. This illustrates the close spacing of the particles in practical ER and MR liquids.

Table 5.27.1 ER and MR liquid densities

Electrorheological liquid density				
		Solid	Oil	Total
Density	kg/m³	2600	800	1520
Volume	litre	40	60	100
Mass	kg	104	48	152
Magnetorheological liquid density				
		Solid	Oil	Total
Density	kg/m³	7874	800	2920
Volume	litre	30	70	100
Mass	kg	236.2	56.0	292.2

Table 5.27.2 ER–MR general material properties

Property		Units	Oil	Aluminum silicate	Iron
Density	ρ	kg/m³	800	2600	7874
Compressibility	β	%/MPa	0.04	—	—
Volume expansion	α	ppm/K	1000	10	36
Specific heat	c_P	J/kg K	2500	1050	450
Thermal conductivity	k	W/m K	0.14	5.0	81.0
Viscosity	μ	mPa s	40	—	—

The compressibility of pure light mineral oil is very small at about 0.04 %/MPa, so pure ER and MR liquids are about 0.02–0.03 %/MPa. This is insignificant unless increased by gas emulsification.

A light mineral oil has a high volumetric expansion coefficient of 1000 ppm/K (parts per million per kelvin), aluminium silicate is low at about 10 ppm/K, and iron is also relatively low at 36 ppm/K (three times the linear value to make it volumetric). The solids therefore tend to reduce the expansion of the mixture. Nevertheless, the expansion may be important. For the pure oil between −40 and 130°C the volume change is about 17%. For practical ER materials, the operating temperature range is limited to +10 to +90°C in DC operation. AC operation is somewhat better at −25 to +125°C, but this is still less than the vehicle manufacturer's desired range.

A typical light mineral oil has a specific thermal capacity c_p of 2500 J/kg K, aluminium silicate is about 1050 J/kg K, and pure iron is lower at 450 J/kg K. The solids therefore reduce the thermal capacity significantly. Volumetrically, the oil specific thermal capacity is 2.0 J/cm³ K, aluminium silicate is 2.73 J/cm³ K and iron is 3.51 J/cm³ K, so the mixture thermal capacity variation by volume is relatively small, and increases with solid content.

For a typical mineral oil, the thermal conductivity is 0.14 W/m K, for aluminium silicate it is 6 W/m K, and for pure iron it is 81 W/m K. The solids will therefore increase the conductivity, particularly in the case of iron, although the solid content and greater viscosity will reduce thermal currents and consequent free convection where this is significant. With the magnetic field on, the particles are organised into the fibrils along the field, so the material becomes isotropic with, *inter alia*, thermal and electrical conductivities being dependent on direction.

The viscosity even at zero field is difficult to analyse from first principles. The solid content will increase the effective viscosity compared with the pure oil, but this is hard to quantify. Obviously the solid material internal properties are not relevant, but the volume present and the particle size distribution are important. The particle geometry, smooth sphere or irregular and jagged, may also be significant.

The performance of ER and MR materials has been characterised by the quality factor

$$Q = \frac{\mu_0}{\tau_Y^2}$$

which is the zero-field viscosity divided by the square of the field-on yield stress. ER materials achieve up to 10^{-7} s/Pa whereas MR achieves a figure 1000 times that.

An MR device can operate with much less fluid than can an ER device. For example, a car damper could operate at the required forces with an active liquid volume of 0.3–0.5 cm³, much less than the usual passive-damper 70 cm³ of oil, whereas an ER damper would require the usual amount. However, long life of the MR liquid is really the limiting factor (damage associated with total energy dissipated per liquid volume), so ample total liquid is still required, even though the field-controlled volume is small at any one time.

Costs are always important in manufacturing. ER fluids would be of similar price to simple damper oil. However, MR liquids are expensive. Current price runs at 100 to 200 $/litre for bulk purchases. For a 70 cm^3 damper this is 7–14 $/damper, 28–56 $/vehicle material cost. This is a considerable additional cost in relation to the initial cost of a conventional damper.

Colloidal Ferrofluids

MR fluids should be distinguished from so-called ferrofluids, which are colloidal, having very small particles, typically magnetite (Fe_3O_4) or manganese–zinc ferrite. The particle size is 5–10 nm, 1/1000 of the diameter of MR liquid particles, 10^{-9} times the individual particle mass. On the application of a magnetic field, ferrofluids remain liquid, they do not form particle fibrils, and so do not generate a yield stress. The ferrofluid experiences a body magnetic force which attracts it to regions of high field strength.

6

Valve Design

6.1 Introduction

This chapter considers valve design, and the relationship between the volumetric flow rate Q and the pressure difference P for a given valve. Valves are fitted into the piston and the body of the damper—the piston valves and foot valves respectively. For a given speed of damper action, fluid is displaced through a valve at volumetric flow rate Q. The valve resistance to flow requires a pressure difference across the valve to produce this flow rate. This pressure, acting on the piston annulus area or rod cross-sectional area, will create a force resisting damper motion. Hence the $F(V)$, force–velocity, characteristic of the damper is intimately related to the $P(Q)$, pressure–flowrate, characteristic of the valves. The relationship between the individual valve characteristics and the complete damper characteristic is investigated in the next chapter.

In ride quality studies, using analytical or computer simulations of complete vehicles, the dampers are usually modelled as linear, implying a linear valve characteristic. This can be achieved by a valve with a viscous pressure drop, such as a simple tube. However, viscous losses are too temperature sensitive, so more elaborate valves with dynamic losses are used. This also allows the $P(Q)$ characteristic to be controlled to a desired nonlinear form, within limits.

Practical dampers, then, are based on using energy dissipation primarily by turbulence, usually by allowing the liquid to pass through a small hole, giving a turbulent exit jet which dissipates in a bulk of liquid. Viscosity continues to have some effect, as seen in the dependence of the discharge coefficient on Reynolds number, but the viscosity sensitivity is much reduced from that of laminar flow. With the dynamic loss kind of valve, the pressure loss for a given volumetric flow is more dependent upon the fluid density than on viscosity. However, the density is also dependent on temperature, so although the temperature sensitivity is reduced, it is certainly not eliminated.

The dynamic loss type of valve introduces a new problem: the pressure loss is now dependent upon the square of the exit velocity. This means that for a simple orifice of fixed area the pressure loss depends on the square of the volumetric flow rate, which will give a damper force proportional to damper velocity squared. This is completely unacceptable. However, unlike the problem of viscosity variation with temperature for flow through a tube, this has an entirely practical solution—the valve area can be made to vary to produce a desired characteristic. All dampers have this area variation in a passive form, with a larger pressure difference forcing the valve to open to a larger area, giving a moderated fluid exit speed. Nowadays some dampers also have area variation by manual intervention (i.e. adjustable) or by automatic control. These modern refinements are considered in Chapter 8. A recent further innovation is control of the liquid characteristics (ER and MR Dampers, Chapter 9).

Because the damper characteristics are asymmetrical with direction, either because the designer feels that this is better, or cheaper to implement, the individual valves are usually one way only, that is

---

The Shock Absorber Handbook/Second Edition John C. Dixon
© 2007 John Wiley & Sons, Ltd

they do not allow flow in the reverse direction. This is not strictly true when there is a simple leak path, but this can be treated as an independent one-way leak path for each separate valve.

The following simple algebraic analysis may be made to investigate the required valve exit area. Given a piston geometry with a known piston annulus area A_{PA}, and a required damping coefficient C_D, the force at speed V_D is $C_D V_D$ and the pressure required is

$$P = \frac{C_D V_D}{A_{PA}}$$

The fluid exit speed, by Bernoulli's equation, is then

$$U = \sqrt{\frac{2P}{\rho}}$$

The volumetric flow rate is

$$Q = U A_V = A_{PA} V_D$$

The required valve exit area may then be deduced to be

$$A_V = \sqrt{\frac{\rho A_{PA}^3 V_D}{2 C_D}}$$

Forming the nondimensional ratio of areas A_V/A_{PA} leaves a function of a second nondimensional group on the right:

$$\Pi_D = \frac{\rho A_{PA} V_D}{2 C_D}$$

This could be considered to be a nondimensionalisation of the damper velocity, by dividing by a characteristic speed

$$V_{CH} = \frac{2 C_D}{\rho A_{PA}}$$

with a value of around 12 km/s. The small actual damper speed relative to this high characteristic speed (with a square root) indicates the small valve exit area relative to the piston annulus area.

In all cases, the valve exit area is only a few square millimetres. An example numerical calculation may be made as follows:

				Normal	Survival
Damper speed	m/s	V_D	=	0.100	2.00
Damping coefficient	Ns/m	C_D	=	2500	2500
Damper force	N	F_D	=	250	5000
Piston annulus area	cm^2	A_{PA}	=	5.00	5.00
Pressure	MPa	P	=	0.50	10.0
Flow rate	cm^3/s	Q	=	50	1000
Density	kg/m^3	ρ	=	900	900
Valve exit velocity	m/s	U_E	=	33.3	149.1
Area ratio	—	f_A	=	333	74.5
Valve exit area	mm^2	A_E	=	1.50	6.71
Equivalent diameter	mm	D_E	=	1.38	2.92

The area ratio is the ratio of valve exit area to piston annulus area, and follows directly from the ratio of speeds. The high exit velocity is also notable. This high energy turbulent jet must be dissipated safely without creating noise. The typical exit area of 1.5 mm^2 is important in indicating the care needed with detail design and manufacture. For survival, the pressure at the design limit speed must be acceptable. Even then, a relatively small area is adequate, but the value must be carefully observed.

6.2 Valve Types

Valves can be arranged to be responsive to many factors. The obvious ones are: (1) position; (2) velocity; (3) acceleration. The damper is essentially a device for dissipating energy and as such the velocity sensitivity is the basic one, i.e. the damper relationship $F(V)$ and the corresponding valve relationship $P(Q)$. All other valve sensitivities are just variations on the basic theme. Other designs, described below, include pressure rate sensitivity, stroke length sensitivity, and so on, and particular other nonmechanical methods of implementation, e.g. piezoelectrically operated valves.

Basic mechanical damper valves may be classified conveniently by the design configuration used in the variability of area. There are numerous possible forms, but the three basic ones are:

(1) disc valve;
(2) rod valve;
(3) spool valve;
(3) shim valve.

These are illustrated in Figure 6.2.1. The photographic plates also illustrate some of the valves to be found in passenger car dampers, e.g. plate pages 8 and 9. Shim disc valves are convenient for dampers

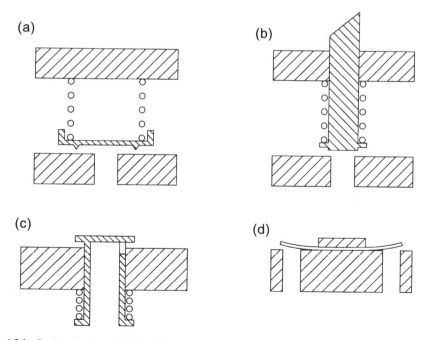

Figure 6.2.1 Basic valve types: (a) disc valve supported by coil spring; (b) rod valve with coil spring; (c) spool valve with coil spring; (d) shim valve.

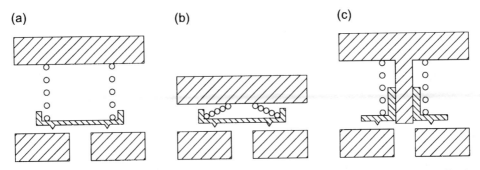

Figure 6.3.1 Disc valves: (a) disc with coil spring; (b) with spiral spring (when free); (c) guided disc with coil spring.

that need to have their characteristics changed, and are therefore usually favoured for racing, whereas the spool valve type and rigid disc with coil spring are more common on passenger cars.

6.3 Disc Valves

The simple disc with coil spring, Figure 6.3.1(a), is obvious in operation, being sealed until an opening pressure difference is reached, depending upon the active pressure area and the preload force of the coil spring. The active pressure area, that is the effective area of the disc within the seal, may be much larger than the cross-sectional area of the flow passage. The flow area is the circumference of the seal times the lift. If the coil spring is of low stiffness (but not necessarily of low preload force) then further opening can occur easily, and large flow areas and flow rates will be possible with little increase of pressure difference. Alternatively, a stiff spring with low preload will give a more gradual increase of area. In practice, because the required flow area is only a few square millimetres, even a small lift of this kind of valve tends to give a 'large' area, so it is difficult to make this type truly progressive. Rather, it acts as a simple blow-off valve with a constant pressure

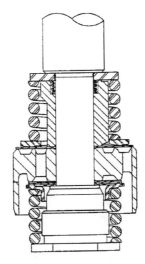

Figure 6.3.2 Piston with coil-spring disc valves for both compression and extension.

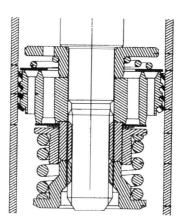

Figure 6.3.3 Piston with coil-spring disc valves for compression and conical spring for compression. The extension disc in some cases acts as a shim at low pressures. Reproduced from Duym (2000) Simulation tools, modelling and identification, for an automotive shock absorber in the context of vehicle dynamics, *Vehicle System Dynamics*, 33, pp. 261–285, with permission from Taylor and Francis Ltd, www.informaworld.com

characteristic. Nevertheless, this type may be suitable in some cases, particularly for a low-preload valve to constrain flow to one direction only with quite a low forward pressure drop, as in a foot valve. Unless the disc is guided, it will probably open asymmetrically because of imperfect symmetry of the spring, which may actually make the individual valve more progressive, although possibly inconsistent one to another. Also, production tolerances make it difficult to achieve consistent preload.

Variations of the disc valve include one using a conical spiral spring, Figure 6.3.1(b). This may have several flow holes disposed circumferentially. This is conveniently compact axially, and suitable for low preloads, so it is a likely choice for a foot valve.

In a third variation, Figure 6.3.1(c), the mobile disc with a sleeve slides on the guiding rod, restrained by a coil spring.

If a disc with coil spring is used on both sides of the piston, the total length may be disadvantageous, Figure 6.3.2. A conical coil is more compact, Figure 6.3.3.

6.4 Rod Valves

To obtain a progressive pressure drop for a disc type of valve requires a small circumference to allow a worthwhile lift, leading to a rod valve, in which the fluid is controlled by a flat rod end on a hole, Figure 6.4.1(a). The hole may be as small as convenient manufacturing permits. This allows a relatively large progressive valve lift, but a lift of $h = \pi R^2 / 2\pi R = R/2$ makes the exit area equal to the hole area, so there is a limit to the effective flow area. However, a hole diameter of say 3 mm with area 7 mm² allows a useful progressive lift of 1.5 mm or more. The spring must be designed to give the desired force increase over this distance, so the stiffness will still be high.

To allow a greater lift for a given flow area, a tapered rod, may be used, Figure 6.4.1(b). When the taper is long, this is known as a needle valve. A straight taper is the most practical to manufacture, but more elaborate tapers are possible. Taper needles were developed in great detail and produced in quantity for use in SU carburettors, shaped by grinding. The rod valve is rather long axially for use on the piston body, but lends itself to effective use positioned in the end of the main shaft controlling extension flow.

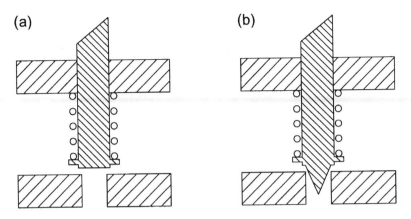

Figure 6.4.1 Rod valves: (a) basic flat-ended; (b) tapered.

6.5 Spool Valves

The spool valve or bobbin valve, shown in Figure 6.2.1 (c), has automatic radial positioning, usually opening against a coil spring although a leaf spring would also be a possibility. A simple form of spool valve cross-drilled with four round exit holes is a common choice for an extension foot valve. The exit hole(s) in the spool may be given a suitable profile to provide a flow area varying with spool position in any required way. The spool valve is particularly adaptable in this respect. A linear $Q(P)$ characteristic is obtained with a flow slot that reduces in width at the new flow point as the valve opens, so that the total flow area increases as the square root of the opening length. In principle, any required characteristics may be designed in, limited by the practical precision of the exit hole profile and the costs of manufacture. An accurately shaped hole could be made by broaching. With a single-sided exit, as shown in Figure 6.2.1(c), the spool will react laterally to the fluid exit momentum, requiring a considerable lateral support force, possibly several tens of newtons, creating friction, and possible hysteresis or inconsistency in the spool position for a given pressure difference. This may be prevented, at some extra cost, by the use of two or more symmetrically positioned exit holes.

A piston spool needs to be compact to fit between the rod and the piston outer diameter. Four may be used, two acting each way. For low enough flow resistance at maximum speed the bore needs to be about 2 mm. The bearing can be 3.2 mm in diameter, with a flange diameter of 5 mm. The spring stiffness needs to be in the range 50–100 kN/m, possibly best achieved by a Belleville washer stack or

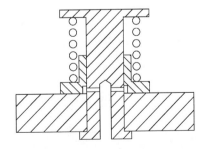

Figure 6.5.1 Spool valve variation with slider.

by a cantilever spring bearing on a raised spot on top of the spool. Assembly is required to achieve the double flange. A circlip is suitable if subsequent disassembly is envisaged, as on a racing damper. Peening or rolling over would probably be preferred for a standard production item. The narrower the slots in total, the greater the axial movement required, the softer the spring, and the less critical is assembly.

In an interesting variation of the spool valve, Figure 6.5.1., the spool is fixed to the base and a slider moves along it, controlled by a coil spring. This may be considered instead to be a variation of the disc valve, but the decisive feature is that the controlled exit flow point is radial.

6.6 Shim Valves

The third main type of valve is the shim valve with basic principle as shown in Figure 6.2.1(d). In practice a pack of shims is used with varying diameters, a system particularly common on racing dampers, partially because the characteristics can be changed easily, Figure 6.6.1. On passenger cars, the shim valve shows to advantage because it is relatively easy to set up accurately with consistent results. This is because the flat shims sit naturally on the piston without problems of manufacturing dimensional inconsistencies affecting the preload.

Figure 6.6.1 shows the usual configuration of one pack on each side of the piston, which will typically contain six holes, three for fluid motion in each direction. Sometimes six holes are used for compression. Hole A is one of the three holes for compression flow, with free entry, and the exit limited by the upper pack of shims. In extension, hole B is one of three active holes, with the lower pack providing resistance.

The shim thickness is 0.2–0.5 mm, and the piston surface is sometimes coned at 0.5–2° to give a preload, possibly just to prevent a leakage path. Generally, there is also a small parallel hole. The shim pack comprises up to six shims of reducing diameter. This gives a controllable stiffness, with greater strength where the bending moment is large, and adds some shim-on-shim friction which may help to prevent valve flutter.

The valve opening height is only a fraction of a millimetre, so the flow path is roughly two-dimensional. With three holes of diameter 6 mm, the exit circumference is nominally 57 mm, so an exit area of 3 mm^2 requires a mean lift of only about 0.05 mm. Here the discharge coefficient will be sensitive to Reynolds number, and also to radiusing or chamfering of the corner at the entry below the shims. There does not seem to be any published information on detailed investigations of flow through such geometry, although Mughal (1979) has reported discharge coefficients for reed type valves. Because of the small valve lift, the characteristics are rather sensitive to burring or small damage at the valve seat, caused by foreign particles being forced through the valve at low lifts. Of course, a very small piece of foreign matter jammed beneath a valve can cause a considerable reduction of resistance.

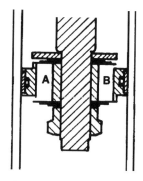

Figure 6.6.1 Complete double-acting shim valve as used in practice.

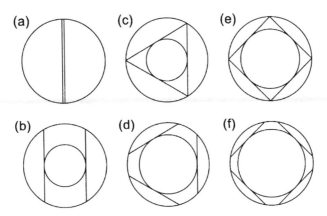

Figure 6.6.2 Shim valve bending modes.

In a more complex variant of the shim pack valve, further support is provided by one or more supplementary shims in a way that makes the preload force adjustable. Also, the shims may be spaced by small diameter discs, so that support and extra stiffness is introduced progressively with increasing deflection.

Even a thin shim is very stiff against compound curvature, so it prefers to bend with planar curvature. This means that there is a limit to the bendable distance available, governed by the number of bending sections on the shim, as in Figure 6.6.2. In part (a), the shim has a small central support, and can bend in two wings, each over a distance $D/2$. With a solid central support up to $D/2$, ($D \cos 60°$), there are still two wings, as in (b), forming shorter cantilevers that are stiffer. At this point there is a transition, and a three-wing mode becomes possible for a support diameter exceeding $D/2$, as in (c). Although a two-wing mode is still possible, the mode that will occur depends on the fluid hole positions. At a support radius $0.71\ D$ ($D \cos 45°$), bending with four wings is possible, as in Figure 6.6.2(e). For consistent and predictable valve behaviour, the number and position of flow holes must be compatible with the preferred bending mode of the shim and its support. For example, with a rigid support of diameter $0.4\ D$, two-wing bending will be more compliant and more likely than three-wing bending which would be based away from the support, so using three fluid holes could be problematic in such a case.

Figure 6.6.3 shows the typical designs used in practice, with three holes, or six holes in three pairs, for fluid flow. The design intent is for a three-wing bending mode. A support smaller than $0.5\ D$ in diameter would make the wings fight each other for space, and permit the more compliant two-wing bending, with possibly erratic results.

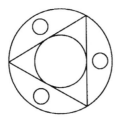 

Figure 6.6.3 Shim valve flow hole positions as typically used for three-wing bending mode.

When shim valves are to be used for large volume flows (not dampers), radial slots are used in the shim to facilitate greater bending, e.g. 10 or 12 sections, with a small rigid central support to give a long cantilever length. These are called petal valves.

6.7 Valve Characteristics

Some mention has been made of the overall valve area–pressure relationship, and the pressure–flowrate relationship, but this is not the only important aspect of the valve. In fact the following qualities, *inter alia*, are important:

(1) steady state pressure–flowrate;
(2) friction and hysteresis;
(3) transient response (flutter, overshoot);
(4) temperature sensitivity;
(5) cavitation;
(6) wear;
(7) fatigue;
(8) consistency in production;
(9) required precision of manufacture;
(10) economy of manufacture.

In a complete valve, the variable-area component above will normally be combined with an orifice in parallel to give some flow even with the valve fully closed. Also, the valve will be limited in its maximum area, or there will be a series orifice to control the flow at very high pressure. These various factors, being the areas of series and parallel holes, the maximum area and the valve pressure–area characteristic, are all juggled to obtain the desired, or at least best available, complete valve characteristic. These valve characteristics may be studied to a useful extent by analysing a basic valve model with variable area, without concern for the actual physical implementation.

Considering an ideal valve free of mechanical friction, there will normally be a unique pressure for any flow rate, and vice versa, under steady-flow conditions at least. For some designs, the mechanical friction is inherently low, e.g. a single shim valve, for others it may be high, e.g. an asymmetrical spool valve. With friction, for a given pressure or flow rate the valve position is indeterminate within the friction band, depending upon the recent history of operation, primarily whether the flow rate is increasing or decreasing. For moderate friction this is not necessarily problematic. Indeed, some friction may be useful to prevent valve overshoot or valve oscillation due to positional instability, although this is better solved by addressing the fundamental problem, usually due to having a flow area varying too rapidly with pressure. Valve friction is one cause of hysteresis in the $F(V)$ characteristic of the damper, and is likely to be associated with valve wear.

Because of the small flow areas, even a small amount of wear of the valve or seals can dramatically affect the overall damper characteristic. Production costs are always important, of course, and a valve design which is insensitive to production variations is valuable for providing greater consistency, easier manufacture and lower costs. In practice, variation of individual performance and rapid loss of function due to seal and piston wear is a significant weakness of low-cost dampers.

Valve flutter is basically a simple mechanical vibration at the valve natural frequency, depending upon valve mass m and spring stiffness k. In practice, the mass of the oil moving with the valve must also be considered (the fluid 'added mass'). Where the mass m and a linear stiffness k are identifiable,

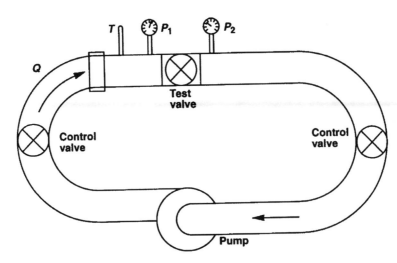

Figure 6.7.1 Valve test circuit.

the undamped natural frequency is

$$f_N = \frac{1}{2\pi}\sqrt{\frac{k}{m}} \quad [\text{Hz}]$$

In the case of shim valves, a cantilever analysis is required.

Flutter only occurs when the valve motion damping is low, so the flutter damping will not affect the frequency significantly. Damping of such oscillations may be provided through the presence of the oil, or by mechanical friction. Flutter and valve overshoot show up under transient conditions, making the $F(V)$ curve inconsistent.

The basic steady-state pressure–flowrate valve characteristic may be tested by an arrangement such as that in Figure 6.7.1, comprising pump, volumetric flow rate measurement such as an orifice plate or venturi meter, temperature gauge, and pressure measurements, resulting in the $P(Q)$ characteristic such as in Figure 6.7.2, possibly with some friction-band hysteresis. For a common passenger car damper,

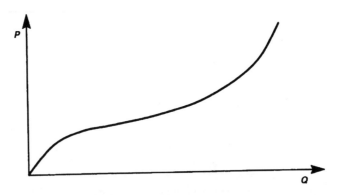

Figure 6.7.2 Typical valve test result.

volumetric flow rates are normally around 0.1 litre/s, going up to 1 litre/s *in extremis*, with corresponding pressure drops of around 1 MPa normally, going up to 10 MPa. The actual power dissipation in the valve test, at the valve itself, is given by

$$\dot{W} = PQ$$

and may be around 50 W for a normal test but up to 5 kW in a severe test. Hence a full test is quite demanding of the equipment in terms of pump power, and generally robustness, but may be less demanding of the valve and oil than in the damper because the oil circulation is so good in the steady-state test, and localised heating is less. Any sort of continued severe testing will require a substantial oil reservoir and cooling arrangements to maintain satisfactory test conditions. A good alternative, particularly for high flows and pressures, is to fit the valve into a purpose-made damper and to test it by constant velocity stroking.

Testing of the valves in isolation from the damper serves to validate, or otherwise, theory of valve behaviour, to evaluate the accuracy of specific mathematical models of valves, and to give reliable valve characteristics for use in complete damper models, permitting the evaluation of other aspects of damper performance such as mechanical friction of the piston and seals. Hence valve testing is an adjunct to complete damper testing, not a substitute for it. The advantage of testing the valve in isolation is that uncertainties regarding the effect of the other aspects of the damper are eliminated, such as piston seal leakage.

6.8 Basic Valve Models

Mathematical (algebraic) modelling of the valves provides performance prediction for new designs of valve, and also provides design insight and understanding of valve behaviour possibly leading to new concepts. In practice, such analytical work is limited in its ability to deal with nonlinear behaviour (e.g. spool valves with difficult area functions) and is therefore supplemented by computer-simulation numerical models. Numerical models are, however, less good at providing design insight than simple algebraic models.

A damper valve will normally be considered as a combination of flow orifices, with, in practice, one of the areas being dependent upon the pressure across it. The primitive valve is an orifice of fixed area A. Simplistically, by Bernoulli's equation, the pressure difference gives an ideal flow velocity U_T. In practice, because of viscosity and frictional losses in the length of pipe leading to the exit, the flow rate is somewhat less, allowed for by the discharge coefficient C_d. Hence, as developed in Chapter 5

$$Q = C_d A U_T = C_d A \sqrt{\frac{2 P_{1-2}}{\rho}} \tag{6.8.1}$$

The discharge coefficient actually arises from an area factor due to an area reduction (the *vena contracta* being smaller than the orifice itself) and a velocity factor from mean velocity deficit.

Rearranging Equation (6.8.1) gives the pressure drop P_{1-2} as

$$P_{1-2} = \frac{1}{2}\rho\left(\frac{Q}{C_d A}\right)^2 \tag{6.8.2}$$

Equations (6.8.1 and 6.8.2) provide the basis for damper valve analysis, frequently with roughly estimated values of C_d, which are likely to be around 0.7 for practical damper orifices.

Consider an orifice with variable effective area. For example, the variational relationship $A(P)$ may often be represented conveniently by an exponent relationship with index n:

$$A = C_{AP} P^n \tag{6.8.3}$$

Using Equation (6.8.2) gives

$$P = \frac{1}{2}\rho \left(\frac{Q}{C_d C_{AP} P^n} \right)^2$$

$$P^{1+2n} = \frac{1}{2}\rho \left(\frac{Q}{C_d C_{AP}} \right)^2 \tag{6.8.4}$$

$$P = C_1 Q^{2/(1+2n)} \tag{6.8.5}$$

For the case of $n = 0$, there is then no area variation and a correspondingly normal quadratic pressure drop. With $n = 1/2$, i.e. with the orifice area proportional to the square root of the pressure, then the pressure drop is directly proportional to Q, giving a linear characteristic. With $n = 1$, area proportional to pressure, the pressure is proportional to flow rate to the power 2/3, the $P(Q)$ curve developing the knee shape. Further increase of n gives further reduction in the power of Q in variation of P, as seen in Figure 6.8.1. The higher values of index n therefore exhibit a regressive form of characteristic with a knee, generally considered desirable.

The variation of flow area with pressure difference across the valve,

$$A \equiv A(P)$$

can in fact be broken down into variation of area with valve position $A(x)$ and variation of valve position with pressure difference $x(P)$, so

$$A \equiv A\{x(P)\}$$

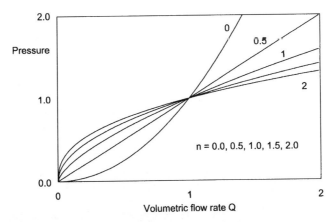

Figure 6.8.1 Basic variable-area valve $P(Q)$ characteristic with pressure index n.

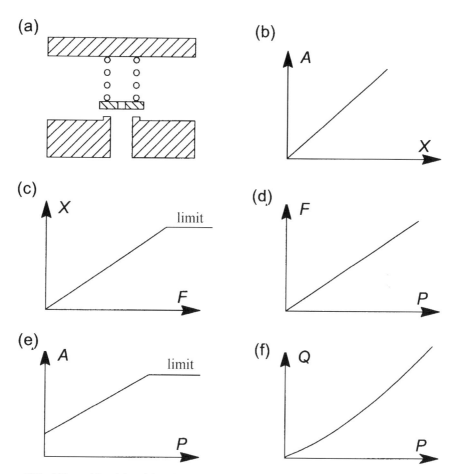

Figure 6.8.2 Valve $A(x)$, $x(F)$, $F(P)$, $A(P)$ and $Q(P)$ characteristics: (a) valve physical configuration; (b) area vs displacement; (c) displacement vs force; (d) force vs pressure; (e) area vs pressure; (f) resulting flowrate vs pressure.

The required variation of area with pressure could therefore be achieved by attention to either $A(x)$ or $x(P)$, that is by the area variation with position or by position variation with pressure. The opening force is related to the pressure difference by a function $F(P)$, and the displacement to force by function $x(F)$, so ultimately the area depends on the displacement which depends on the force which depends on the pressure:

$$A \equiv A\{x[F(P)]\}$$

The required variation $A(P)$ may therefore be achieved by attacking any of the three underlying relationships $A(x)$, $x(F)$ or $F(P)$. In practice the last of these is usually fairly linear, and the required variation is achieved through $A(x)$ or $x(F)$. For a linear damper,

$$A \propto \sqrt{x}$$

or

$$x \propto \sqrt{F}$$

or some suitable combination of these.

A suitable area variation $A(x)$ is possible by correct design of a spool valve, with a profiled orifice. Using other kinds of valve, with area proportional to displacement, a variable increasing stiffness is required. This may be achieved by progressively bringing into action a series of supplementary springs. A progressive single spring can be used, for example a coil spring with a coil of progressive pitch. The closest coils collapse together first, leaving fewer active coils and greater stiffness. However such springs are not easily and cheaply manufactured with good consistency for the small deflections required. A shim pack with separation is a more practical method of achieving this result.

In practice, any one valve in the damper is somewhat more complex than previously described because there may be an upper area limit with the valve fully open, and also the variable area component may be coupled with additional series and parallel orifices. A method frequently used to obtain an approximately correct $F(x)$ is by use of a linear spring with a maximum area and a parallel hole. For a constant width of orifice, this gives an area function as in Figure 6.8.2, where (e) shows the resulting $A(P)$ relationship and (f) shows $Q(P)$.

Assuming basic linearity for simplicity, for a valve lift X and exit width B the valve exit area is $A = BX$, and the lift is $X = F/K$. The force is $F = A_F P$ where A_F is the effective force–pressure area. With a leak area A_L the total valve exit area is then

$$A_{VL} = \frac{BA_F P}{K} + A_L$$

The volumetric flow rate is then

$$Q = \left(\frac{BA_F P}{K} + A_L\right)\sqrt{\frac{2P}{\rho}}$$

6.9 Complete Valve Models

Figure 6.9.1 shows a possible general combination of orifices. A_P is the effective area of the parallel orifice (i.e. in parallel with the primary variable area component), A_S is the effective area of the series orifice, and A_L is a leak area, possibly deliberately added or possibly to allow for failure of the valve to seat perfectly, or other leakage. Figure 6.9.2 shows the idealised relationship for the valve area A_V as a function of the pressure difference $(P_2 - P_3)$, although this may be more complex in some cases as previously discussed. An actual leak area could be incorporated into this A_V function, but in any case should normally be quite small for a damper in good condition. A parallel hole A_P or leakage area A_L is often provided by design to ensure prompt pressure equalisation at zero or very low damper speed and to give a desired characteristic. If the series hole A_S is large, which is often but not always the case, then A_P and A_L are equivalent. Wear of the damper seals, a frequent cause of deterioration of damper performance, can frequently be considered as an increased value of A_P. The series hole A_S may be provided as a limitation on flow through the variable area, in effect alternative to the upper area limit A_M.

The variable area is considered to become active at the pressure P_{vfc} (pressure for valve fully closed), and reaches the maximum area A_M at P_{vfo} (pressure for valve fully open).

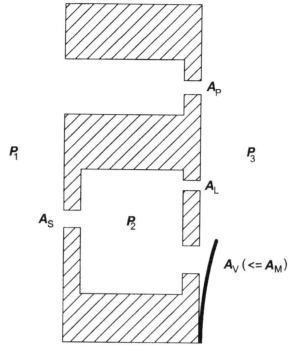

Figure 6.9.1 Valve areas and pressures.

Inclusion of a series orifice is analytically difficult. For a linear A_V (P) the explicit equation for P_2 becomes a cubic, the solution of which is unwieldy by hand. Hence with such a series hole, computer simulation is necessary in practice.

For basic understanding of valve behaviour, the effect of A_S is not so different from a valve area limit A_M. Consider therefore a somewhat simplified version of the valve, as in Figure 6.9.3, with a parallel orifice and a variable area of characteristic as in Figure 6.9.2, rising to a maximum effective area A_M, from a pressure difference P_{vfc} (valve fully closed) to P_{vfo} (valve fully open).

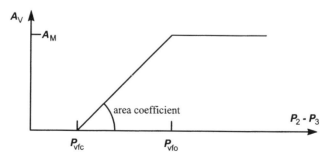

Figure 6.9.2 Variable area characteristic.

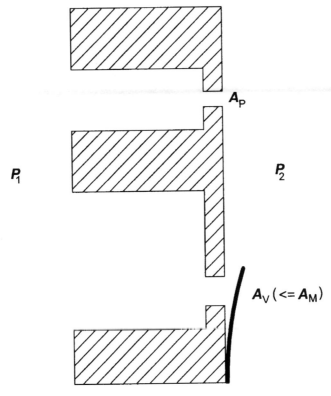

Figure 6.9.3 Basic valve model.

The resulting $P(Q)$ curve appears in Figure 6.9.4. The three parts of the curve are called stages:

Stage 1: valve A_V closed, $A_V = 0$, flow in parallel orifice A_P only, $P \propto Q^2$;
Stage 2: valve partially open, $A_V > 0$;
Stage 3: valve fully open, $A_V = A_M$, $P \propto Q^2$ again.

Curve A applies to the fully closed valve, allowing flow in the parallel orifice only. Curve B is for a fully open valve allowing flow in both orifices at constant area. In the transition region, with a partially open valve, there is a corresponding transition curve. These do not always emerge exactly as expected. The one shown is for a linear $A_V(P)$ function. As may be seen, the complete resulting curve is not so badly removed from a linear $P(Q)$ over a good range, to somewhat beyond Q_{vfo}. Hence, this approach can give good results, with careful choice of parameters, using simple low-cost reproducible linear components. Also, where some nonlinearity is deemed desirable, it may be possible to design this in. The transition around first opening of the valve and the initial part of the regressive characteristic is called the knee of the curve, and is considered important for handling qualities.

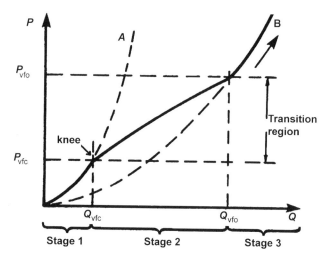

Figure 6.9.4 Transition $P(Q)$ characteristic.

The particular analysis is as follows. Curve A applies to the fully closed valve, allowing flow through the parallel orifice only. From Equation (6.8.1), by Bernoulli

$$Q_P = A_P \sqrt{\frac{2P}{\rho}}$$

where A_P is taken as the effective area, allowing for C_d. With the valve fully open,

$$Q_M = A_M \sqrt{\frac{2P}{\rho}}$$
$$Q = Q_P + Q_M$$

shown as curve B. Within the intermediate range, there is a transition according to the $A(P)$ function of the valve. For a linear $A(P)$, the total Q is a linear proportion between curves A and B. In this transition region, for a linear valve the variable area is proportional to the pressure above P_{vfc}. The variable valve area is

$$A_V = f_{VO} A_M$$

where f_{VO} is the valve area fraction open. Given that this is in the range 0–1,

$$\left.\begin{array}{c} A_V = f(P)A_M = \left(\dfrac{P - P_{vfc}}{P_{vfo} - P_{vfc}}\right) A_M \\ Q_V = A_V \sqrt{\dfrac{2P}{\rho}} \\ Q = Q_P + Q_V \end{array}\right\} \qquad (6.9.1)$$

In summary then, the complete $Q(P)$ function is

$$Q = \{A_P + A_V\}\sqrt{\frac{2P}{\rho}}$$

and more specifically

$$\left. \begin{array}{ll} Q = \{A_P\}\sqrt{\frac{2P}{\rho}} & P \leq P_{\text{vfc}} \\ Q = \left\{A_P + \left(\frac{P - P_{\text{vfc}}}{P_{\text{vfo}} - P_{\text{vfc}}}\right)A_M\right\}\sqrt{\frac{2P}{\rho}} & P_{\text{vfc}} < P < P_{\text{vfo}} \\ Q = \{A_P + A_M\}\sqrt{\frac{2P}{\rho}} & P \geq P_{\text{vfo}} \end{array} \right\} \quad (6.9.2)$$

Once the valve is fully open, then it behaves once again as a plain orifice, and further increase of flow rate requires a quadratic pressure increase. In practice this may be useful in providing a hydraulic bump-stop action, although the very high pressures may cause damage to the damper, especially the seals. In some cases the seals, by accident or design, act to limit the high pressure by blowing open.

Often there is no preload on the variable area, as in the case of shims on a flat piston, in which case the above equations obviously simplify, with $P_{\text{vfc}} = 0$. Then, within the variable area range

$$Q = \{A_P P^{1/2} + K_A P^{3/2}\}\sqrt{\frac{2}{\rho}}$$

Greater insight into the influence of valve parameters on the $P(Q)$ curve may be obtained by studying the figures presented in Chapter 8, Adjustables.

The three-stage $P(Q)$ curve of Figure 6.9.4 may be partially characterised by the ratios $P_{\text{vfc}}/P_{\text{vfo}}$ and $Q_{\text{vfc}}/Q_{\text{vfo}}$. The transition pressure ratio k_{TP} is

$$k_{\text{TP}} = \frac{P_{\text{vfc}}}{P_{\text{vfo}}} \quad (6.9.3)$$

and is controlled entirely by the variable area valve. The transition volume flow ratio k_{TV} is given by

$$k_{\text{TV}} = \frac{Q_{\text{vfc}}}{Q_{\text{vfo}}} \quad (6.9.4)$$

Using Equations (6.9.2) for the opening and closing points

$$k_{\text{TV}} = \left(\frac{A_P}{A_P + A_M}\right)\left(\frac{P_{\text{vfc}}}{P_{\text{vfo}}}\right)^{1/2} = \left(\frac{A_P}{A_P + A_M}\right)k_{\text{TP}}^{1/2}$$

where these are, of course, effective areas, but actual areas will suffice if the discharge coefficient is effectively equal for all orifices.

The parallel hole area is usually small compared with the maximum valve area, giving

$$k_{\text{TV}} \approx k_{\text{TP}}^{1/2}\left(\frac{A_P}{A_M}\right)$$

Valve Design

In the case of a zero-preload valve with flow area proportional to pressure, it is possible to obtain an analytical estimate for a good parallel orifice. From Equation (6.9.4) with $n = 1$, for the variable orifice

$$P = \left(\frac{\rho}{2C_d^2 C_{AP}^2}\right)^{1/3} Q_V^{2/3}$$

or

$$Q_V = P^{3/2} \left(\frac{2}{\rho}\right)^{1/2} C_d C_{AP}$$

For the parallel fixed orifice A_P

$$Q_P = A_P \sqrt{\frac{2P}{\rho}}$$

For good mutual compensation, set the volumetric flow rates equal at the required mid-range pressure P_{MR}, resulting in

$$A_P = P_{MR} C_d C_{AP}$$

i.e. the parallel area should be approximately equal to the variable area value at the mid-range pressure. For example, for a valve with area 2 mm²/MPa (2×10^{-12} m²/Pa) at a mid-range pressure of 2 MPa the parallel area required is around 4 mm².

6.10 Solution of Valve Flow

For the general model valve of Figure 6.9.1, the following equations apply. These are the constitutive equation of the variable area, equations of continuity at constant density, and the application of Bernoulli's equation across each orifice. The orifice areas are effective areas allowing for discharge coefficients. The pressures are measured relative to the downstream reservoir pressure after the valve.

$$A_V = k_A(P_2 - P_{vfc}) \leq A_M \quad (1)$$

$$\tfrac{1}{2}\rho Q_P^2 = A_P^2 P_1 \quad (2)$$

$$\tfrac{1}{2}\rho Q_S^2 = A_S^2(P_1 - P_2) \quad (3)$$

$$\tfrac{1}{2}\rho Q_L^2 = A_L^2 P_2 \quad (4)$$

$$\tfrac{1}{2}\rho Q_V^2 = A_V^2 P_2 \quad (5)$$

$$Q_S = Q_L + Q_V \quad (6)$$

$$Q = Q_P + Q_S \quad (7)$$

The known values are $A_P, A_S, A_L, k_A, A_M, P_{vfc}$ and one of P_1 and Q. The unknown values are $P_2, A_V, Q_P, Q_S, Q_L, Q_V$, and one of Q or P_1. This makes seven unknowns, appropriate to the seven equations. Generally, a numerical solution is required. However, some limited cases have useful analytical solutions. The following assumes that the variable area valve is within its varying range. Outside this range, the solutions are easy.

In the highly simplified case of the variable valve area alone, the relevant equations are:

$$A_V = k_A(P_1 - P_{vfc}) \tag{1}$$

$$\tfrac{1}{2}\rho Q^2 = A_V^2 P_1 \tag{2}$$

To obtain Q given P_1, is easy, solving directly for A_V. To solve for P_1 given Q, as is often required for a damper, requires solution of a cubic obtained by eliminating A_V:

$$P_1^3(k_A^2) + P_1^2(-2k_A^2 P_{vfc}) + P_1(k_A^2 P_{vfc}^2) + (-\tfrac{1}{2}\rho Q^2) = 0$$

In the simplified case of a large series hole A_S of insignificant resistance, the leak area A_L is equivalent to the parallel hole A_P, so both A_S and A_L may be neglected. Also, P_2 is no longer relevant. The appropriate equations are then:

$$A_V = k_A(P_2 - P_{vfc}) \tag{1}$$

$$\tfrac{1}{2}\rho Q_P^2 = A_P^2 P_1 \tag{2}$$

$$\tfrac{1}{2}\rho Q_V^2 = A_V^2 P_1 \tag{3}$$

$$Q = Q_P + Q_V \tag{4}$$

The unknowns are A_V, Q_P, Q_V, and one of P_1 or Q. Given P_1, the valve variable area and the volumetric flow rates follow easily. Given Q, a cubic may be obtained in S, the square root of P_1 by eliminating Q_P and Q_V:

$$S^3(k_A) + S(A_P - k_A P_{vfc}) + \left(-Q\sqrt{\tfrac{1}{2}\rho}\right) = 0$$

Usually there is only one real solution, to give a real solution for P_1. If there are more real solutions, there should be only one with a root within the physically meaningful range.

In the simplified case of no parallel hole A_P, the following equations apply:

$$A_V = k_A(P_2 - P_{vfc}) \tag{1}$$

$$\tfrac{1}{2}\rho Q_S^2 = A_S^2(P_1 - P_2) \tag{2}$$

$$\tfrac{1}{2}\rho Q_L^2 = A_L^2 P_2 \tag{3}$$

$$\tfrac{1}{2}\rho Q_V^2 = A_V^2 P_2 \tag{4}$$

$$Q_S = Q_L + Q_V \tag{5}$$

$$Q = Q_S \tag{6}$$

The knowns are A_S, A_L, k_A, P_{vfc} and P_1 or Q. The unknowns are P_2, A_V, Q_L, Q_V, Q_S and Q or P_1. To solve for P_1 given Q, the pressure drop through A_S is known directly. A cubic may be obtained in $S = \sqrt{P_2}$:

$$S^3(k_A) + S(A_L - k_A P_{vfc}) + \left(-Q\sqrt{\tfrac{1}{2}\rho}\right) = 0$$

Then A_V and P_1 follow easily. To solve for Q given P_1, again solve first for P_2, this time with the cubic:

$$P_2^3(k_A^2) + P_2^2(2A_L k_A - 2k_A^2 P_{vfc})$$
$$+ P_2(A_S^2 + A_L^2 - 2A_L k_A P_{vfc} + k_A^2 P_{vfc}^2) + (-A_S^2 P_1) = 0$$

The other values then follow easily.

In the completely general case, solving for the flow rate Q given the pressure P_1 requires solution of P_2 by the same cubic immediately above. Then all else follows easily. Solution of the pressure P_1 given Q involves an elaborate sextic, and may be solved equally well by iteration on the original equations. A simple binary search takes about 1–2 μs per iteration, some 40 μs in total. A subroutine using a cubic solution of Q via P_2 for a given P_1 may be used in the loop, iterating with various P_1 until Q is accurate enough. The curve is well behaved, so a better iterative method would be much faster, but may have difficulties very near to the end points (valve very close to opening or reaching maximum area).

Appendix E offers an explanation of the solution of the cubic equation by analytical means. Iteration is also effective.

6.11 Temperature Compensation

As mentioned in Chapter 5, Fluid Dynamics, the energy losses producing the pressure difference may arise from viscosity in laminar flow, or from turbulence. The former gives a pressure loss proportional to the volumetric flow rate, the latter gives one proportional to volumetric flow rate squared. The former is basically more desirable because it gives a damper force proportional to velocity.

Experimental dampers have achieved this by having valves comprising simply long small-bore pipes, which, under a suitable Reynolds number, have a pressure loss governed by the Hagen–Poiseuille equation. Unfortunately, however, the viscosity of a liquid is highly sensitive to temperature, so this method has proved unsatisfactory in practice. Although synthetic oils have been developed with less sensitivity to temperature than mineral or vegetable oils, the ideal of zero viscosity variation has not even been approached. The future of such a design of valve therefore depends upon the invention of a suitable fluid with a viscosity that either exhibits little or no sensitivity to temperature, or has a viscosity having some other dependence which could be used to compensate for the temperature variations. For example, some liquids have a viscosity (and a shear stress) dependent upon the electric or magnetic field strength. This can be used to provide not just temperature compensation, but actual damper adjustability, at extra cost, as described in Chapter 9.

The viscosity and density variations can also be overcome by a thermally controlled orifice size. There are four main possibilities:

(1) differential expansion—longitudinal;
(2) differential expansion—bending;
(3) fluid volumetric expansion;
(4) solid volumetric expansion;
(5) solid volumetric phase change.

Differential Expansion—Longitudinal

It is possible in principle simply to have a centre rod of a high expansion solid material within a hollow main rod, but the differences of common metal expansion coefficients are not conveniently large. For example, 24 ppm/K for aluminium against 12 ppm/K for steel. Over a length of 100 mm and a temperature change of 100 °C, the differential expansion is 0.12 mm, usable with good design not really large enough for convenient and economic application to production. Use of an Invar rod would double the available movement if used in an aluminium tube, or *vice versa*. With a rod of plastic, this concept may certainly be attractive. Some example plastic linear expansion coefficients are ABS 90 ppm/K, PTFE 90 ppm/K, HD polyethylene 110 ppm/K, Nylon 6/6 90 ppm/K, ethyl cellulose 200 ppm/K, polyethylene vinyl acetate 162–234 ppm/K.

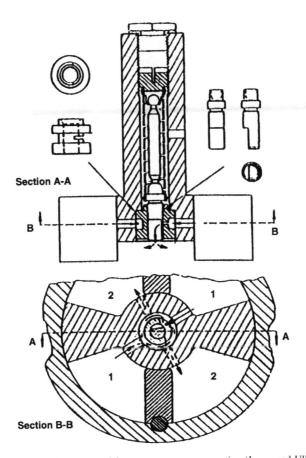

Figure 6.11.1 Bimetallic strip used for temperature compensation (James and Ullery, 1932).

Differential Expansion—Bending

Use of a bimetallic strip in bending for temperature compensation is an old idea, for example as proposed by James and Ullery (1932), as shown in Figure 6.11.1. Note that the design must be such that the fluid pressure does not distort the spiral bimetallic strip excessively, this lacking rigidity. The bimetallic strip acts in effect as a lever, giving a large movement with, consequently, only a small force. The deflection of the spiral for a given temperature is inversely proportional to the thickness, but the stiffness in bending is proportional to the fourth power of the thickness. This idea has been used more recently in a more compact single layer spiral form, Figure 6.11.2.

Fluid Volumetric Expansion

A modern implementation of the fluid expansion compensation principle is as shown in Figure 6.11.3. Thermal expansion of the trapped liquid, alcohol, moves the tapered plunger thereby reducing the orifice size. With careful choice of dimensions and materials, this system can work very well. Suitable

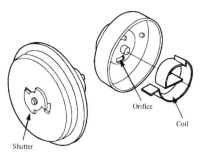

Figure 6.11.2 Single spiral bimetallic strip operating a rotary shutter. Reproduced from *Design News* 10-2-89, p. 195.

liquids have volumetric ('cubical') thermal expansion coefficients in excess of 1000 ppm/K, e.g. methyl alcohol 1400 ppm/K. Hence a 100 K temperature rise can give a volume increase of 14%, making it easy to achieve a position change of 1–2 mm (20 μm/K).

Solid Volumetric Expansion

A low-yield substance, solid not liquid, possibly waxy, could be used in a similar manner to a liquid with some advantages. By changing its shape it can conform to a container and the expansion can therefore be amplified by allowing it to escape from a only small exit area. This is the principle of a liquid thermometer.

Solid Volumetric Phase Change

Another possibility would be to use a wax-expansion variable valve, as on engine cooling thermostats, although this does not seem to have been applied to dampers so far. The special wax has a substantial volumetric expansion at a phase change at a particular temperature. This method would not give smooth compensation over a temperature range, but may be helpful in combating severe fade.

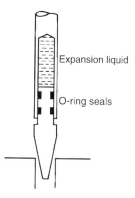

Figure 6.11.3 Liquid expansion method of temperature compensation.

6.12 Position-Sensitive Valves

Sensitivity of the force to position is really achieved by a spring, not a damper, but this is not the intended meaning of the phrase in this context. Position-sensitive valves have been unusual on cars, but have some history of use on motorcycles, and are common on aircraft landing gear. In the last case, they may use a long taper needle entering an orifice in the piston, reducing the effective orifice size as the undercarriage compresses. Figure 6.12.1 shows some aircraft oleo legs, whilst Figure 6.12.2 shows a taper needle damper for cars. The idea of position sensitivity should be distinguished in principle from the idea of stroke length sensitivity. In the latter case, the force cannot be identified directly with the actual position of the piston in the cylinder.

An alternative method to achieve position sensitivity is to machine one or more longitudinal grooves on the inside of the pressure cylinder. This would probably require a thicker basic cylinder wall than would otherwise be used, as this is usually only about 1 mm. Typically, two three or four grooves are used. The grooves require a suitable sectional area, a few square millimetres, and may be profiled longitudinally to give a progressive effect. Manufacturing would require a shaping operation with attention to the groove edges to pre-empt adverse wear effects on the piston seal. The basic concept is to introduce a leak area bypassing the piston valves when the damper is near to its central position, softening the damping for normal driving. Figure 6.12.3 (p. 242) shows the groove concept. Figure 6.12.4 (p. 243) shows test results for the $F(X)$ loop where the effect of the groove on the damper force is apparent. The resulting curves shown are theoretical ones, but experiments were in good agreement. Figure 6.12.5 (p. 243) shows a four-groove design.

There are some weak aspects of the bypass channel method. In so far as it acts as an orifice it has a non-ideal characteristic with very little damping at very low speed. To the extent that this is ameliorated by viscosity it will be temperature sensitive. As a position-sensitive method rather than a stroke-sensitive method, the low-damping action does not adjust to different ride height for different vehicle loads. These objections also apply to the taper-rod system.

6.13 Acceleration-Sensitive Valves

Figure 6.13.1 (p. 244) shows the original 'inertial valve' proposed by Kindl (1933).

Figure 6.13.2 (p. 244) shows an inertial valve proposed by Speckhart and Harrison (1968). In this system the orifice area depends upon the valve position governed by the acceleration of the piston. The upper part of the piston is fixed to the rod, only the lower part can move independently. When the rod and piston, attached to the wheel, are suddenly struck upwards by a bump in the road, the spring-loaded inertial lower section of the piston is left behind by inertia, opening the bypass port to soften the damping force. The damper force is therefore acceleration sensitive.

The floating sub-piston has a mass m_F. The spring preload is F_P. The valve is designed so that a fluid pressure differential across the valve does not affect the free piston position, otherwise the spring must be very stiff and the floating piston mass too great. The floating piston will begin to lift when $A_P m_F = F_P$.

Such systems have not found commercial success so far, but the concept has been developed by other more expensive methods, such as electronic control. Also, the later pressure-rate valve is similar in intent. The design as shown may not be very practical for ordinary dampers. A close fit is needed at the port to give a good seal at low accelerations, so alignment is critical. Also, the rod must be attached to the wheel, whereas in most installations the rod attaches to the vehicle body. On a passenger car, the rod would need to emerge from the bottom of the damper. A pressure-balanced spool valve may then be more suitable. Some struts have a double-tube arrangement to carry the side loads, with the piston driven by a rod from beneath. These struts are obvious candidates for application of the inertial valve.

To overcome the difficulty that the rod must emerge from the bottom of the damper, it would be much better to have the inertia valve on the body of the damper, connected from the top of the pressure

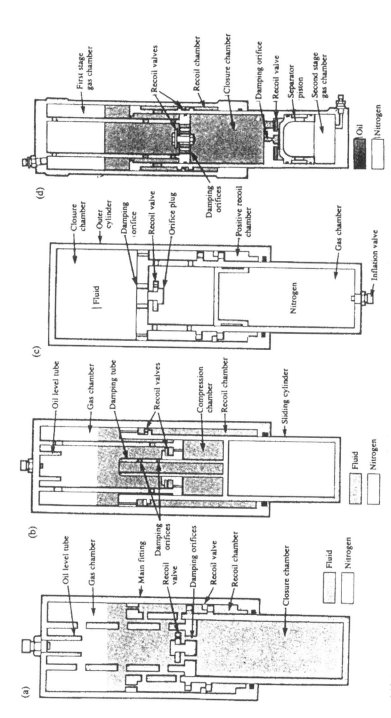

Figure 6.12.1 Aircraft oleo dampers: (a) unseparated oleo with positive recoil control; (b) unseparated oleo with three level damping position dependent; (c) separator type oleo; (d) two-stage damper oleo (Young, 1986).

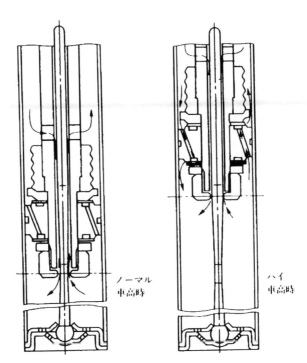

Figure 6.12.2 Position-sensitive damper using tapered rods. Reproduced from Komamura and Mizumukai (1987) History of Shock Absorbers, *JSAE*, 41(1), pp. 126–131.

chamber to the bottom, in parallel with the piston. On a double-tube damper, it would be fitted within the outer tube. In the condition of striking a bump, the compression chamber below the piston is at high pressure, controlled by the foot valve, with reduction back to low pressure through the piston. Therefore it is not sufficient simply to bypass the foot valve with a sliding ring around the lower part of the pressure tube, or cavitation will occur in the extension chamber. The extension and compression chambers really need to be connected directly together via the acceleration valve, whether this is in the piston or not.

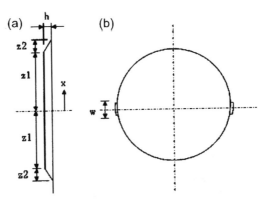

Figure 6.12.3 Position-sensitive pressure cylinder design: (a) longitudinal groove section; (b) cross-section (Lee and Moon, 2005).

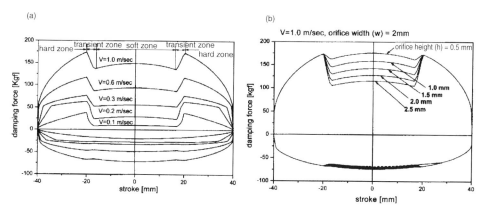

Figure 6.12.4 Grooved cylinder damper F(X) force loops: (a) various speeds; (b) various groove depths (Lee and Moon, 2005).

6.14 Pressure-Rate Valves

Seeking to achieve similar results to the Speckhart and Harrison inertial valve and the Shiozaki piezoelectric valve in a simpler manner, it is possible to design a passive mechanical valve that is responsive to the rate of change of pressure in one chamber. In Figure 6.14.1 (p. 245), the centre element A functions as a normal valve. It seats on part B. With a slow rate of pressure rise, B follows A. With a rapid rate of pressure rise, the lift of B is limited by the constricted hole, so A will lift from B relieving the pressure. This pressure-rate relief concept may be used in parallel with a conventional valve.

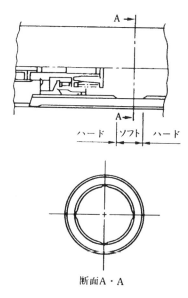

Figure 6.12.5 Cylinder with four shorter grooves. Reproduced from Komamura and Mizumukai (1987) History of Shock Absorbers, *JSAE*, 41(1), pp. 126–131.

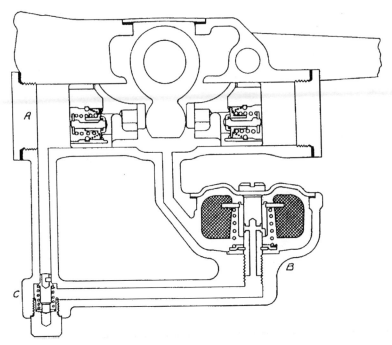

Figure 6.13.1 Inertial valve for opposed-piston damper (Kindl, 1933).

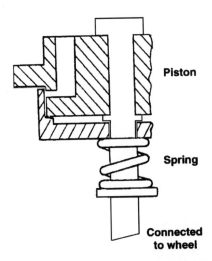

Figure 6.13.2 Inertial valve (from Speckhart and Harrison, 1968, modified for clarity of function).

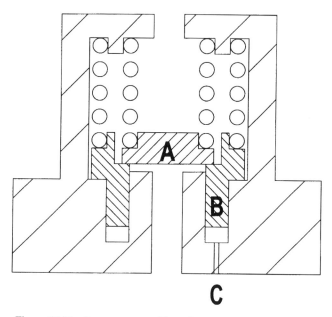

Figure 6.14.1 Pressure-rate-sensitive valve (Potas, 2004, 2005, 2006).

6.15 Frequency-Sensitive Valves

Some dampers have been advertised as having a frequency-sensitive force characteristic, but it is not immediately obvious what this is intended to mean. At a given stroke, frequency sensitivity would just be sensitivity to velocity — in other words a normal damper. At a given velocity amplitude, frequency sensitivity would imply dependence of force on the stroke, and would be better called stroke sensitivity. In a laboratory test, the frequency is well defined, and usually constant. The motion history is known, and also the motion that is to come. On the road, the damper motion is semi-random, dominated by body resonance and wheel hop, and it is not clear what frequency should be attributed to the motion at any instant, given only the history of the motion. Therefore it is unclear exactly what the frequency f is, how the variable f is intended to affect the valve resistance, and how this could be expressed in a practical implementation.

6.16 Stroke-Sensitive Valves

A stroke-sensitive valve is one that has a pressure drop that increases with physical displacement, even though this occurs at constant flow rate. This is claimed, with justification, to give a better combination of ride and handling by increasing the damping for large body motions such as roll at corner entry, and for severe ride motions, whilst allowing less damping for ordinary ride motions and small bumps. The same nonlinear effect occurs with the use of rubber bushes in series with the damper, but at a smaller stroke scale. Also, stroke dependence could be achieved indirectly by the method of position dependence, e.g. a taper rod reducing an orifice, or piston bypass channels, or by a piston that can slide on the rod.

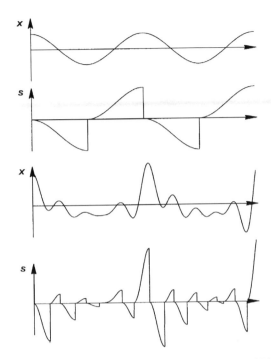

Figure 6.16.1 Instantaneous stroke derived from displacement: top sinusoidal, bottom random.

A stroke-sensitive valve is different in principle from an amplitude-sensitive one. The amplitude of a damper is only clear when tested in the laboratory, when it is normally a constant, and certainly under control. On the road, the damper does not have a clear amplitude at any instant. There is only a motion history with a semi-random nature. In contrast, the idea of the instantaneous stroke can be clearly defined for practical conditions. The instantaneous stroke is the damper displacement since the last reversal of direction. This applies with clarity not only to laboratory sinusoidal tests, but also to real on-the-road semi-random motions. Note that this is to be distinguished from the common use of the term stroke to mean the maximum stroke, twice the amplitude in a sinusoidal test. For a sinusoidal motion of constant amplitude, the instantaneous stroke is then a continuous variable with a somewhat sawtooth type of character, increasing until it suddenly drops to zero at a reversal, as shown in Figure 6.16.1. For a semi-random motion, the maximum value of stroke achieved varies at each reversal.

The piston itself can hardly sense the stroke directly, but the fluid displaced through the piston at any point during the stroke since reversal is just $Q_S = A_{PA} S_I$, so it is a practical possibility to sense Q_S at the piston, and therefore to have a passive mechanical valve that is responsive to this value, and hence to the instantaneous stroke. This is not likely to be functionally better than an actively controlled valve with good sensors, but is a cost-effective way to achieve a useful improvement over a standard passive damper.

Fukushima *et al.* (1983, 1984) advocated stroke-sensitive damping, and proposed a vortex valve to achieve it, with a successful demonstration unit. Figure 6.16.2(a) shows the relationship of the damper characteristic to the vortex valve, whilst (b) shows the valve geometry, and (c) the damper structure with vortex piston and foot valves. Part (d) shows the damper characteristic actually achieved. The basic idea is that for small strokes the vortex character of the chamber is ineffective, and it has a resistance simply the sum of those of the inlet and outlet orifices. For long strokes, the vortex can develop, and there is a centrifugal pressure resistance, which can be analysed as that of a free vortex. The vortex development stroke can be adjusted by the chamber volume.

Valve Design

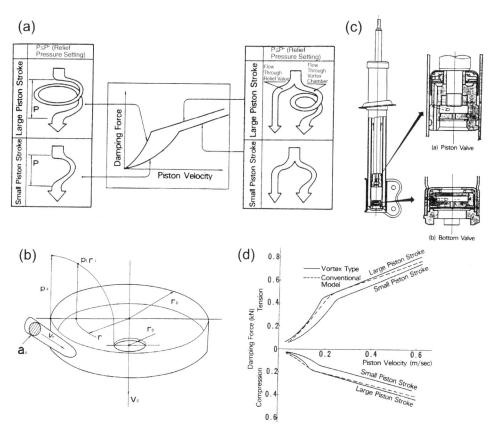

Figure 6.16.2 Vortex valve for stroke sensitive damping: (a) vortex effect on resistance pressure; (b) vortex chamber geometry; (c) damper structure; (d) experimental damper characteristics. Reproduced from Fukushima, Idia and Hidaka (1984) Development of an automotive shock absorber that improves riding comfort without impairing steering stability, *Proc. 20th FISITA Conference*, pp. 218–223.

Figure 6.16.3 shows again the action of the vortex, with the flow almost directly into the central hole for short stroke, but forming a full vortex for long stroke.

The vortex is basically of the free type, although viscosity is a significant factor. The centrifugally generated vortex pressure gradient (Chapter 5) is

$$\frac{dP}{dR} = \frac{\rho V_T^2}{R}$$

where V_T is the tangential velocity. The total vortex pressure drop is easily found by integration over the radius. Application of Bernoulli's equation, with inlet nozzle and outlet orifice discharge coefficients, and the pressure drop across the free vortex when present, relates the flow rate to the pressure. The outlet orifice size needs to be small to generate the most vortex pressure because much of the effect occurs near to the centre, but it is desired to make it large to keep the short stroke non-vortex pressure loss low. Optimising the ratio of pressures vortex-active/vortex-inactive gives a quadratic in $(R_O/R_N)^2$, where R_O is the outlet diameter and R_N is the inlet nozzle diameter, dependent on the vortex chamber diameter and the discharge coefficients. For practical values, the optimum R_O is close to R_N, and the theoretical ratio of resistances, including realistic discharge coefficients, but excluding viscous

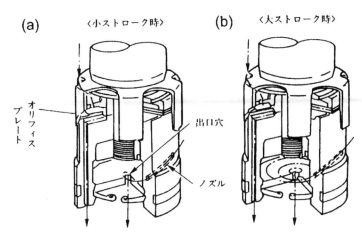

Figure 6.16.3 Vortex valve: (a) short stroke radial flow; (b) long stroke spiral flow. Reproduced from Komamura and Mizumukai (1987) History of Shock Absorbers, *JSAE*, 41(1), pp. 126–131.

effects in the chamber, can be quite high. A factor of 4 would give excellent results, although the original Fukushima test damper ratio was rather smaller at about 1.4, but still showing favourable test results on the vehicle.

The time constant, or stroke constant, is important. The vortex will develop over a stroke sufficient to provide enough oil to fill the vortex chamber. Normally the chamber diameter will be nearly as large as that of the piston, so the development of the vortex will occur over a stroke roughly equal to the chamber axial length. Fukushima's original chambers were short, only about 0.15 times the piston diameter, and also smaller in diameter than the piston, so a greater effect may be achievable than in the original tests, if desired.

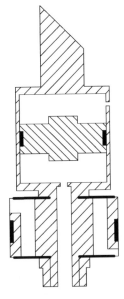

Figure 6.16.4 Use of a floating secondary piston in a chamber, to give a stroke-sensitive parallel flow and force.

Perhaps a simpler way to obtain much the same effect as a vortex valve is to have a floating secondary piston, as in Figure 6.16.4. For short strokes the piston can move, and there is effectively a free flow through the two orifices, in and out of the chamber, in parallel with the main valve. For long stroke, the secondary piston reaches the end of its free motion and closes one of the orifices. This is the principle of the 'DampMatic', which includes elastomer impact cushions on the piston. Another possibility would be to give the piston some central location tendency by springs.

So-called FSD (frequency-selective damping) valves are really stroke-sensitive valves within the above definition. There is an extra 'FSD' valve in parallel with the main piston valve. The extra part opens easily at short stroke, softening the damping force, but with persistence of the stroke the pressure is transmitted to the rear of the valve, forcing it to close hard, eliminating the bypass effect. This is a true stroke dependence, in contrast to the position dependence of the bypass channel method.

6.17 Piezoelectric Valves

Piezoelectrically operated variable-flow valves have been proposed, and prototypes have been demonstrated, although they have not been used for production vehicles. As actuators, they have the advantages of rapid response and production of large forces, but have the disadvantages that the distance of actuation is very small and that large control voltages are required (hundreds of volts).

The small distance of actuation for a single unit, typically only 1 μm (0.001 mm) can be amplified up to a useful valve displacement of 1 mm by two techniques. The piezo elements can be stacked, with ten in series (or even more) giving 0.010 mm movement. Hydraulic amplification by straightforward means at a diameter ratio of 10 then gives an area ratio of 100 with a motion ratio of 100. The total effective gearing is then 1000, or even more if desired. The mechanical advantage is correspondingly low, but this is not a problem because the valve actuation forces are quite small and large piezo forces are available.

Figure 6.17.1 shows details of a design of damper of this type by Shiozaki *et al.* (1991). They intended to take advantage of the high speed of response by using a normally hard set-up with the valve relieving the force when the wheel hits a bump causing a rapid positive rate of change of the force (a similar intent to the inertial valve of Speckhart and Harrison, 1968). The piezo sensing element detects the rate of change of damper force with an output of about 2.5 μVs/N (i.e. 2.5×10^{-6} V per N/s). One problem that arose was that the sharp valve actuation caused audible clicking noises. Various valve edge profiles were investigated to alleviate this. Although technical feasibility was clearly demonstrated, overall it seems that the manufacturing complications and the high control voltage have been too much of a handicap, and commercial acceptance has so far eluded this design concept.

6.18 Double-Acting Shim Valves

A single shim can be used for both compression and extension, as illustrated in Figure 6.18.1. The shim cones in the same direction for both flows, but opens on its outer edge for compression and its inner edge for extension, giving more resistance in extension according to the ratio of inner to outer radii. The coning or balling behaviour in compression may be considered suspect in that it seems likely that there would instead be distinct planar bending wings, as for a conventional shim valve. However, this could not happen for extension, although there could still be asymmetrical distortion. Lee (1997) investigated the concept, attributing it to the de Carbon company, and found good agreement between theory and experiment on the basis of axial-symmetrical deflections. The coning is nonlinear, so dimensional analysis and finite-element analysis were used.

The parts count is low and the piston is axially compact. Tuning the characteristics may be more difficult, although Lee's test results showed satisfactory behaviour. Shim stresses may be high, but the

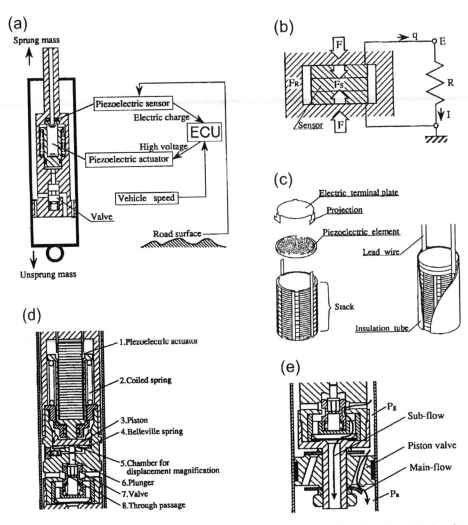

Figure 6.17.1 Piezoelectric sensing and actuating damper valve: (a) general configuration; (b) piezoelectric sensor; (c) stack of elements; (d) construction of actuator including hydraulic amplification; (e) detail of flow through main valve and controlled supplementary parallel valve (sub-flow) (Shiozaki *et al.*, 1991).

concept seems attractive from the economics of manufacturing. The one critical dimension is the depth of the shim seat in the piston relative to the inner piston seat, which must equal the shim thickness for zero preload, or be changed from that value as required.

6.19 Rotary Adjustables

A convenient way to achieve some adjustment is to have a rotatable barrel with various holes in it. This may be manually or electrically controlled, Figure 6.19.1. Various examples of the valve arrangement

Valve Design

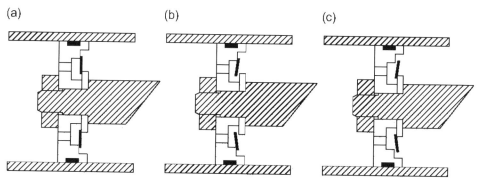

Figure 6.18.1 Example configuration of a double-acting shim valve, showing: (a) closed; (b) extension flow opening; (c) compression flow opening. Radial location of the shim is required at three points, not shown.

itself are given in Figures 6.19.2–6.19.4. The sequence of resistances must be carefully considered. Usually the variable hole is in parallel with the main valve, so it acts as a leak. For uniform steps of pressure resistance at a given flow rate

$$P = NP_1$$

with integer N. Using Bernoulli's equation for flow from an orifice

$$P = \tfrac{1}{2}\rho \left(\frac{Q}{A}\right)^2 = \tfrac{1}{2}\rho \left(\frac{4Q}{\pi D^2}\right)^2$$

$$D = \sqrt{\frac{4Q}{\pi}} \sqrt{\frac{\rho}{2P}}$$

from which the sequence of diameters to give equal increments of pressure may be deduced.

Alternative designs use several holes in series, or taper or needle valves moving into an orifice.

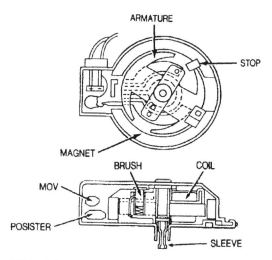

Figure 6.19.1 Solenoid actuator for adjustable barrel valve (Soltis, 1987).

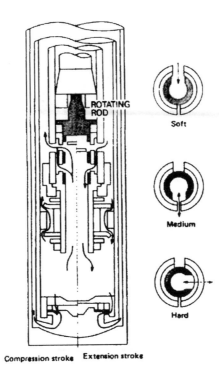

Figure 6.19.2 Three setting rotary valve (Sugasawa *et al.*, 1985).

6.20 Bellows Valves

Figure 6.20.1 shows a double-tube damper foot valve with a sealed bellows containing gas, possibly pre-pressurised, which permits a flow of limited volume in parallel with the main foot valve.

The bellows concept may be better applied to the piston, where, with a suitable bellows stiffness and strength, it would effectively allow flow between the compression and extension chambers for small strokes, giving the desired sensitivity of damper force to stroke.

If the flow resistance into the bellows chamber is small, and the bellows has a pressure/volume relationship dP/dV then it acts similarly to a series stiffness

$$K_B = \frac{dP}{dV} A_{PA}^2$$

Because of the piston rod, the bellows would be mounted only beneath the piston, but could flow oil for both compression and extension, each through its own valve. The motion range of the bellows may need to be limited by mechanical stops to allow suitable stiffness without overstressing.

6.21 Simple Tube Valves

The viscous tube valve damper is a simple displacement damper with a tube bypass along the outside of the cylinder, so that all liquid displaced by the piston annulus passes through the tube. Experimental ones have been built, Figure 6.21.1, but the great sensitivity of viscosity to temperature makes them unsuitable for use when the temperature of the oil may vary significantly, through ambient temperature

Valve Design

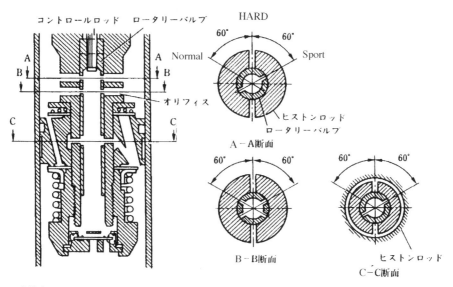

Figure 6.19.3 Rotary adjustable valve. Reproduced from Komamura and Mizumukai (1987) History of Shock Absorbers, *JSAE*, 41(1), pp. 126–131.

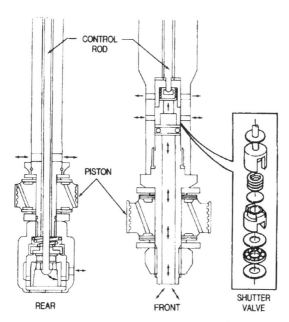

Figure 6.19.4 Rotary barrel valve, flow shown in soft position (Soltis, 1987).

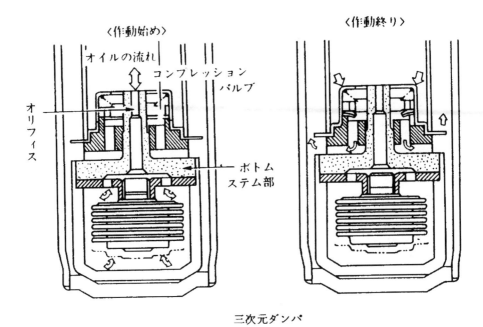

Figure 6.20.1 Foot valve with bellows: (a) the bellows contracts to accommodate a limited volume of fluid from the extension chamber in the bellows chamber, or vice versa; (b) at higher pressure or longer stroke the parallel shim valve operates normally. Reproduced from Komamura and Mizumukai (1987) History of Shock Absorbers, *JSAE*, 41(1), pp. 126–131.

or by hard working. Therefore they are unsuitable for automotive use, but are analysed here for interest and to illustrate their operation.

The piston diameter D_P and rod diameter D_R give a piston annulus area

$$A_{PA} = \frac{\pi}{4}\left(D_P^2 - D_R^2\right)$$

At a damper velocity V_D, the volumetric flow rate through the tube is

$$Q = A_{PA} V_D$$

The tube inner diameter is D_T and length L_T. The viscous pressure drop, by the Hagen–Poiseuille equation, is

$$P_V = \frac{128 \mu L_T Q}{\pi D_T^4}$$

The mean fluid flow velocity V_T in the tube is

$$V_T = \frac{Q}{A_T}$$

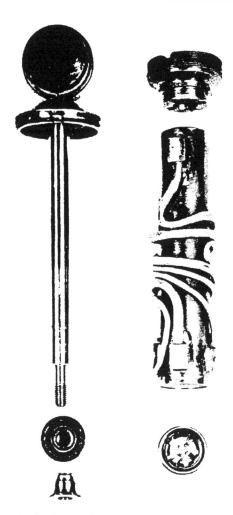

Figure 6.21.1 This experimental tube damper showed the advantages of tuning the $F(V)$ curve, and led to improvements in the earlier hard blow-off telescopics. Reproduced from Peterson (1953) *Proc. National Conference on Industrial Hydraulics*, 7, 23–43.

where A_T is the tube cross-sectional area. The Hagen–Poiseuille equation is applicable only for laminar flow, requiring

$$Re = \frac{\rho V_T D_T}{\mu} \leq Re_{max} = 2000$$

The Reynolds number may also be expressed as

$$Re = \frac{4\rho Q}{\pi \mu D_T}$$

The viscous pressure drop is applied to the piston annulus, giving a damping coefficient

$$C_D = \frac{128\mu A_{PA}^2 L_T}{\pi D_T^4}$$

For the flow to remain laminar, the tube diameter has a minimum given by

$$D_{Tmin} = \frac{4\rho A_{PA} V_{Dmax}}{\pi \mu Re_{max}}$$

If the flow becomes turbulent, then the pressure drop will increase sharply, by a factor of about 2, and subsequently will be quadratic with speed.

The tube length to give a desired damping coefficient C_D is

$$L_T = \frac{\pi D_T^4 C_D}{128\mu A_{PA}^2}$$

For a given design, the maximum damper speed for laminar flow in the tube is

$$V_{Dmax} = \frac{\mu Re_{max} A_T}{\rho A_{PA} D_T} = \frac{\pi \mu Re_{max} D_T}{4\rho A_{PA}}$$

Another expression for the length is

$$L_T = \frac{2 A_{PA}^2 C_D}{\pi^3 \mu^5} \left(\frac{\rho V_{Dmax}}{Re_{max}} \right)^4$$

To have a reasonable tube length, a high-viscosity oil must be used. This value may be deduced from the above equation inverted to make μ the subject.

There is also a quadratic damping term because of the high exit velocity from the small diameter tube. This is given by the basic value

$$C_Q = \frac{\alpha \rho A_{PA}^3}{2 A_T^2}$$

which could be reduced by detailed design if required, e.g. an exit diffuser. The damper force is then given by

$$F_D = C_D V_D + C_Q V_D^2 \quad V_D > 0$$
$$F_D = C_D V_D - C_Q V_D^2 \quad V_D < 0$$

Realistic values are:

$$C_D = 2.000 \, \text{kNs/m}$$
$$D_P = 28.00 \, \text{mm}$$
$$D_R = 12.00 \, \text{mm}$$
$$V_{Dmax} = 2.000 \, \text{m/s}$$
$$Re_{max} = 2000$$
$$\rho = 900.0 \, \text{kg/m}^3$$

$$L_T = 200.0\,\text{mm}$$
$$A_{PA} = 5.027\,\text{cm}^2$$
$$Q = 1.005\,\text{L/s}$$
$$\mu = 160.7\,\text{mPas}$$
$$D_T = 3.600\,\text{mm}$$
$$A_T = 10.10\,\text{mm}^2$$
$$f_A = 49.75$$
$$V_T = 99.49\,\text{m/s}$$
$$Re = 1999$$
$$C_D = 2001\,\text{N s/m}$$
$$\alpha = 2.000$$
$$C_Q = 1120\,\text{N s}^2/\text{m}^2$$

6.22 Head Valves

As described elsewhere, when a double-tube damper is in extension, the pressure in the extension chamber causes some leakage flow through the rod bearing. This leakage passes out to the reservoir. It is used to advantage to circulate the oil, improving cooling. This concept is developed further in some adjustable double-tube dampers, where a valve in used in the damper body head. This valve allows oil to pass from the top of the extension chamber out into the reservoir. Opening this valve lowers the extension chamber pressure, so the piston valving is set for the maximum $P(Q)$ required. The head valve is unidirectional, so in compression the volume requirement of the extension chamber is met entirely by the piston compression valve. The advantage of this system is that the valve is external, on the body instead of in the piston, so adjustment accessibility is very good, and the valve itself is easily given the desired characteristics. Care must be taken that the efflux from the valve passes down into the reservoir oil without entraining air. If the oil take-off point is below the top of the extension chamber then there is some hydraulic bump-stop effect at the top when only the piston extension valve can pass oil.

6.23 Multi-Stage Valves

The usual valve is three-stage as described. However, sometimes it is useful to have additional stages. This may be done in either direction, that is by the addition of compliance or stiffness at some point of partial valve opening. Two systems are common:

(1) On a shim valve, a support shim is given small separation, by a small diameter spacer shim, from the main shim, so that the main shim opens somewhat, e.g. 0.1 mm, and then receives additional support from the backing shim(s).
(2) On a conventional piston extension valve, the coil spring operates against a rigid disc which partially supports a shim. The valve then operates as a shim valve until a high pressure is reached, at which point the coil spring operates allowing the entire shim to lift, giving a blow-off pressure relief effect.

7
Damper Characteristics

7.1 Introduction

The damper is characterised by:

(1) general dimensional data;
(2) force characteristics;
(3) other factors.

Dimensional data include the stroke, the minimum and maximum length between mountings, diameters, mounting method, etc. Force characteristics indicate how the force varies with compression and extension velocities, production tolerances on these forces, any effect of position, and so on. Other factors include limitations on operating temperature, power dissipation, cooling requirements, etc.

In considering the required characteristics of dampers it is desirable, if possible, to express their complex behaviour in a few simple parameters that can be correlated with subjective ride and handling quality. Thus the complexity of the force–speed characteristic might be reduced, albeit imperfectly, to the following parameters:

(1) overall mean damping coefficient C_D;
(2) asymmetry, the transfer factor e_D;
(3) $F(V)$ shape, the progressivity factor λ.

The most fundamental parameter is the total average damping coefficient. This results in a damping ratio for the vehicle which varies considerably according to the type of vehicle and the philosophy of the particular vehicle manufacturer. Typical overall damping ratios are 0.2–0.4 for a passenger car, and 0.4–0.8 for a performance-oriented passenger car or competition car. Considering the variation in vehicle mass and spring stiffness, the required damping coefficient per wheel varies typically from 1 to 5 kN s/m for passenger cars, and higher of course for commercial vehicles.

The second fundamental parameter is the asymmetry, the relative amounts of bump and rebound damping, which on passenger cars tends to be around 30/70, although not narrowly constrained, varying between 20/80 and 50/50. On motorcycles it seems to be even more asymmetric, perhaps from 20/80 to 5/95.

The third parameter, the shape of the force against speed curve, may be represented by the progressivity factor λ. With intelligent choice of valve parameters it is possible to achieve a wide range of force–speed graph shapes and progression factors. In general the preference is for a force that

---

The Shock Absorber Handbook/Second Edition   John C. Dixon
© 2007 John Wiley & Sons, Ltd

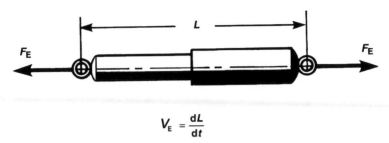

Figure 7.1.1 Positive velocity and force in extension. The damper is in tension, with its length increasing.

increases rather less than proportionally with speed (i.e. a reducing damping coefficient) within the normal operating range, so that the damping ratio is higher at low damper speed. This is to provide good control of handling motions whilst avoiding unacceptable harshness on bad bumps. The actual curve itself should of course be smooth.

As far as the cyclic characteristic is concerned, it is important for the $F(X)$ loop to be smooth in shape. Basically this means having smooth valve characteristics and avoiding cavitation in the usual range of operation.

The force exerted by a damper depends on its velocity, and also on its recent history of operation which influences the temperature and fluid properties. For a normal damper the effect of position is secondary, although this is not always the case, for example for motorcycle forks which include hydraulic buffering near to the full compression position, or for a combined spring/damper unit. In the latter case, the force is merely the sum of damper and spring forces, which act in parallel.

The basic characteristic of a complete damper is represented by a graph of force against velocity. The damper extension velocity is

$$V_{DE} = \frac{dL}{dt}$$

where L is the length between mounts, Figure 7.1.1. The compression velocity is then defined as

$$V_{DC} = -V_{DE} = -\frac{dL}{dt}$$

Essentially this means that any velocity, compression or extension, can be expressed as a compression or extension velocity with appropriate sign. Normally, of course, the one giving a positive sign to the motion of immediate interest is used, i.e. only one of the two velocity variables V_{DC} and V_{DE} is in use at any one time. For drawing $F(V)$ graphs it is often convenient to plot the forces against absolute velocity.

The normal installation of a suspension damper is such that suspension bump causes damper compression, so bump velocity is a common alternative term for compression velocity, and rebound velocity for extension velocity. However bump and rebound may be inappropriate terms for other types, e.g. steering dampers. The plain term 'damper velocity' may refer to any of the above, according to context, or to the absolute value of the velocity, then always being positive.

For forces, various conventions are in use. A positive extension damper force F_{DE} is one exerted on the damper causing extension. The damper is in tension. A positive extension force in this sense is therefore one also pulling the mounting points on the vehicle towards each other. Compression forces are opposite to the above. As in the case of velocities, it may be convenient to define compression and extension forces that simply have a sign reversal, $F_{DC} = -F_{DE}$, and to use whichever one is positive.

The weight, mass and acceleration of the damper are generally fairly small with a small mA product. Often, therefore, the forces on the two ends can be considered to be equal for practical purposes.

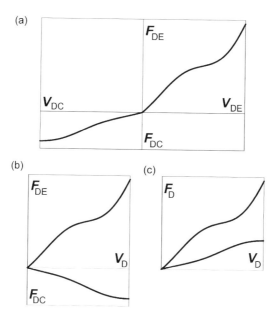

Figure 7.1.2 Damper characteristics: (a) force vs velocity; (b) force vs absolute velocity; (c) absolute force vs absolute velocity.

The relationship between the damper $F(V)$ characteristic and the $P(Q)$ characteristics of the various valves is in principle quite straightforward. For incompressible flow, a given damper speed results in a volumetric flow rate through the relevant valves. With the valve $P(Q)$ curves this permits solution of the various chamber pressures, from which the forces may be deduced. Analytically, it may be more convenient to begin with a pressure difference across a valve, leading to a consequent valve opening, valve flow rate and hence damper velocity. Analysis of linear valves or other simple valves gives considerable design insight, whilst computer simulation provides analysis of complex valves.

Figure 7.1.2 shows the form of a typical basic $F(V)$ characteristic, with the extension (rebound) force upwards, and the compression (bump) force downwards. The abscissa in (b) is the magnitude of

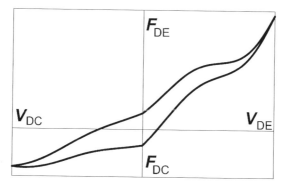

Figure 7.1.3 Example damper $F(V)$ loop resulting from stiffness.

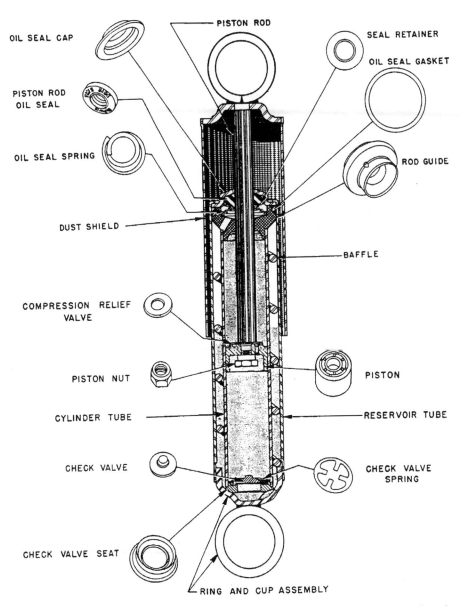

Figure 7.1.4 The Chrysler Oriflow damper introduced in 1951 was an attempt to move away from the simple abrupt hydraulic blow-off characteristic to something nearer to a linear $F(V)$ curve. Reproduced from Peterson (1953) *Proc. National Conference on Industrial Hydraulics*, 7, 23–43.

velocity, i.e. its absolute value, condensing the graph into less space. This is sometimes also done for the forces, if the compression and extension forces are sufficiently distinct to avoid confusion, as in (c). This is normally not desirable for adjustable dampers, or where other parameter variation is included in the graph.

The graph of Figure 7.1.2 assumes that the position of the damper along its stroke is not important, i.e. the force is to be dependent on velocity rather than position. This is approximately true for a conventional damper, but ceases to be true for a combined spring–damper unit. Any positional dependence of force turns the sinusoidal-test $F(V)$ line into a loop, Figure 7.1.3, because the speed is zero at the two extreme positions.

The characteristics may also be investigated through an $F(X)$ plot in sinusoidal motion, which reveals some additional features. This is discussed later. For a spring the $F(X)$ plot is a line. Any damping makes the $F(X)$ plot into a loop.

As discussed in Chapter 11, Testing, the energy dissipated per cycle and the mean power dissipation in sinusoidal motion are simply related to the linear damping coefficient. It is sometimes useful to express the energy or power dissipation of a real nonlinear damper as an equivalent linear damping coefficient:

$$C_{\text{Deq}} = \frac{2P_m}{\omega^2 X_0^2} = \frac{P_m}{2\pi^2 f^2 X_0^2}$$

$$C_{\text{Deq}} = \frac{2E_C}{\pi \omega X_0^2} = \frac{E_C}{2\pi^2 f X_0^2}$$

where

$$E_C = P_m T = \frac{P_m}{f} = \frac{2\pi P_m}{\omega}$$

7.2 Basic Damper Parameters

The most basic damper parameters have been stated to be:

(1) mean damping coefficient C_D;
(2) asymmetry e_D;
(3) mid-range shape of $F(V)$, λ.

The basic damper $F(V)$ graph may be calculated from the valve characteristics on the basis of steady state analysis. Testing, covered later, is generally by sinusoidal oscillation, with the peak force related to the peak velocity. Although this is not truly steady state, agreement is usually quite good.

Mean Damping Coefficient

For an approximately linear damper the force may be modelled as proportional to speed. This has the considerable merit of simplicity, and is often used for ride and handling analysis. However dampers are usually asymmetric in operation, Figure 7.1.2, so different coefficients may be used for the two directions (the bilinear model). Neglecting the small Coulomb-type friction, with a gas-pressure static damper compression force F_{DG} (positive compression force), the forces in extension and compression are:

$$F_{DE} = -F_{DG} + C_{DE} V_{DE}$$
$$F_{DC} = F_{DG} + C_{DC} V_{DC}$$

The mean damper coefficient is

$$C_D = \tfrac{1}{2}(C_{DC} + C_{DE})$$

Asymmetry Coefficient

In terms of C_D, the damper asymmetry factor (force transfer factor) e_D is

$$e_D = \frac{(C_{DE} - C_{DC})}{2C_D} = \frac{(C_{DE} - C_{DC})}{(C_{DE} + C_{DC})}$$

The directional coefficients are:

$$C_{DC} = C_D(1 - e_D)$$
$$C_{DE} = C_D(1 + e_D)$$

Zero asymmetry is $e_D = 0$ with $C_{DC} = C_{DE}$, of course. An asymmetry of $+100\%$, $e_D = +1$, corresponds to pure extension damping, with zero compression damping. The common C_{DC}/C_{DE} ratio of 30/70% yields an e_D value of 0.40.

Progressivity Factor

The force on a damper is not really exactly proportional to speed. Indeed, in Stage 1 and Stage 3 of the curve it is proportional to V^2. In the usual operating range of Stage 2, it does not follow any exact power of V, but does deviate to some extent, by accident or design, from direct proportionality, typically being regressive, with index < 1. It is of interest therefore to introduce a progressivity parameter to express this deviation from proportionality.

One method of developing such a progressivity factor is to consider a best fit or model curve

$$F = C_\lambda V^\lambda$$

over the speed range of interest (i.e. through the greater part of Stage 2). Neglecting Stage 1 and Stage 3, this forms a reasonable model for basic ride simulation studies, somewhat more realistic than a simple linear model.

Progressivity values are as follows:

$\lambda = 0$ constant Coulomb – type friction
$\lambda = 1$ Linear viscous – like friction
$\lambda = 2$ dynamic quadratic friction

The simplest method of fitting is to two suitable points, giving

$$F_1 = C_\lambda V_1^\lambda \qquad F_2 = C_\lambda V_2^\lambda$$

from which

$$\lambda = \frac{\log(F_2/F_1)}{\log(V_2/V_1)} \qquad C_\lambda = \frac{F_1}{V_1^\lambda}$$

Where some degree of progressivity (or regressivity, $\lambda < 1$) is deemed desirable, as is usual, this provides a basis for quantification of the desired effect.

It was shown earlier that a valve characteristic $P(Q)$ depending on Q^n will result in a damper force depending on V^n. Defining a valve progressivity

$$\lambda_V = \frac{\log(P_2/P_1)}{\log(Q_2/Q_1)}$$

then the valve progressivity required will equal that chosen for the damper.

With progressivity, i.e. nonlinearity, the damper coefficient is not constant, and where a summary mean value is required it will normally be evaluated in the middle of Stage 2, possibly at the root mean square of the speed limits of the range $V = \sqrt{V_1 V_2}$.

7.3 Mechanical Friction

Mechanical friction arises from the following:

(1) piston oil seal;
(2) piston side force (struts);
(3) rod oil seal;
(4) rod side force (struts).

A damper will exhibit measurably greater damping than is supplied by the valves alone because the piston and seal friction and their increase with pressure add to the coefficients. Nevertheless, simple theory gives a useful guide to the valve coefficients required for a known damping coefficient, assuming linear behaviour. The pressure-dependent mechanical friction may add 10–20% to the fluid forces, but being related to the pressure this friction is not objectionable. It could even be used deliberately instead of the fluid force apart from the problem of mechanical wear which is largely absent from fluid-based forces. Figure 7.3.1 illustrates such a design of damper.

Friction forces in a conventional damper depend on the operating conditions. For a truly axial load there are no additional radial reactions, although the rod seal has a preload pressure. The piston friction depends on the seal design. For a plain piston it will generally be negligible, provided that loads are essentially axial. However a seal will be fitted, e.g. a piston ring, or a sliding seal—a PTFE-based ring preloaded by an O-ring. In this case the friction force is difficult to calculate because although the pressures forcing the seal against the cylinder wall may be known, the lubrication conditions between the sealing ring and cylinder are unknown. Radial preload of the seal gives the damper a Coulomb friction term which is objectionable to ride quality, and is minimised.

A simple analysis of piston seal friction may be made as follows. When the pressures on the two sides of the piston (P_{CC} and P_{EC}) are equal, there will be no pressure friction term. The seal ring axial length is L_S. The seal outer surface area is $A_S = \pi D_P L_S$. Consider the case of $P_{EC} > P_{CC}$. Typically there is an O-ring half-way down the sealing ring, so half of the seal area has an extra pressure $P_{EC} - P_{CC}$ pressing out, causing a normal force at the seal outer face against the cylinder:

$$F_{NS} = \tfrac{1}{2} A_S (P_{EC} - P_{CC})$$

The actual conditions here are uncertain because of unknown wear of the seal profile and canting of the seal. However, as an estimate simply use a plausible Coulomb friction coefficient μ_M on the normal force, giving a piston friction force

$$F_{FP} = \mu_M F_{NS} = A_{PF}(P_{EC} - P_{CC})$$

where A_{PF} is an effective piston friction area:

$$A_{PF} = \tfrac{1}{2}\pi \mu_M D_P L_S$$

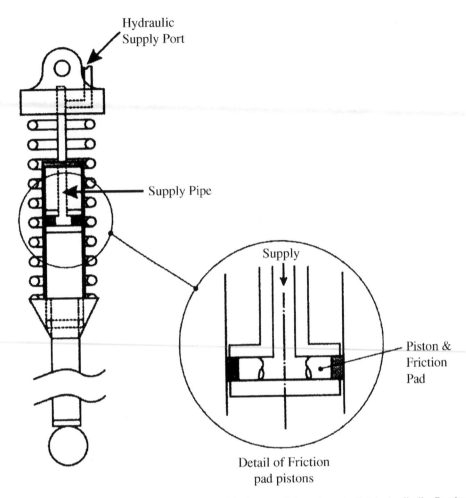

Figure 7.3.1 A dry friction telescopic damper in which the normal force is controlled hydraulically. Reprinted from *Control Engineering Practice*, 12, Guglielmino and Edge, A controlled friction damper for vehicle applications, pp.431–443, Copyright 2004, with permission from Elsevier.

This area can conveniently be nondimensionalised against the piston area to give a piston friction area coefficient k_{PFA}:

$$k_{PFA} = \frac{A_{PF}}{A_P} = \frac{2\mu_M L_S}{D_P}$$

As example values, assuming a Coulomb friction coefficient of 0.3 and a seal length of 6 mm on a piston of diameter 28 mm, then $k_{PFA} = 0.129$. The piston friction area is then $0.792\,\text{cm}^2$, and for a pressure difference of 3 MPa the friction force is 238 N, compared with a fluid resistance force of about 1500 N, a friction addition of 16%, in accord with practical testing experience.

Where the damping function is incorporated into a strut, there are significant internal side forces, with resulting additional friction, Figure 7.3.2. The vertical and lateral force at the tyre is reacted at the lateral arm via the bottom ball joint, leaving a moment applied to the strut. Although the vertical force has a small moment arm about the ball joint, it is a large force and is always present. The moment is reacted by the damper rod and piston as a force couple over the length from the rod bearing to the

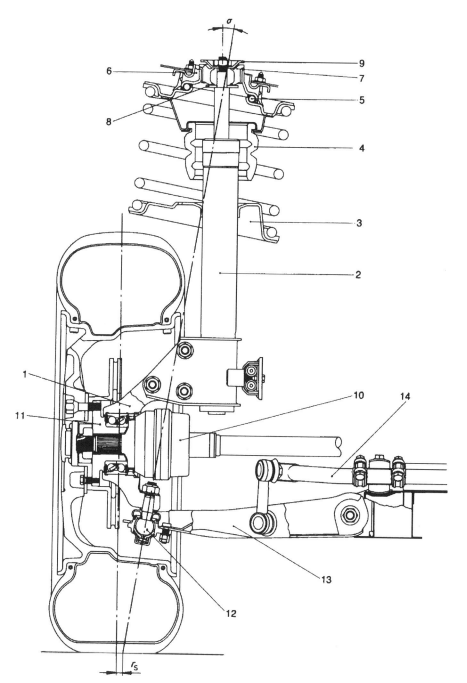

Figure 7.3.2 An example front strut, illustrating how the vertical and cornering forces at the base of the tyre give a moment about the bottom ball joint applied to the strut, reacted at the piston and rod bearing (Lancia).

piston. This creates rod and piston side forces with increased friction and wear. The longitudinal tyre shear forces have a similar effect. As a result, these two bearing points within the strut take significant side loads, sometimes exceeding 1 kN, adding to friction and wear. In more complex struts, at extra expense, additional bearings may be used to protect the piston seal from the side loads. Longitudinal tyre forces have a similar detrimental effect.

7.4 Static Forces

Although the main damper forces are related to velocity, there are also some static characteristics to be considered. When traversed very slowly, to eliminate fluid dynamic forces, a pressurised damper will exert:

(1) a force produced by pressurisation times rod area;
(2) a stiffness from pressure rise due to rod insertion;
(3) a static (Coulomb-type) friction arising from rod and piston friction.

These will all appear on a very-low-speed $F(X)$ curve, Figure 7.4.1. Unpressurised (double-tube) dampers exhibit only the third of these to any degree.

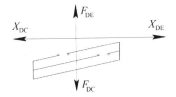

Figure 7.4.1 The force loop for creeping motion reveals the gas pressure force and friction forces only.

The damper parameters related to the above are:

(1) static (compression) force F_{SC} at central position;
(2) stiffness K_D through the range;
(3) Coulomb static damper friction force F_F.

Creeping the damper in and out at the central position requires damper 'static' compression and extension forces

$$F_{DSC,in} = F_G + F_F \quad \text{(in)}$$
$$F_{DSC,out} = F_G - F_F \quad \text{(out)}$$

where F_G is the gas pressure force. For a double-tube damper, the friction exceeds the gas force and a small tension force is required to creep the damper out. For a pressurised damper, the gas force is larger, and the damper may be allowed to creep out at an applied compression force.

The difference between the two creeping forces is twice the friction force, so

$$F_F = \tfrac{1}{2}(F_{DSC,in} - F_{DSC,out})$$

The mean value is the gas force:

$$F_G = \tfrac{1}{2}(F_{DSC,in} + F_{DSC,out})$$

The change of gas force over the range of motion gives the effective compressive stiffness:

$$K_D = \frac{dF_G}{dX_{DC}}$$

This is only significant for a pressurised damper. The basic damper static gas force arises from static compression chamber pressure times the rod area:

$$F_G = P_{CC} A_R \approx P_G A_R$$

The static stiffness arises from increase of the internal pressure due to rod insertion, and therefore depends upon rod area, stroke, initial pressure and gas volume, and is not strictly constant. Also the compression could be considered to be adiabatic for rapid compression, or isothermal for slow compression or for highly emulsified gas in small bubbles. The static force is likely to increase by perhaps 20% over the whole stroke, with an associated stiffness of perhaps 200 N/m, which will be a few percent of the suspension spring stiffness according to the particular installation. In some cases the stiffness is higher, and the increase may be 50% or more. For such dampers, it may be desirable to consider the nonlinear aspect of the gas spring effect, and its temperature sensitivity.

$$K_D = A_R \frac{dP_G}{dX_{DC}} = \frac{A_R^2 P_{G0}}{V_{G0}}$$

The pressure may be around 1 MPa (145 psi), which with a rod diameter of 12 mm gives a force of 113 N (25 lbf). The Coulomb friction depends on the seals, but can be quite small with good design, 5 N or less (1 lbf). The stiffness may be around 200 N/m (1 lbf/in).

7.5 Piston Free Body Diagram

The basic piston and rod geometry is shown in Figure 7.5.1. The piston diameter is D_P giving piston end area A_P. The rod diameter is D_R giving cross-sectional area A_R. The piston annulus area, apparent on the rod side of the piston, is

$$A_{PA} = A_P - A_R$$

Figure 7.5.2 shows the free body diagram of the piston and rod for compression. The six piston forces are as follows:

(1) F_{DC}: the mechanical compression force exerted externally on the damper rod, i.e. the damper compression force.
(2) $F_{P,PCC}$: the force exerted on the lower surface of the piston by compression chamber pressure P_{CC}:

$$F_{P,PCC} = A_P P_{CC}$$

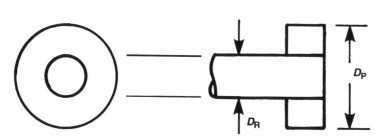

Figure 7.5.1 Piston geometry.

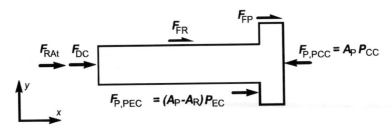

Figure 7.5.2 Free body diagram of piston and rod.

(3) $F_{P,PEC}$: the force exerted on the piston annulus on the rod side of the piston (distributed around the annulus) by expansion chamber pressure P_{EC}:

$$F_{P,PEC} = A_{PA} P_{EC}$$

(4) F_{FP}: the piston mechanical friction force exerted on the piston by the piston/cylinder seal, opposing the direction of motion, distributed around the piston side.
(5) F_{FR}: the rod friction force exerted by on the rod by the rod seal, opposing the direction of motion, distributed around the rod.
(6) F_{RAt}: the force exerted on the rod cross-sectional area by atmospheric pressure.

$$F_{RAt} = P_{At} A_R$$

If all pressures are measured as gauge pressures relative to atmospheric, the F_{RAt} force disappears. In any case, it is relatively small. The mechanical friction forces are generally dependent upon the liquid pressure (as, in the case of a piston-ring type of seal).

The pressure differential across the piston effectively acts on the annular area only, $A_{PA} = A_P - A_R$, and therefore the piston force $F_{P,PCC}$ may usefully be thought of as two forces one on the annular area and one on the rod area:

$$F_{P,PCC} = P_{CC} A_{PA} + P_{CC} A_R = F_{A,PCC} + F_{R,PCC}$$

In this context it may be useful to think of the rod passing through the piston as in Figure 7.5.1. Physically this is often the case. Retaining nuts and washers, the presence of valve discs and so on do not alter this basic concept.

In the free body diagram of Figure 7.5.2, it is assumed that the applied damper forces are axial; substantial nonaxial forces (e.g. as on a MacPherson strut) will add radial forces at the piston and at the rod bearing, giving additional friction forces.

The total force along x is

$$\sum F_X = F_{DC} + P_{At} A_R + P_{EC} A_{PA} - P_{CC} A_P + F_{FR} + F_{FP} = ma_X \approx 0$$

The combined mass of the piston and rod is about 200 g for an average car damper. For sinusoidal motion at amplitude X_0 and frequency f, giving radian frequency $\omega = 2\pi f$, the peak acceleration is $X_0 \omega^2$ and peak speed $X_0 \omega$. Even at an acceleration of 100 m/s², the acceleration force is 20 N and negligible compared with the damping force. Also the acceleration force is out of phase with the peak speed, and hence for sinusoidal motion has little or no effect on the peak force.

7.6 Valve Flow Rates

Figure 7.6.1 shows a general damper configuration. The packaging may be different from this, but the operating principle is generally the same. The figure shows a remote reservoir with a free piston separating gas and liquid, but the gas may be accommodated in an alternative way, e.g. around the main cylinder in the case of the double-tube damper. The piston has two valves, PE (piston extension) and PC (piston compression). Entry or exit of the rod causes fluid displacement through one of the two foot valves, FC (foot compression) and FE (foot extension) respectively.

Consider the damper to be displaced by a small distance X_{DC} inwards. The volume of the compression chamber (Chamber 2) is reduced, requiring an increased pressure there during the motion to force liquid out, some going to the expansion chamber (Chamber 3) and some to the foot chamber (Chamber 1).

A volume Λ_{FC} (lambda) of liquid is displaced by the rod and is moved through the foot valve:

$$\Lambda_{FC} = A_R X_{DC}$$

The volume moved through the piston compression valve is

$$\Lambda_{PC} = (A_P - A_R) X_{DC} = A_{PA} X_{DC}$$

For a damper compression velocity V_{DC}, the volumetric flow rates (assuming correct operation and incompressible liquid) are therefore

$$Q_{FC} = A_R V_{DC}$$
$$Q_{PC} = A_{PA} V_{DC}$$

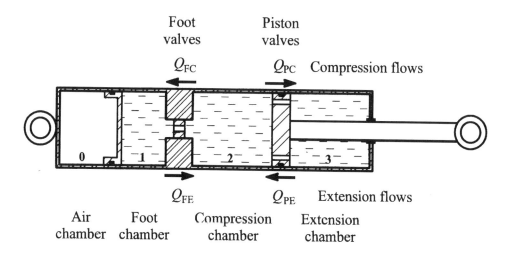

Figure 7.6.1 General damper configuration.

For damper extension velocity V_{DE} the foot extension valve and the piston extension valve are operative, giving, with correct operation,

$$Q_{FE} = A_R V_{DE}$$
$$Q_{PE} = A_{PA} V_{DE}$$

From the above analysis the volumetric flow rates for each valve are known for any piston velocity, under the assumption of normal operation, with no cavitation and no compressibility.

7.7 Pressures and Forces

From the valve characteristics, the pressure drop across a valve may be calculated for a given flow rate. Beginning at the gas reservoir, the pressure of which may be altered by rod insertion in the case of a pressurised damper, all the chamber pressures may then be calculated.

Normally for this calculation the valves will be nonlinear and represented by a complex model, with computer simulation. However, by way of a simple example for analytical purposes, consider linear valves such that the pressure drops are

$$P_{FC} = k_{FC} Q_{FC} = k_{FC} A_R V_{DC}$$
$$P_{PC} = k_{PC} Q_{PC} = k_{PC} A_{PA} V_{DC}$$
$$P_{FE} = k_{FE} Q_{FE} = k_{FE} A_R V_{DE}$$
$$P_{PE} = k_{PE} Q_{PE} = k_{PE} A_{PA} V_{DE}$$

where k_{FC} and so on are the linear valve resistances (Pa/(m^3/s) = Pa s/m^3 = N s/m^5). Only two of these equations are operative at any one time, of course.

In compression then, the chamber pressures P_{CC} and P_{EC}, neglecting friction of the free separator piston, are

$$P_{CC} = P_G + P_{FC} = P_G + k_{FC} A_R V_{DC}$$
$$P_{EC} = P_{CC} - P_{PC} = P_{CC} - k_{PC} A_{PA} V_{DC}$$

The compression force F_{PC} on the piston-rod assembly due to fluid pressure is therefore

$$F_{PC} = P_{CC} A_P - P_{EC} A_{PA}$$

The piston area can be resolved into rod and annulus areas, giving

$$F_{PC} = P_{CC}(A_{PA} + A_R) - P_{EC} A_{PA}$$
$$= P_{CC} A_R + (P_{CC} - P_{EC}) A_{PA}$$
$$= P_{CC} A_R + P_{PC} A_{PA}$$

so

$$F_{PC} = P_G A_R + P_{FC} A_R + P_{PC} A_{PA}$$

This shows how the foot valve pressure drop P_{FC} acts on the rod area A_R, and how the piston valve pressure drop P_{PC} acts on the annular area A_{PA}.

Substituting for the linear valve pressures,

$$F_{PC} = P_G A_R + k_{FC} A_R^2 V_{DC} + k_{PC} A_{PA}^2 V_{DC}$$
$$= P_G A_R + (k_{FC} A_R^2 + k_{PC} A_{PA}^2) V_{DC}$$
$$C_{DC} = k_{FC} A_R^2 + k_{PC} A_{PA}^2$$

By similar analysis for extension,

$$P_{CC} = P_G - P_{FE} = P_G - k_{FE} A_R V_{DE}$$
$$P_{EC} = P_{CC} - P_{PE} = P_{CC} - k_{PE} A_{PA} V_{DE}$$

The piston force in extension (sign convention tension positive) is

$$F_{PE} = P_{EC} A_{PA} - P_{CC} A_P$$
$$= k_{PE} A_{PA}^2 + k_{FE} A_R^2 - P_G A_R$$

This analysis leads to several conclusions regarding damper force production:

(1) The gas reservoir pressure P_G always acts on the rod area A_R to give a static force (independent of velocity).
(2) The foot valves give a pressure drop which acts directly on the rod area A_R in both compression and extension.
(3) The piston valves give a pressure drop which acts on the annulus area A_{PA} for both compression and extension.
(4) Linear valve $P(Q)$, i.e. constant k, gives linear force $F(V)$.
(5) Forces are proportional to the areas squared and hence to D^4. This is because the area produces the volume flow rate which produces the pressure which then acts on the area (for linear valves).

The foregoing analysis is easily repeated for nonlinear valves with

$$P = kQ^n$$

from which it emerges that the first three of the previous conclusions remain valid. However the forces produced by each valve become proportional to V^n and to the relevant area to the index $1+n$. Evidently then, although transformed, the basic form of the valve $P(Q)$ curve appears in the damper $F(V)$ curve.

If the valve characteristic is represented as a polynomial, e.g.

$$P_{FC} = k_{FC1} Q_{FC} + k_{FC2} Q_{FC}^2 + \cdots + k_{FCr} Q_{FC}^r + \cdots = \sum_{r=1}^{n} k_{FCr} Q_{FC}^r$$

and similarly for P_{PC}, then the piston force F_{PC} becomes

$$F_{PC} = P_G A_R + \sum_{r=1}^{n} (k_{FCr} A_R^{1+r} + k_{PCr} A_R^{1+r}) V_{DC}^r$$

7.8 Linear Valve Analysis

For linear valves, in extension, the damper force is

$$F_{DE} = (k_{FE} A_R^2 + k_{PE} A_{PA}^2) V_{DE} - P_G A_R$$

For a free-flowing extension foot valve, as is normally required, $k_{FE} = 0$, and

$$C_{DE} = \frac{dF_{DE}}{dV_{DE}} = k_{PE}A_{PA}^2$$

Hence, the required piston extension valve resistance k_{PE} is

$$k_{PE} = \frac{C_{DE}}{A_{PA}^2}$$

For example, to achieve 2 kNs/m with an annulus area of 5 cm² will require

$$k_{PE} = \frac{2000}{(5 \times 10^{-4})^2} \frac{\text{N s}}{\text{m.m}^4} = 8\,\text{GPa s/m}^3 = 8\,\text{MPa}/(\text{L/s})$$

This gives a good initial estimate for the valve, which may be reduced somewhat to allow for mechanical friction.

For compression

$$F_{DC} = (k_{FC}A_R^2 + k_{PC}A_{PA}^2)V_{DC} + P_G A_R$$

$$C_{DC} = \frac{dF_{DC}}{dV_{DC}} = k_{FC}A_R^2 + k_{PC}A_{PA}^2$$

In this case, the piston valve and foot valve are both active, although the latter is relatively ineffective in force generation because of the small rod area. It is used to facilitate use of the piston valve, by preventing cavitation behind the piston. Further design requires a choice of the ratio k_{FC}/k_{PC}. For example, if this is made just sufficient to prevent cavitation (Section 7.9), even in the absence of general pressurisation,

$$\frac{k_{FC}}{k_{PC}} = \frac{A_{PA}}{A_R}$$

Using this in the previous equation for C_{DC} to eliminate k_{FC} gives

$$C_{DC} = k_{PC}A_{PA}A_R + k_{PC}A_{PA}^2 = k_{PC}A_{PA}A_P$$

For a known desired C_{DC} and geometric dimensions this then gives an estimate of the valve coefficients k_{PC} and k_{FC}:

$$k_{PC} = \frac{C_{DC}}{A_{PA}A_P}$$

$$k_{FC} = \frac{C_{DC}}{A_R A_P}$$

To achieve a damping coefficient of 1000 Ns/m with a rod area of 1 cm² and a piston area of 6 cm² requires $k_{PC} = 3.33$ GPa s/m³ and $k_{FC} = 16.7$ GPa s/m³.

7.9 Cavitation

Cavitation occurs when the oil vapour pressure exceeds the local static pressure. This is analogous to boiling, but happens because of the pressure reduction rather than by temperature and vapour pressure increase. Nevertheless, at high temperature the oil vapour pressure is higher, and cavitation is somewhat easier to create. When cavitation happens, numerous pockets of oil vapour are created throughout the oil. A small increase of pressure can easily turn this back into liquid, with a severe slam shock, causing bad noise and possible damage to the damper internals. Even at a high damper temperature (service limit 130°C) the oil vapour pressure will be less than 100 kPa. Cavitation may be avoided by correct design, such that the pressure at all points in the damper exceeds the vapour pressure.

For a pressurised single-tube damper, cavitation may occur in the extension chamber during compression when the pressure drop through the piston exceeds the pressure in the compression chamber, i.e. exceeds the gas pressure. This places a limit on the damper compression speed and compression force. For a linear damper, the expansion chamber pressure during compression is

$$P_{EC} = P_{CC} - P_{PC} = P_G - k_{PC} A_{PA} V_{DC}$$

To maintain this positive,

$$V_{DCmax} = \frac{P_G}{k_{PC} A_{PA}}$$

The maximum compression force achievable is

$$F_{DCmax} = P_G A_P$$

With nonlinear valves, the graph of piston compression pressure drop may be inspected to obtain the speed at which the pressure drop equals the gas pressure. Obviously, to obtain large compression forces, large gas pressure is needed.

If the free piston is omitted from a single-tube damper and the gas is emulsified into the liquid, then the previously well-defined occurrence of cavitation may be blurred, and certainly the collapse shock is softened, but the basic operating limit remains.

For a double-tube damper to avoid cavitation above the piston, i.e. on the rod side, during compression, the foot compression valve must produce a pressure drop not less than that of the piston compression valve, which, however, passes much more liquid. For linear valves

$$P_{EC} = P_{CC} - P_{PC} = P_G + k_{FC} A_R V_{DC} - k_{PC} A_{PA} V_{DC}$$

The gas pressure in a double-tube damper is small, so we require

$$k_{FC} A_R V_{DC} > k_{PC} A_{PA} V_{DC}$$

$$k_{FC} > \frac{A_{PA}}{A_R} k_{PC}$$

For nonlinear valves, inspect the $P(Q)$ curves. Scale the foot valve $P(Q)$ by the ratio of areas to compare the pressures effectively as $P(V)$.

In extension of a double-tube damper, cavitation may occur in the compression chamber, which has pressure

$$P_{CC} = P_G - P_{FE}$$

Then it is required that $P_{CC} > P_{vap}$, so

$$k_{FE} A_R V_{DE} < P_G - P_{vap}$$

Therefore, k_{FE} must be rather small. Allowing a pressure drop of 20 kPa at a maximum speed of 2 m/s, and a rod area of 1 cm², the maximum k_{FE} is 100 MPa s/m³. Considered as a simple orifice, the effective area needs to be not less than 30 mm². This is one hole of diameter 6.2 mm, or, more likely, a ring of six holes each of diameter not less than 2.5 mm with a very softly sprung disc valve.

7.10 Temperature

Damper forces and coefficients reduce as the oil temperature increases, a phenomenon known as damper 'fade'. According to the design, the reduction may be up to 2% per degree Celsius, 0.02 /K. By minimising the effect of viscosity, the effect can be reduced to around 0.002 /K, about 20% over 100 K temperature change.

For a given orifice and volumetric flow rate, the pressure drop is

$$P = \frac{\frac{1}{2}\rho Q^2}{C_d^2 A^2}$$

The oil temperature affects the discharge coefficient C_d through viscosity variation, and affects the density directly. The oil variational properties are typically

$$\frac{1}{\mu}\frac{d\mu}{dT} = -0.02/K$$

$$\frac{1}{\rho}\frac{d\rho}{dT} = -0.001/K$$

For a good basic design of damper, the variation of damping coefficient may be

$$\frac{1}{C_D}\frac{dC_D}{dT} = -0.002/K$$

indicating that the viscosity variation has largely been dealt with. At this level, the damper is acceptable for general use. In more performance critical, and less cost-critical, applications, temperature compensation may be used, as described in Section 6.11. The basic method is to reduce an orifice area to offset the density and viscosity reduction. In principle this can completely eliminate the problem, but this is difficult in practice because it is only practical to vary a parallel hole.

Temperature compensation can also be achieved by electrically controlled variable-orifice dampers, and by electrorheological and magnetorheological dampers.

7.11 Compressibility

Vigorous activation of the damper generates high pressures, and the compressibility effects may become significant. The compressibility of pure damper oil is low, and strain of the damper cylinder adds only a little effective compressibility. Importantly, however, the inclusion of even a small amount of gas as in the liquid greatly increases the compressibility. This can occur through severe agitation of the damper on rough roads. The gas could be in a single pocket, but is more likely to be in a finely divided emulsion. Figure 7.11.1 shows experimentally observed effects of compressibility, possibly with some cavitation.

If the damper is moving at constant velocity, then the pressures will be constant. In this case, although the fluid may be compressed, the density will be constant, so there will be no very obvious effect of compressibility. The effects of compressibility will only be seen in transient operation, and in particular will be related to acceleration of the piston.

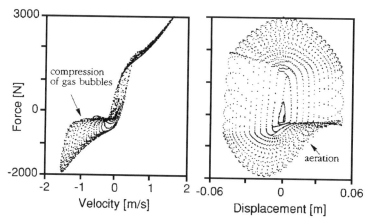

Figure 7.11.1 The effect of compressibility and cavitation on the damper $F(V)$ curve, due to bad valve settings, as measured experimentally (from Duym et al., 1997).

This may be demonstrated mathematically as follows. Representing volume by Λ, the compressibility β_E of the oil–gas emulsion is defined as

$$\beta_E = -\frac{1}{\Lambda}\frac{d\Lambda}{dP} = -\frac{1}{\Lambda}\frac{d\Lambda}{dt}\frac{dt}{dP}$$

The rate of change of volume in a pressure chamber due to compressibility coupled with changing pressure will therefore be

$$\frac{d\Lambda}{dt} = -\Lambda \beta_E \frac{dP}{dt}$$

The pressure is related to the damper velocity through the characteristics of the valves, so the rate of change of pressure is related to the damper acceleration dV/dt. For a linear valve

$$P = kAV$$
$$\frac{dP}{dt} = kA\frac{dV}{dt}$$

so

$$\frac{d\Lambda}{dt} = -\Lambda \beta_E kA \frac{dV}{dt}$$

Hence acceleration creates a compressible volume change which causes a deficiency of flow through the valves, which reduces the force compared with the incompressible case. In short then, compressibility creates a relationship between damper force and damper acceleration.

In extension, when dV_E/dt is positive the extension chamber pressure is increasing. Compressibility then reduces the flowrate through the piston, reducing the piston pressure drop and the consequent damper force. Writing the damper force in extension as

$$F_{DE} = -F_G + K_D X_{DE} + C_{DE} V_{DE} + (m + C_A)\frac{dV_{DE}}{dt}$$

where m is the appropriate inertia for the moving part, the residual acceleration coefficient C_A due to compressibility may be shown to be approximately

$$C_A = -k_{PE}^2 A_{PA}^2 \beta_E \Lambda_{EC}$$

Example values for an emulsified oil might be $k_{PE} = 8\,\text{GPa s/m}^3$, $A_{PA} = 5\,\text{cm}^2$, $\beta_E = 0.1/\text{MPa}$, $\Lambda_{EC} = 50\,\text{cm}^3$, giving $C_A = -80\,\text{N s}^2/\text{m} = -80\,\text{kg}$ equivalent mass.

It will be apparent that for a sinusoidal activation the greatest accelerations occur at the ends of the stroke, where the velocity is zero. Hence the effect of compressibility is to introduce hysteresis into the sinusoidal $F(V)$ curve, with greatest spread of force at $V = 0$. Hence, when tested at various stroking frequencies for a given peak velocity, the higher frequencies will tend to exhibit greater hysteresis. This is certainly observed in practice, and is known as damper lag.

When testing a damper sinusoidally, the acceleration is proportional to the displacement, so a force proportional to acceleration has similar results to a stiffness force proportional to displacement. The latter, however, is not dependent on frequency.

Some compressibility is probably advantageous overall, because it will reduce the transmission of higher frequencies (NVH—noise, vibration and harshness). In this respect it is probably similar in result to the inclusion of rubber bushes in the end fittings. Some dampers have been designed to take advantage of this by working on fully emulsified oil rather than separating out the air.

7.12 Cyclical Characteristics, $F(X)$

Although the conventional damper produces a force essentially independent of position, in sinusoidal testing the force can be plotted against either velocity or position, the latter sometimes giving some useful additional insight. Of course, the $F(X)$ graph does not imply that the force actually depends causally upon the position X, rather the force happens to have the values shown at the values of X, whilst the forces are actually determined by other factors, e.g. the velocity. The $F(X)$ graph is easily observed during testing, e.g. on a storage oscilloscope, with suitable sensors of course.

Figures 7.12.1–7.12.13 show a series of examples of pairs of $F(X)$ and $F(V)$ plots where it must be emphasised that the motion is a sinusoidal one.

Considering a sign convention in which the displacement, velocity and force are positive in the same direction, Figure 7.9.1 is for a linear spring $F = -kx$ with a restoring stiffness, which has a clockwise-going loop in $F(V)$. Note that although a graph of $F(V)$ is drawn, the force is not actually controlled by V, it is only $F = -kX$. The linear damper of Figure 7.9.2 has $F = -CV$ with damping force opposing the velocity. This has an anticlockwise loop in $F(X)$, indicating power dissipation. Note that although a graph of $F(X)$ is drawn, the force is not actually controlled by X, it is only $F = -CV$. Combining the above two gives the double loop of a spring-damper unit, where each loop is sheared.

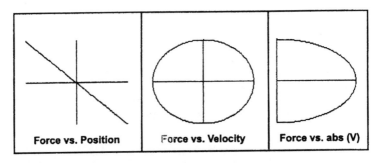

Figure 7.12.1 $F(X)$, $F(V)$ and $F(abs(V))$ for a sinusoidal test. Force type: $-kx$ (linear spring).

Damper Characteristics 279

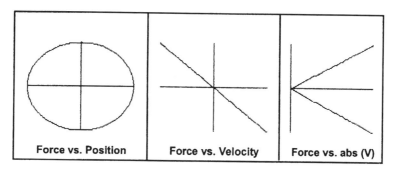

Figure 7.12.2 $F(X)$, $F(V)$ and $F(\text{abs}(V))$ for a sinusoidal test. Force type: $-kV$ (linear damper).

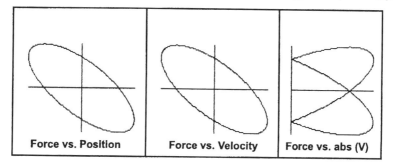

Figure 7.12.3 $F(X)$, $F(V)$ and $F(\text{abs}(V))$ for a sinusoidal test. Force type: $-0.5(kx + cV)$ (linear spring + damper).

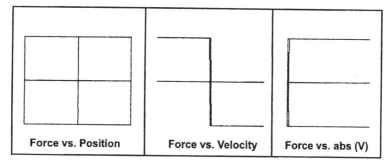

Figure 7.12.4 $F(X)$, $F(V)$ and $F(\text{abs}(V))$ for a sinusoidal test. Force type: $-c\,\text{sgn}(V)$ (Coulomb friction).

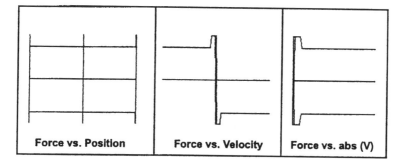

Figure 7.12.5 $F(X)$, $F(V)$ and $F(\text{abs}(V))$ for a sinusoidal test. Force type: stiction.

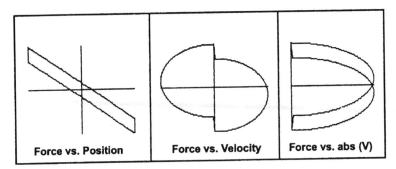

Figure 7.12.6 $F(X)$, $F(V)$ and $F(\text{abs}(V))$ for a sinusoidal test. Force type: linear spring + Coulomb friction.

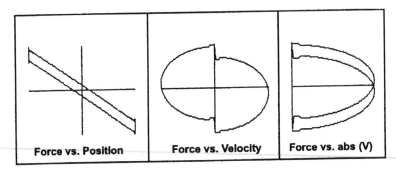

Figure 7.12.7 $F(X)$, $F(V)$ and $F(\text{abs}(V))$ for a sinusoidal test. Force type: linear spring + stiction.

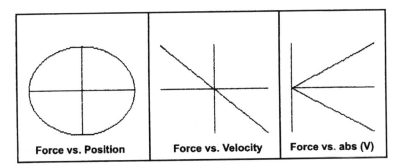

Figure 7.12.8 $F(X)$, $F(V)$ and $F(\text{abs}(V))$ for a sinusoidal test. Force type: $-cV$ (linear damper), repeated.

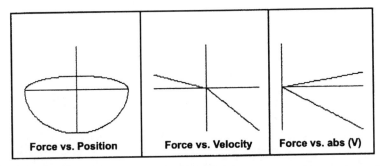

Figure 7.12.9 $F(X)$, $F(V)$ and $F(\text{abs}(V))$ for a sinusoidal test. Force type: bilinear damper (25/75).

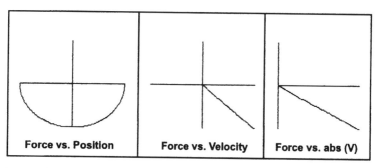

Figure 7.12.10 $F(X)$, $F(V)$ and $F(\text{abs}(V))$ for a sinusoidal test. Force type: unidirectional linear damper.

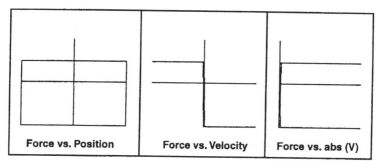

Figure 7.12.11 $F(X)$, $F(V)$ and $F(\text{abs}(V))$ for a sinusoidal test. Force type: asymmetrical blow-off valves.

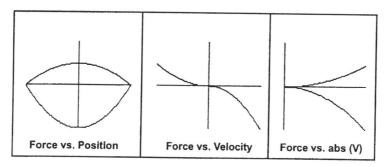

Figure 7.12.12 $F(X)$, $F(V)$ and $F(\text{abs}(V))$ for a sinusoidal test. Force type: asymmetrical fixed orifices, $F = -cV\text{abs}(V)$.

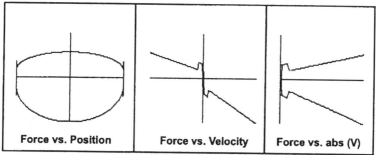

Figure 7.12.13 $F(X)$, $F(V)$ and $F(\text{abs}(V))$ for a sinusoidal test. Force type: bilinear + stiction.

Basic Coulomb friction has a constant magnitude, in opposite direction to velocity

$$F = -C\text{sgn}(V)$$

resulting in the rectangular $F(X)$ loop of Figure 7.12.4. With stiction, i.e. extra frictional force at zero or very low velocity, the $F(X)$ loop acquires protuberances, as in Figure 7.12.5.

Combining a linear spring with Coulomb friction gives Figure 7.12.6, and a spring with stiction gives Figure 7.12.7.

Figures 7.12.8 onwards show curves more obviously relevant to basic damper behaviour. The bilinear damper of Figure 7.12.9 has an asymmetric loop, reaching the unidirectional extreme of Figure 7.12.10. A preloaded valve with no parallel hole can give a blow-off characteristic similar to Coulomb friction, but possibly with asymmetry, resulting in the asymmetrical rectangular loop of Figure 7.12.11. For a fixed orifice, the force is given by

$$F = -CV\text{abs}(V) = -\text{sgn}(V)CV^2$$

resulting in the pointed bi-parabolic loop of Figure 7.12.12.

Finally, Figure 7.12.13 shows a bilinear damper with stiction resulting in the addition of protuberances to the basic bilinear loop from Figure 7.12.9.

7.13 Extreme Cyclic Operation

During normal operation, the pressures in a damper do not fall low enough to result in significant cavitation or de-solution of gas. However, in vigorous operation this may occur. It may also occur at low speed, of course, if the damper is incorrectly selected or set, with improper balance of foot and piston valves, or inadequately pressurised. If cavitation occurs in cyclic operation then it is necessary to consider various possible 'modes' of operation explicitly, Table 7.13.1.

Conceptually there are eight possible modes. The two motion states are compression and extension, each combined with four possible cavitation states (two states at the rod end times two states at the foot end). The individual cavitation states are, of course, cavitation or no cavitation. Hence from the table, for example, State 6 denotes motion in extension, with cavitation at the rod end only.

In Table 7.13.1 a distinction is drawn between the four states that can arise steadily (S) (or cyclically), and the four that can occur only cyclically (C). For example, in compression, cavitation will not begin at the foot end, it will only exist if carried over from the previous half cycle. Hence this can only be a 'cyclical' phenomenon. Those modes denoted as 'steady' could occur during damper motion

Table 7.13.1 Damper cavitation mode numbers

Cavitation state	Mode number	
	In compression	In extension
Nil	1, S	5, S
Rod end only	2, S*	6, C*
Foot end only	3, C	7, S
At both ends	4, C	8, C

S (steady) or can occur at steady velocity
C (cyclic) arises in cyclic operation only
*Double valve P–V solution

at constant speed. This is important from the point of view of a computer numerical simulation, although of course all real damper motions are effectively cyclical.

The method of numerical solution of the pressures depends on the mode. With rod-end cavitation (modes 2 and 6) (asterisked) the foot-end pressure affects the volume flow rate through both foot-end piston valves, further influencing the method of solution.

The conditions of the onset of cavitation and its relation to valve characteristics is discussed earlier in this chapter.

Cavitation is certainly to be avoided in normal operation. Whilst the initiation of cavitation is not of itself too detrimental, although affecting the forces somewhat, the collapse of cavitation creates extreme pressures and stresses with potential damage, plus severe noise. In some cases, it has been known to cause very rapid erosion of the piston and valves.

7.14 Stresses and Strains

The cylinder wall must be designed to withstand low-stress fatigue and occasional higher stresses, plus handling. For single-tube dampers, impact damage of the working cylinder is a serious hazard.

Consider an idealised simple circular cylinder of inner diameter D_P and thin wall of thickness w, mean diameter $D_C = D_P + w$, with internal pressure P resulting in axial and hoop stresses σ_A and σ_H respectively. There are fixed ends on the cylinder, and no external forces. Considering the free body above a transverse section, longitudinal equilibrium of forces requires

$$A_P P = \pi D_C w \sigma_A$$
$$\sigma_A = \frac{D_P^2 P}{4 D_C w} \approx \frac{D_P P}{4w} \qquad (7.14.1)$$

Considering a partial longitudinal section on a diameter, forming a free body of length L, there are no shear forces on the ends, so lateral equilibrium requires

$$D_P L P = 2 w L \sigma_H$$
$$\sigma_H = \frac{D_P P}{2w} \qquad (7.14.2)$$

Hence, the hoop stress is about twice the axial stress. Considering a bore diameter of 30 mm and a wall thickness of 1.5 mm, the hoop stress is ten times the working pressure. Hence in most cases, for ordinary dampers the minimum wall thickness is likely to be governed by practical requirements for rigidity, general handling strength, resistance to accidental damage by dropping, or in the case of single tube dampers by resistance to stone impacts, rather than by fatigue or yield failures. Struts are different, carrying significant transverse suspension loads.

The axial and hoop strains also depend on Poisson's ratio v (typically 0.29 for steel, 0.33 for aluminium) according to

$$\varepsilon_A = \frac{\sigma_A - v\sigma_H}{E}$$
$$\varepsilon_H = \frac{\sigma_H - v\sigma_A}{E}$$

where E is the Young's modulus of elasticity (about 206 GPa for steel, 70 GPa for wrought aluminium alloys).

For comparison with pressure sensors, the internal pressure may be measured experimentally by strain gauging the cylinder, although caution must be exercised over thermal strains due to radial temperature gradients. These can be reduced by thermal insulation over the gauge. For experimentally observed axial and hoop strains, by rearrangement of the earlier simultaneous equations for the strains

$$\sigma_A = \frac{(\varepsilon_A + \nu\varepsilon_H)E}{(1 - \nu^2)}$$

$$\sigma_H = \frac{(\varepsilon_H + \nu\varepsilon_A)E}{(1 - \nu^2)}$$

The approximately two-to-one relationship of cylinder stresses and Poisson's ratio combine to result in much higher hoop strains than axial strains, e.g. for a plain cylinder in the ratio

$$\frac{\varepsilon_H}{\varepsilon_A} = \frac{\sigma_H - \nu\sigma_A}{\sigma_A - \nu\sigma_H} = \frac{2-\nu}{1-2\nu}$$

which takes a value of 4.07 for steel (at $\nu = 0.29$) and 4.91 for aluminium alloy (at $\nu = 0.33$). Hence the hoop strain should be considered the primary indicator.

Application of stress analysis to a real damper requires consideration of the forces acting, as shown in Figure 7.14.1. The wall cross-sectional area is

$$A_W = \pi(D_P + w)w$$

The wall tensile forces in the two chambers are

$$F_{ECwall} = P_{EC}A_{PA} + F_{FR}$$
$$F_{CCwall} = F_{ECwall} + F_{FP}$$

where F_{FR} and F_{FP} are the friction forces at the rod and piston respectively. The axial stresses for the two chambers follow directly from these axial forces. Note that the axial stress in the expansion chamber is reduced by the presence of the rod which reduces the area upon which P_{EC} acts. As a result, with a high Poisson's ratio the axial strain may be negative.

The hoop stresses are as for a simple cylinder (even for the extension chamber containing the rod)

$$\sigma_{H,EC} = \frac{P_{EC}D_P}{2w}$$

$$\sigma_{H,CC} = \frac{P_{CC}D_P}{2w}$$

Working in the other direction, starting with experimentally measured strains, it is easy to deduce the axial and hoop stresses. The hoop stresses alone are sufficient to deduce the chamber pressures. The axial stresses give the axial forces in the walls. The piston friction force is then

$$F_{FP} = F_{CCwall} - F_{ECwall}$$

The rod friction force is

$$F_{FR} = F_{ECwall} - P_{EC}A_{PA}$$

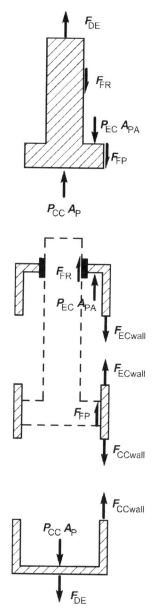

Figure 7.14.1 Wall forces for a single-tube damper in extension motion (wall force actually distributed circumferentially).

However, this is not a well-conditioned calculation, being the subtraction of two similar values.

Surprisingly, strain of the pressure cylinder may have a greater effect on the volume that pure oil compressibility. Consider a circular steel cylinder of inner diameter $D_P = 28$ mm, wall thickness $w = 1.2$ mm, filled with oil initially at negligible pressure. The oil has initial density $\rho = 860 \text{ kg/m}^3$, bulk modulus $K = 1.5$ GPa. The metal has Young's modulus $E = 207$ GPa, and Poisson's ratio $\nu = 0.29$. The pressure is raised to $P = 10$ MPa.

The relative reduction of oil volume is $-P/K = -6.67 \times 10^{-3} = -0.67\%$. Using the above formulae for stress and strain, the steel cylinder has $\sigma_H = 116.7$ MPa, $\sigma_A = 58.3$ MPa, $\varepsilon_H = 482$ μstrain, $\varepsilon_A = 118$ μstrain. The relative volume increase of the cylinder is $2\varepsilon_H + \varepsilon_A = 1082 \times 10^{-3}$ (1.08%), greater than the effect of the oil. The sum of these effects is 1.75%, which is not normally worth considering, from an engineering viewpoint. Distortion of the cylinder ends, or valves in some cases, could add something to the above figures. However, the total is likely to remain unimportant with the exception of the presence of free gas.

7.15 Damper Jacking

When an asymmetrical damper is actuated, it produces a mean force through the cycle. For normal dampers this is a tension force, pulling the suspension down, lowering the ride height over rough roads, an undesirable effect. In some cases, exactly the opposite may be needed. The damper jacking stiffness is defined as the mean jacking force per unit of displacement amplitude in sinusoidal actuation. Upward jacking is defined as positive, because link and spring jacking are then positive, so the damper jacking stiffness is negative for the usual greater force in extension.

Consider sinusoidal actuation, with amplitude Z at frequency f, of a linear asymmetrical damper with damping coefficients C_{DC} and C_{DE}, having transfer factor e_D:

$$C_{DC} = (1 - e_D)C_D$$
$$C_{DE} = (1 + e_D)C_D$$

The sinusoid period is $T = 1/f$. The stroke is $2Z$ each way in time $T/2$, with mean velocity $V = \pm 4Z/T = \pm 4fX$. The average directional forces are therefore

$$F_{DC} = 4C_{DC}fZ$$
$$F_{DE} = 4C_{DE}fZ$$

The mean jacking force, over a complete cycle, is therefore

$$F_{DJ} = \tfrac{1}{2}(F_{DC} - F_{DE})$$
$$= 2(C_{DC} - C_{DE})fZ$$
$$= -4e_D C_D fZ$$

The damper jacking stiffness for frequency f is

$$K_{DJ} = \frac{F_{DJ}}{Z} = -4e_D C_D f \quad \text{(N/m)}$$

The underlying damper property is the damper jacking coefficient:

$$C_{DJ} = -4e_D C_D \quad \text{(N/m Hz)}$$

giving

$$K_{DJ} = C_{DJ} f$$

With a suspension stiffness (wheel rate) K_S, an amplitude of Z therefore results in a mean suspension jacking distance Z_{DJ}:

$$Z_{DJ} = \frac{K_{DJ} Z}{K_S} = \frac{C_{DJ} f Z}{K_S}$$

The damper jacking ratio R_{DJ} is

$$R_{DJ} = \frac{Z_{DJ}}{Z} = \frac{K_{DJ}}{K_S} = \frac{C_{DJ} f}{K_S}$$

Example values at each wheel are: $C_D = 2.5$ kN s/m, $e_D = 0.6$, giving $C_{DJ} = -6$ kN/m Hz and $K_{DJ} = -9$ kN/m at 1.5 Hz. A motion amplitude of 20 mm will result in a jacking force of -180 N. With a suspension stiffness of 26 kN/m this gives a jacking distance of about -7 mm, and a damper jacking ratio of $R_{DJ} = -0.35$, a considerable effect.

It is easily shown that R_{DJ} at the undamped heave natural frequency, R_{DJN}, is given by

$$R_{DJN} = -\frac{e_D}{\pi^2 f_N}\left(\frac{C_{DT}}{m_S}\right)$$

where C_{DT} is the vehicle total damping coefficient. The 1-dof heave equation of motion is

$$m_S \ddot{Z} + C_{DT} \dot{Z} + K_{ST} Z = 0$$

so $C_{DT}/m_S = -2\alpha$ where α is the damping factor. Then

$$R_{DJN} = -\frac{2e_D}{\pi}\left(\frac{C_{DT}}{m_S \omega_N}\right)$$

$$= -\frac{2e_D}{\pi}\left(\frac{-2\alpha}{\omega_N}\right)$$

so R_{DJN} may be expressed as

$$R_{DJN} = \frac{4}{\pi} e_D \zeta_H$$

Typically for a passenger car $\zeta_H = 0.4$ giving $R_{DJN} = 0.3$. The effect will be even greater for sports cars.

7.16 Noise

Ideally, dampers should work quietly, but bad design can lead to many noises from fluid dynamics to friction screeching. Also, the transmission of noise from the suspension to the vehicle body is always under scrutiny for improvement. Noise is really frequencies above 30 Hz, beyond the ride and handling regime, which effectively stops somewhat above the wheel hop frequency.

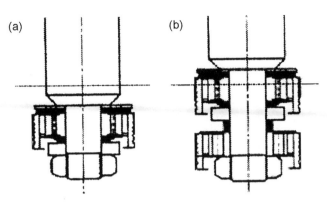

Figure 7.16.1 A damper piston with two bearing areas (Yamauchi *et al.*, 2003).

One of the problems is stick–slip friction at the piston. To overcome this, a longer piston bearing area, or two well-separated areas, helps, reducing the normal forces when the piston moves out of alignment, Figure 7.16.1.

General damper properties in the range 30–500 Hz have been studied, Figure 7.16.2, but will not be examined in detail here.

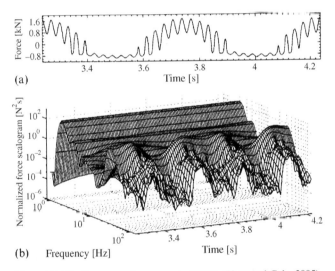

Figure 7.16.2 Damper scalogram up to 100 Hz (Yung and Cole, 2005).

8
Adjustables

8.1 Introduction

Suspensions may be classified by type and speed of response as:

(1) Passive   – slow (manual adjustment)
(2) Adaptive   – slow (roughness and speed)
               – fast (individual bumps)
(3) Active   – very slow (load levelling)
               – slow (roughness and speed)
               – fast (individual bumps)

The fast adaptive type of damper is sometimes called 'semi-active' but this is a misleading term. An active suspension requires a large power input from a pump. An adaptive suspension requires power for the valves only. The basic purposes of adjustment are:

(1) To optimise damper characteristics for varying conditions of road roughness and driving style,
(2) To compensate for wear.

Various forms of damper adjustment are possible:

(1) manual after removal from the car,
(2) manual *in situ*,
(3) remote manual from the driving seat,
(4) automatic (adaptive).

Self-levelling requires a slow increase of static force, rather than of damping coefficient, but happens to be done via the damper unit. This will not be considered further here.

Where removal of the damper from the car is required to make the adjustment (possibly only one end of the damper), this is to permit relative rotation of the two ends. With this kind of adjustment, the damper is fully compressed to engage the internal adjuster so that relative rotation can, for example, alter the piston valve preload. If the damper is not fully compressed, rotation has no effect. This type normally adjusts the extension force only.

Adjustment *in situ* is made by rotation of a knob or lever on the damper body or rod. Often adjustments for compression and extension are available independently, but sometimes one adjustment has an effect in both directions.

For remote adjustments, for example by a switch in the vehicle control panel, the information is transmitted electrically, with a stepper motor positioning, by rotation, a barrel with several holes, so that the parallel hole size is changed. This is the most practical form of electrical adjustment because only a small activating torque is required, and the adjustment can readily be made whilst the damper is functioning.

The next stage is for the adjustment to be made automatically according to operating conditions, i.e. more damping for high speed or when there are large suspension excursions. In this case, the driver will normally be able to select either strong damping, weak damping or automatic control. Automatic systems may be slow acting in the sense of time response of the adjustment in relation to the vehicle heave and pitch natural frequency. Driving conditions normally change relatively slowly, so fast response is not necessary, e.g. a response time of 3 s. The basic rule for damping selection is to have low damping only when the vehicle is travelling at moderate speed in a fairly straight line. Any deviation from this placid condition calls for the high damper setting. Specifically, high lateral or longitudinal accelerations, high speeds, or rough roads all require more damping. Interestingly, very low speeds may benefit from high damping, which is then not detrimental to ride, and the zero speed condition is better with very high damping to reduce rock when passengers enter or leave the vehicle.

In contrast, a fast adaptive (semi-active) system seeks to obtain many of the advantages of a fully active system, but at low cost, by optimising the energy removal with a response which is fast relative to the body ride motions, with a time constant substantially less than 1 s.

Adjustments are made to the valve, and a sharp distinction must be made between the foot valve and the piston valve in this respect. The foot valve is mounted in a fixed part of the damper and is therefore relatively accessible and amenable to adjustment. However, except for highly pressurised dampers the foot valve is only active in compression, and invariably its pressure drop is active for force production only on the rod area. Functionally, it is highly desirable to adjust the piston valves. However the piston is not very accessible. One system has been mentioned which involves compressing and rotating the parts. Alternatively the piston rod may be made hollow to contain an internal shaft which can perform the adjustment. In this case the shaft usually screws in or out to change the position of a taper needle in a hole, thereby altering the effective parallel hole size. This is likely to be in series with a one-way valve, hence acting only in extension. Obviously the possibility exists for the adjuster shaft to influence some other aspect of the valve, although this is not as easy to implement as a simple parallel hole system. Also, there is a case for allowing the piston valve parallel hole variation to apply to both compression and extension; this may be achieved simply by removing the one-way valve.

Racing dampers nowadays often have independent adjustment for high and low speed ranges separately for bump and rebound. At least this is stated, but in some cases the adjustments have not done all that is claimed.

It must be appreciated from the outset that adjustable dampers vary considerably in the way that the $P(Q)$ curve of the valve and the associated $F(V)$ of the damper are altered by the adjustment. As will be shown, this depends greatly on the type of adjustment used, i.e. on the particular valve parameter that is varied. It should also be appreciated that so-called adjustable dampers have been marketed on which the adjustment has no measurable effect over the normal operating range of the damper. Also, because it is easier to arrange for adjustment of a foot valve than a piston valve, adjustments are frequently rather limited in effect. Because foot valve adjustments produce pressure acting on the rod area only, relatively large pressures must be generated to produce useful forces.

8.2 The Adjustable Valve

Figure 8.2.1 (a repeat of Figure 6.9.1 without A_L) shows the basic conceptual valve. It is defined by the following features:

(1) a parallel hole of area A_P;
(2) a series hole of area A_S;

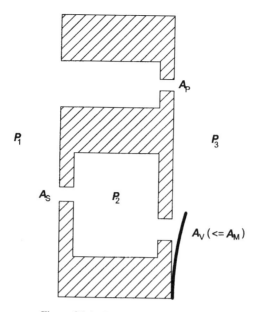

Figure 8.2.1 Valve areas and pressures.

(3) a variable-area hole with area coefficient k_A;
(4) a valve fully closed pressure, P_{vfc};
(5) a maximum area for (3), A_M.

For a simple linear variable hole the area is proportional to pressure difference with

$$A_V = k_A(P - P_{vfc}) \leq A_M$$

where k_A is the area coefficient, P is the pressure difference, and P_{vfc} is the valve fully closed pressure. This could equally well be called the valve just opening pressure. The valve fully open pressure difference is

$$P_{vfo} = P_{vfc} + \frac{A_M}{k_A}$$

For an adjustable valve, or in the design of a nonadjustable valve, any of the above parameters may be varied. For simplicity it is likely that only one parameter will be adjustable, but this is not universal or even particularly desirable from the point of view of performance.

The parallel or serial hole (where present) may easily be varied by use of a taper needle in the orifice. Alternatively, one of a series of holes of various sizes may be selected, for example by rotating a barrel. The area coefficient may be varied by altering the valve stiffness, in one case by shortening a cantilever arm, or possibly by varying the width of passage by rotation of a barrel in the case of a spool valve, or even rotating the spool itself. The initial pressure P_{vfc} can be varied by altering the preload of the valve spring. With a nonlinear spring this would also increase the stiffness for a given opening. Finally, the maximum valve area may be varied by moving a stop.

Figures 8.2.2–8.2.6 show the effect of these variations, the curves being obtained by numerical solution following the theory in Chapter 6 (Valve Design). These graphs are explained further in subsequent sections. Comparison of the graphs shows how different forms of adjustment alter the valve

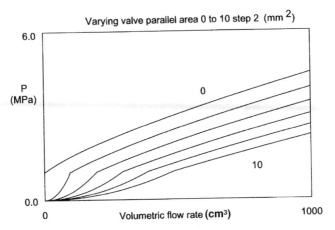

Figure 8.2.2 Effect of adjustment to parallel area.

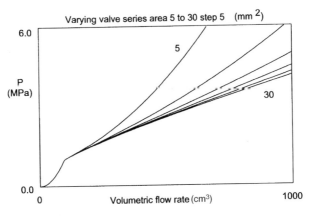

Figure 8.2.3 Effect of adjustment to series area.

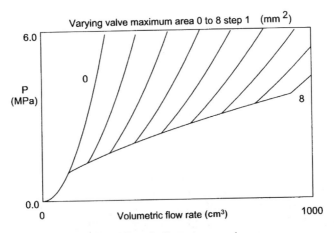

Figure 8.2.4 Effect of adjustment to maximum area.

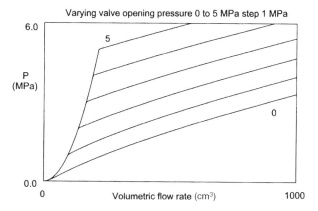

Figure 8.2.5 Effect of adjustment to initial pressure P_{vfc}.

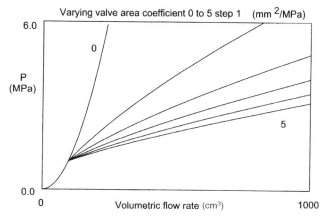

Figure 8.2.6 Effect of adjustment to area coefficient.

characteristics in totally different ways, and in different speed ranges. Therefore, in the use of adjustable dampers it is essential to know the type of adjustment on that damper. The baseline design of these figures includes some valve preload. This is not necessarily desirable, but helps to illustrate the effect of the various adjustabilities more clearly. The baseline specification is

$$A_{\text{P}} = 2\,\text{mm}^2$$
$$A_{\text{S}} = 50\,\text{mm}^2$$
$$A_{\text{M}} = 10\,\text{mm}^2$$
$$P_{\text{vfc}} = 1\,\text{MPa}$$
$$k_{\text{A}} = 3\,\text{mm}^2/\text{MPa}$$

The foregoing is based primarily on the idea of a manually adjusted valve, but also applies to electrically controlled valves that are varied in a similar way with a slow response. However, with an electrically controlled valve of fast response the damper characteristic can, within limits, be controlled entirely by the software, according to the sensor data available, and in that case the characteristics may be more-or-less any shape that is desired, and also dependent on factors other than damper velocity. This is implemented to some extent in semi-active dampers, considered later.

8.3 Parallel Hole

Because of its ease of implementation, variation of parallel hole area is the most common form of damper adjustability. The parallel hole increases the flowrate for any given pressure difference, so this variation basically displaces the existing curve to greater or less flow rates (left and right on the graph), having an effect throughout all three stages of valve operation. A less satisfactory aspect is that the effect is proportionately much greater at low speed, and the damping of low speeds may become unsatisfactorily low.

If this method is implemented by choosing one of a series of holes, then the sequence of hole diameters required is highly nonlinear. Equal increments of force adjustment for a given speed require equal increments of pressure for a given fluid volumetric flow rate. This requires equal increments of u^2, and hence equal increments of orifice diameter to the fourth power. Equal steps of diameter give highly unequal adjustability effects, and are unsuitable. For the same reason, area variability by screwing of a tapered needle may have a very nonlinear effect. To some extent, in that case the nonlinear adjustment effect may be ameliorated by the effect of viscosity on controlling the flowrate, more influential than in the case of a simple orifice, especially for a fine taper, but then the viscosity makes the valve more temperature dependent.

8.4 Series Hole

Variation of a series hole has its effect at the middle and upper end of the speed range, as in Figure 8.2.3, with a progressive increase totally unlike the effect of parallel hole variation. The effect is proportionately greatest at high speeds. Low speed, Stage 1, is unaffected. The effect is highly nonlinear with area, and the hole sizes need to be chosen with care.

Series hole adjustment has been used on at least one type of racing damper. It has some advantages in ease of implementation in common with parallel hole variation. It may be that the two could be combined with good effect.

8.5 Maximum Area

The effect of variation of maximum valve area, as in Figure 8.2.4, is primarily a high-speed adjustment, since by definition it affects Stage 3 only. However it also has a well-defined effect on the flowrate at which Stage 3 begins, so for a small maximum area the effects are seen at quite low flowrates. The onset is much sharper than that of a restrictive series hole. For flowrate and pressure below that necessary to fully open the valve, there is no effect at all. A valve lift limiter to control the maximum area can be implemented fairly easily.

8.6 Opening Pressure

The initial opening pressure P_{vfc} is governed by valve preload and is a well established method of control, though involving rather complex assembly. Increased preload has no effect within Stage 1, but extends its range. Increased preload therefore defers the beginning of Stage 2; it raises the pressure level within Stage 2, i.e. this part of the graph is shifted vertically. Increased preload extends the upper end of Stage 2, by deferring the valve full opening, but within Stage 3 has no effect at all. A good degree of adjustability is achieved throughout the main speed range, but at high preloads the characteristic is extremely nonlinear.

8.7 Area Coefficient (Stiffness)

The valve area coefficient k_A (practical units mm²/MPa) may be adjusted by a change of spring stiffness, or in the case of a spool valve, possibly by a rotation of an internal part of the spool to alter the effective width of the opening, so that a given axial position of the spool reveals a different effective flow area. Variation of the valve stiffness is hardly practical in the case of a coil spring, but may be achieved in the case of a cantilever by altering the free length. Hence a shim valve with two holes, with the shims supported by a two-lobed cam, will exhibit a stiffness and area coefficient dependent upon the rotational position of the support, as in Figure 8.7.1.

Figure 8.2.6 shows the effect of variation of area coefficient where it may be seen that Stage 1 is unaffected, but Stage 2 has an excellent variation. A smaller area coefficient (greater stiffness) gives the higher pressures and defers the end of Stage 2. Once Stage 3 is reached there is no effect.

This method provides the basis for a damper of excellent linearity retained over a good adjustment range. The author has tested a commercially available damper using this principle (variable stiffness shim valve) which did exhibit this excellent behaviour. The 'cam' position can be controlled by a shaft down the centre of the rod. The profile of the 'cam' offers some control over the distribution of the adjustability.

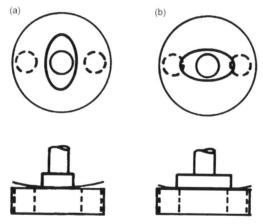

Figure 8.7.1 Variable stiffness valve by rotating a cam, changing the length of the shim cantilever: (a) soft position; (b) hard position.

8.8 Automatic Systems

Suspension systems can be broken down into various classifications:

(1) active;
   (i) very slow (load levelling);
   (ii) slow (speed, road roughness);
   (iii) fast (individual bumps);

(2) adaptive;
  (i) slow adaptive (speed, road roughness);
  (ii) fast adaptive (individual bumps);
(3) passive;

Active suspensions require sensors and large suspension-force power input. Adaptive systems require sensors and control power for valves. Passive systems require no independent control or power input.

The amount of power required by an active system is directly dependent on the speed of action. Load levelling is quite easy, and can even be achieved by energy taken from the dampers themselves (self-powered slow active) when the vehicle is in motion. Slow active will adjust the ride height with speed and road roughness, and of course includes load levelling. It was very good on aerodynamic ground-effect racing cars to control the critical ground clearance with speed and fuel load changes. Fast active has been demonstrated to be very good at both ride and handling, but it is heavy, expensive and sometimes noisy. Other simpler systems have evolved, such as active roll control exerted by an actuator in one or two anti-roll bars, a system which has now become commercially available.

In a fully active suspension, the spring and damper are replaced by or supplemented by a hydraulic actuator which, with a very rapid response and with suitable sensors and logic, can maintain the body level and effectively free of ride motions and also greatly improve road holding on rough surfaces, because the normal force on the tyre can be maintained much more nearly constant. Early research simulations of active suspension and test vehicles clearly demonstrated these advantages. They also showed that fully active systems would be expensive and heavy, have large power flows, and need large pumps. Subsequent use on both passenger cars and racing cars confirmed the benefits and costs. Hence such systems, although very effective, were bound to be expensive in initial cost, and have implications for fuel consumption because of the power requirements.

As a result, attention turned to lower-grade systems in a search for greater cost effectiveness. Various possible paths were identified:

(1) Use a slow response in order to drastically reduce power inputs, pump size, and costs, i.e. use a slow active system. With a frequency response below the wheel hop frequency some of the benefits are lost, but the vehicle can be held level against manoeuvring roll and pitch, and of course against load changes. These were found to be great advantages.
(2) To completely abandon any attempt at active control requiring power inputs. At a stroke, this eliminates all the expensive pumps and actuators and their expense. What is left is control of energy removal, i.e. of energy dissipation, *viz.* an adaptive (damping) system.

Of course, a fast adaptive system has a performance falling short of full active suspension control, but is much cheaper, providing perhaps 50% of the benefit at 5% or less of the cost, and hence being much more cost-effective. This leaves only the remaining 50% of the benefit costing 95% for fully active systems, which are therefore of relatively poor cost-effectiveness.

Remotely controlled and automatic systems have already been briefly described. Figure 8.8.1 shows the block diagram for an automatic system. A system that is only remotely controlled does not require sensors or strategy, which could be rephrased to say that these functions are then fulfilled by the driver. The point of an automatic system is that it requires no input from the driver. Excluded from this definition would be an occasional switching of preferences, such as from best ride to best handling. An automatic system does require information sensors, and actuators for the valves, which are an extra cost.

Adjustables

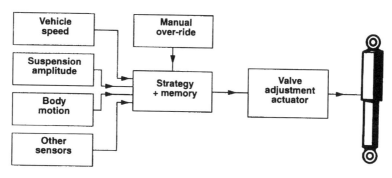

Figure 8.8.1 Automatic adjustment.

As described in Chapter 3, Ride and Handling, on a given road, at higher speed the ride stimulus at the ride natural frequency is much greater, demanding a suitably increased damping for safety and control, whilst at a moderate speed the situation is less critical and a lower damping coefficient may be used for better ride. In rallying, road roughness may actually be the factor that limits speed, through problems with loss of control or reduced traction. For wheel-driven speed record vehicles, even on seemingly very smooth surfaces such as salt flats, the very high speed means that traction and safe control can be major problems.

For passenger cars or other road vehicles, electrical sensors for vehicle speed may already be installed, for example in conjunction with anti-lock brakes. Sensors to reveal suspension amplitude or workspace over a recent time interval are also desirable. An accelerometer on the body may be more reliable because of its less difficult environment. Other factors may also be considered, such as the position or velocity of controls (steering hand wheel, pedals). Figure 8.8.2 shows an ultrasonic height measuring system that has been used successfully.

Depending upon the number of damper settings available, perhaps as few as two (low or high) or effectively continuously variable, the logical control unit will apply its programmed strategy to the input data to determine a preferred damper setting. For an adaptive system, a smoothing time delay

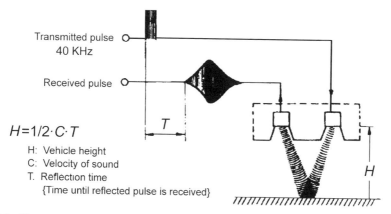

Figure 8.8.2 Ultrasonic height measurement at 40 kHz by detecting the time delay, similar to bats and dolphins (Sugasawa et al., 1985).

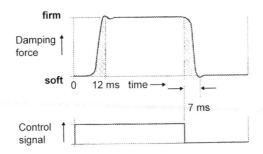

Figure 8.8.3 Response of a switchable damper (Decker et al., 1990).

may be incorporated. For example, a two state low/high system may not switch until the new state has been continuously favoured for a given time, e.g. one second. Alternatively, some hysteresis may be incorporated, so that if the damping required is scored by the logic on a scale of 0 to 1, and greater than 0.5 would favour high damping, then switching up to high damping may only be activated when the score is 0.6 or more for a given period, and switching down when the score is 0.4 or less for a given period. This obviously reduces the amount of switching that occurs when conditions are borderline, reducing wear on the actuator and valve adjustments, and also reducing switching noise.

Fast adaptive damping systems are much more sophisticated than slow adaptive systems. For best results, they need to have a response in milliseconds, in order to continuously maintain a favourable damping according to the more-or-less instantaneous motion of the sprung and unsprung vehicle masses. Actually, switchable dampers may be quite slow in action, with a time constant in tens of milliseconds,

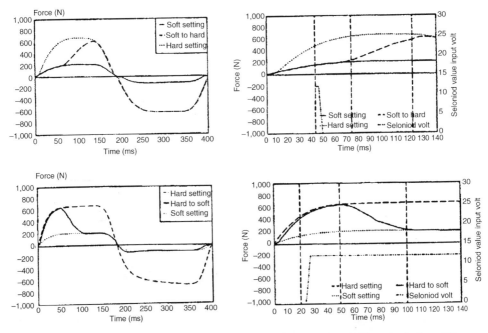

Figure 8.8.4 Delays on a three-state switchable damper. Reproduced from Abd-El-Tawwab, Crolla and Plummer (1998) The characteristics of a three-state switchable damper, Journal of Low Frequency Noise, Vibration and Active Control, 17(2), pp. 85–96.

obviously depending on the resistance and inertia properties and the drive torque available, Figures 8.8.3–8.8.4, a point on which ER/MR dampers may have a significant advantage.

Slow adaptive systems are much easier to implement, and provide a useful benefit at modest cost. The strategy of a slow adaptive system is simple: when driving at modest speed on a fair road and in an approximately straight line use low damping. In all other cases, use high damping.

8.9 Fast Adaptive Systems

Fast adaptive damping requires timely information from sensors on the body and from suspension vertical motions. Several operating strategies have been proposed.

The underlying concept of fast adaptive damping is that the damper is rapidly altered according to whether the force it can produce is deemed desirable or not. Considering a heave-only ('quarter-car') model, the body (unsprung mass) vertical displacement is z_B with velocity $\dot{z}_B$ and acceleration $\ddot{z}_B$, all positive upwards. The wheel has corresponding z_W. The suspension bump deflection is

$$z_S = z_W - z_B$$

with corresponding derivative equations for the suspension velocity and acceleration. Note that suspension deflection z_S is positive for bump, which corresponds to damper compression, i.e. negative extension velocity, for a normal installation.

For an ideal damper with force dependent on velocity only, the direction of damper force (i.e. damper tension or compression) is governed exclusively by the direction of the suspension bump velocity. For positive $\dot{z}_S$, the damper can produce a force upwards on the sprung mass and downwards on the unsprung mass. We may choose to exert such a force or not, according to some sensor information. Because the passenger discomfort is based on vehicle body accelerations, one obvious strategy is to choose to exercise a damper force if it will reduce the body acceleration. Considering a two-state damper with high damping and low (or zero) damping, Table 8.9.1 shows the selection table for the damper. For example, $\ddot{z}_B < 0$, body accelerating downwards, is to be opposed only if the suspension bump velocity is positive, and $\ddot{z}_B > 0$ is to be opposed only if the suspension bump velocity is negative. Otherwise minimum damping is to be selected.

The strategy of Table 8.9.1 can be summarised as:

$$\text{if} \quad \dot{z}_S \ddot{z}_B < 0 \quad \text{then use high else use low}$$

Theoretical and practical studies show that such a strategy can indeed improve vehicle ride. However there may be a deleterious effect on road holding because the favourable body forces are reacted against the unsprung mass in a way which may be a disadvantage for uniformity of tyre vertical force and hence for tyre capability of shear force generation.

If vehicle handling quality is chosen as the criterion, rather than ride, then the damper state selection strategy would be based on making the tyre vertical force as steady as possible, as in Table 8.9.2. In this case, for example, if the tyre vertical force is less than the mean value it is desirable to increase it by exerting a force down on the wheel, so high damping will be selected only if the suspension bump velocity is positive. This illustrates the principle, although from a practical point of view, sensor data on tire vertical force, or deflection, is not cheaply acquired.

The strategy of Table 8.9.2 can be summarised as

$$\text{if} \quad \dot{z}_S (F_V - F_{V,\text{mean}}) < 0 \quad \text{then use high else use low}$$

Table 8.9.1 Selection of adaptive damper state (high/low damping) according to $\ddot{z}_B$ for ride quality

	$\ddot{z}_B < 0$	$\ddot{z}_B > 0$
$\dot{z}_S < 0$ (rebound)	low	high
$\dot{z}_S > 0$ (bump)	high	low

Table 8.9.2 Selection of adaptive damper state (high/low damping) according to tire F_V for handling quality

	$F_V - F_{V,\,\mathrm{mean}} > 0$	$F_V - F_{V,\,\mathrm{mean}} < 0$
$\dot{z}_S < 0$ (rebound)	low	high
$\dot{z}_S > 0$ (bump)	high	low

Table 8.9.3 Semi-active partial realisation of absolute damping

	$\dot{z}_B < 0$	$\dot{z}_B > 0$
$\dot{z}_S < 0$ (rebound)	low	high
$\dot{z}_S > 0$ (bump)	high	low

The above two ride and handling criteria could be combined in a ride/handling compromise, weighted according to their relative importance to the driver at any particular time. With a handling factor f_H in the range 0–1

$$\text{if} \quad \{(1 - f_H)\dot{z}_S\ddot{z}_B + f_H f_E \dot{z}_S (F_V - F_{V,\,\mathrm{mean}})\} < 0 \quad \text{then use high else use low}$$

where f_E is a constant factor equalising the terms when $f_H = 0.5$. Whereas for a simple remotely controlled damper the driver will select high or low damping, or automatic, with an adaptive system he will select 'optimise for ride' ($f_H = 0$) or 'optimise for handling' ($f_H = 1$) or some automatic compromise with a calculated f_H according to conditions.

Other less obvious criteria for damper switching have been proposed, and in fact the early proposals for adaptive dampers were based on consideration of body velocity rather than acceleration. The early work on this (Karnopp, 1983) sought to achieve absolute damping (so-called 'Sky-hook damping') in which the vehicle body is damped according to its absolute velocity rather than the suspension velocity. This would ideally be implemented by producing an actual force proportional to $\dot{z}_B$ rather than $\dot{z}_S$. This would require fully active suspension, but a fast adaptive partial realisation with a two-state damper is possible with

$$\text{if} \quad \dot{z}_S \dot{z}_B < 0 \quad \text{then use high else use low}$$

as shown in Table 8.9.3. Absolute damping has, however been shown to lead to poorly damped wheel motion with deterioration of handling.

Various other concepts have been proposed, for example a 'spring cancellation' strategy in which the springs are effectively softened by using the damper force to oppose any variation of spring force. The damper force switching is then a function of the suspension displacement.

Emerging from the above, it is apparent that the damper state could be made a function of a range of variables, the essential six variables (for a heave-only model) being the position, velocity and acceleration of the sprung and unsprung masses:

$$C = f(z_B, \dot{z}_B, \ddot{z}_B, z_W, \dot{z}_W, \ddot{z}_W)$$

alternatively expressed through the body position z_B and suspension bump position z_S:

$$C = f(z_B, \dot{z}_B, \ddot{z}_B, z_S, \dot{z}_S, \ddot{z}_S)$$

or the wheel and suspension parameters.

For a complete car any one damper may be influenced in general by 24 variables, the six above for each wheel, hence effectively incorporating influences from roll and pitch angles, velocities and accelerations. There is scope here for a wide variety of commercial systems, according to the particular parameters preferred.

8.10 Motion Ratio

As shown in Chapter 4, Installation, the effective damping seen at the wheel depends on the motion ratio of damper to wheel. In some cases it is convenient to adjust the damping at the wheel by altering the motion ratio rather than by adjusting the damper itself. Of course, this is most likely to be suitable on racing cars. Adjustable dampers may be used in addition. When a coil-over-damper arrangement is used, this lends itself well to motion ratio adjustment because the change of ratio affects the stiffness and damping coefficient at the wheel in a similar way.

9
ER and MR Dampers

9.1 Introduction

Electrorheological (ER) dampers and related devices have been under development for many years. This has not led to commercial vehicle damper products. However, since around year 2000 magnetorheological (MR) dampers have come into commercial use on some more expensive passenger vehicles.

Rheology is the science of the deformation of solids and the flow of fluids under stress. ER (electrorheological) and MR (magnetorheological) liquids have properties dependent on the electric field or magnetic field respectively, which is simply a response. ER and MR fluids are sometimes known as 'smart' or 'intelligent' materials, which is, of course, just nonsensical hype. There is no intelligence or information processing capacity. By adjusting the electric or magnetic field as appropriate, the liquid properties are changed, controlling the damping force, which is no longer governed solely by the extension or compression speed. As discussed later, in Sections 9.4 and 9.7, the damper itself is considerably different from conventional design, lacking the usual variable-area valves. Therefore, the damper force must be continuously controlled.

A conventional damper oil is considered to be a Newtonian liquid, in that it has a simple viscosity, albeit temperature dependent. ER and MR liquids have a yield stress and a post-yield marginal viscosity, both dependent on the applied field. Hence, they are basically Bingham plastics, characterised by two parameters, the yield shear stress τ_Y and the subsequent marginal viscosity μ. In practical use, it is the controlled variation of the yield stress that is the main operational parameter. Chapter 5, Fluid Mechanics, gives some background information on the properties of such liquids.

9.2 ER–MR History

Small electrorheological effects have long been known, but large-scale effects with possible practical applications were first studied by Winslow (1947). The first important work on MR is attributed to Rabinow (1951). For forty years significant efforts were made on ER, largely neglecting MR, but since 1990, when work on MR increased, it has become apparent that MR may be much more practical because of the lower operating voltage, lower power requirement, higher shear yield stresses achievable, broader operating temperature range, and greater tolerance of the liquid to contamination, particularly water. Against this must be weighed much greater expense, and some significant hazards in manufacture.

The Shock Absorber Handbook/Second Edition   John C. Dixon
© 2007 John Wiley & Sons, Ltd

March 25, 1947. W. M. WINSLOW **2,417,850**
METHOD AND MEANS FOR TRANSLATING ELECTRICAL
IMPULSES INTO MECHANICAL FORCE
Filed April 14, 1942

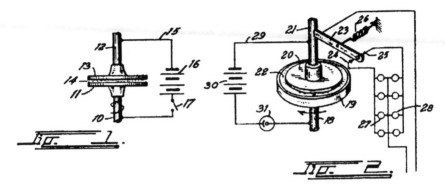

Figure 9.2.1 The original concept of electrofluid control, of a clutch or a rotary relay, W.M. Winslow (1947) US Patent 2,417,850.

Winslow's 1947 patent, with the mysterious title 'Method and Means for Translating Electrical Impulses into Mechanical Force' is really a patent on various configurations of clutches and relays using 'electro-fluids', although the 22nd and last claim is very general:

> The method of instantaneously increasing the viscosity of a force-transmitting fluid composed of a dielectric liquid and a finely divided substance suspended therein; which consists in applying an electric field to the fluid.

This would surely have included 'electrofluid' motor vehicle dampers, had the possibility of such an application been conceived at the time. In Figure 9.2.1, on the left is a controllable clutch for power transmission, in the second case operating relay switches at 24–25 to control loads such as lamps, 27–28. The possibility of a controllable rotary damper is evident to modern eyes.

To quote further from the patent:

> The invention comprises what might be termed an electro-fluid clutch. . . This invention contemplates the use of what is believed to be a novel phenomenon of electricity. I have found that if two plates are separated by certain substantially dielectric fluids containing certain substances the fluid mixture will tend to cause the two plates to act as a unit as long as an electrical potential difference exists between the plates. [In Figure 1] the fluid mixture is held between the plates by capillary attraction. Electrical potential is applied to the discs by means of the closing device 17. Since the fluid is dielectric, very little current will flow through the circuit. Many fluid mixtures have been found to accomplish this result with more or less efficient results. It appears that the fluid must be a dielectric, or substantially non-conducting at all operating electrical pressures, for very little current flows through the fluid between the plates. Therefore a low viscosity non-conducting liquid is preferred as the fluid medium . . . . suitable are light weight transformer oil, olive oil, mineral oil, etc. A pure fluid or oil alone, however, does not act to tie the plates together under the influence of the electric current [*sic*] to any practical extent. However, when an additional substance, in the nature of a finely divided material, is added thereto the tying effect is very pronounced. Such substances as starch, limestone, gypsum, flour, gelatine, carbon, etc, all create the desired effect with more or less efficient results . . . . To date the applicant finds that a pharmaceutical mixture of refined mineral oil and lanolin in which starch granules (approximately 20% by volume) have been placed gives good results.

Just what takes place in the fluid when the electrical potential is impressed upon the plates is not definitely known. It appears, however, from close observation of the mixture in action that there is a tendency for suspended particles to form an infinite [*sic*] number of strings or lines extending between the plates. These strings immediately disappear when the circuit is broken.... It is manifest that the [apparent] viscosity of the fluid is greatly increased in the presence of the electrical field and a homogeneous fluent mechanical linkage, or coupling, is thereby established. This increase in viscosity takes place without a change of temperature.... The effect can be attained with direct or alternating currents of any frequency. There is no permanent change in the mixture as it instantly releases and regrips rapidly and indefinitely.... The body of a person when moving his feet back and forth across a carpet will store an electrostatic charge sufficient to operate the device.

Rabinow's 1951 patent, based instead on the MR principle, referring *inter alia* to Winslow's ER patent, considers various configurations of clutch, including some multiplate ones, and is generally directed at more substantial devices for power transmission, including in one drawing even a cooling circuit for the liquid. Figure 9.2.2 shows the first few of his configurations, including axial-field and radial-field devices.

To quote from the 1951 Rabinow patent:

A principle object of this invention is to provide an electromagnetically-controlled clutch or brake with substantially no wearing parts, capable of locking two relatively movable rotating elements together with great force and with features of advantage over conventional magnetic eddy current clutches or brakes including the ability to lock in with substantially its maximum torque even at its lowest (or zero) relative speed; perfectly smooth and chatterless operation when there is relative motion between the rotating elements; and substantially constant torque at all slipping speeds within a wide range . . . . Other advantages include very fast response to quick changes in control current; operation requiring only a low potential source of electric power such as can be supplied by a storage battery, and ability to operate on either alternating current or direct current.

My intention is based upon the fact that if two slightly spaced surfaces of paramagnetic materials are connected by a mixture of liquid and a large number or a mass of finely divided relatively movable contiguous discrete paramagnetic particles, such as soft iron particles; and a magnetic field is applied so that the particles are included in the magnetic circuit between the surfaces a substantial component of the field will be perpendicular to the surfaces at the areas of contact between the particles and the surfaces because of the fundamental law of physics that the potential energy of any system must be at a minimum; and the contact pressure between the particles *inter se* and between the particles and the surfaces will build up, whereby the surfaces will tend to lock together so as to transmit force between them as long as the magnetic field continues to exist.

When the field is energised the particles are attracted one to the other and in that way the adhesive consistency of the mass of particles is increased so that the resistance is offered to the relative motion of the particles.

In Figure 1, the top shaft 1 has secured thereto a disc 3, while the bottom shaft 2 has secured thereto a cup 4 containing a magnetic fluid mixture consisting a suitable fluid vehicle and a quantity of finely divided paramagnetic particles, such, for example, as commercially available soft iron dust sold as Carbonyl Iron Powder, of which 8 microns average size, has been found satisfactory. The percentage of dust may be varied within fairly wide limits, but I have found a mixture of containing approximately 50% by volume of dust to give satisfactory results . . . . By making shaft 1 of non-magnetic material or else by making the section 5 of non-magnetic material, or otherwise suitably designing the magnetic circuit for efficient operation in accordance with known good practice . . . . Although the clutch is operative without a fluid vehicle using only the iron particles, I have found that the operation is greatly improved by the use of a fluid mixed with the iron particles. I have found a light lubricating oil to be suitable for this purpose, but, in general, any liquid may be used which has suitable mechanical properties to make the mixture act as a rather viscous fluid at all contemplated operating temperatures . . . . These details can be widely varied according to the dictates of the particular design employed . . . . When the coil is energised, it is found that a strong coupling force exists between the two shafts 1 and 2 . . . . The two elements 3 and 4 will be 'locked' together in that they will rotate at the same speed . . . . It is obvious that the same system can also act as a brake if one of the members is held fixed. Under these conditions there will be no effective braking action until coil 6 is energised at which time a retarding torque will be developed which will exert a braking action on the rotating shaft.

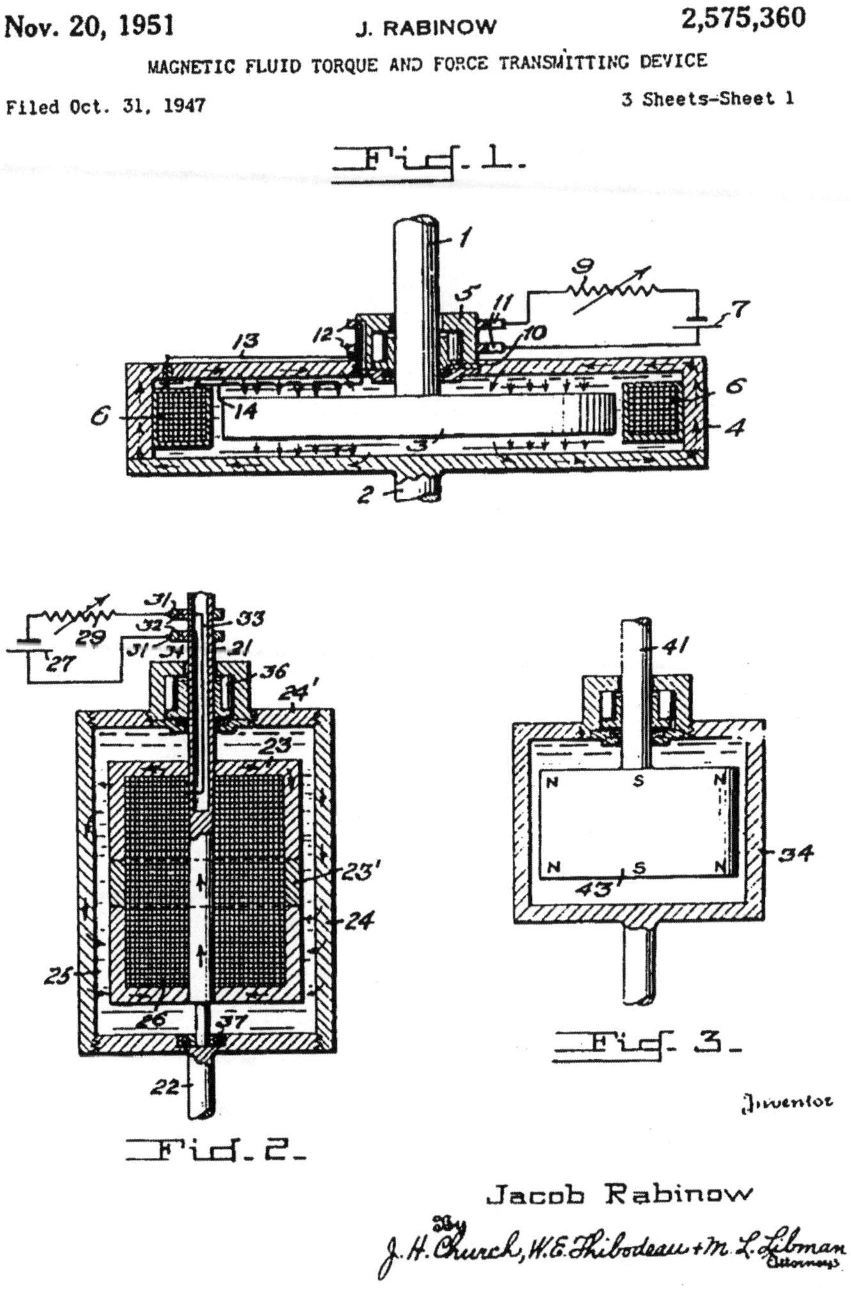

Figure 9.2.2 The original concept of magnetic-fluid control of a clutch (or rotary damper), with axial field or radial field. The third figure is a permanent magnet torque transmitter and limiter, J. Rabinow (1951) US Patent 2,575,360.

The specific patent claims included:

5. A device for controlling the transmission of torque from one element to an adjacent relatively movable element comprising members fixed to said elements respectively and having substantial, opposed, closely spaced surface areas separated by a fixed distance, a mass of contiguous relatively movable discrete paramagnetic particles in the space between the said surface areas and means for creating a magnetic field between said opposed members and including said particles, to produce a coupling effect between said spaced members . . . .
7. The invention according to claim 6 in which said particles are mixed with a non-magnetic fluid.
8. The invention according to claim 7 in which said particles consist of iron powder and said fluid is a light oil.

Here it may be seen that Rabinow emphasises the low operating voltage, large force generation and rapid response. He made public demonstrations with a human supported by an MR link at a stress of 100 kPa, using 90% by mass iron and 10% mass light mineral oil. Again, the possible application to a rotary automotive damper is clear to modern eyes.

Winslow took out a further patent in 1959, on rotary couplings using both ER and MR materials. Figures 9.2.3 and 9.2.4 show two of these, indicating the flavour. Although they were described as electrically controlled clutches for power transmission, the possible application to a controllable rotary damper is obvious.

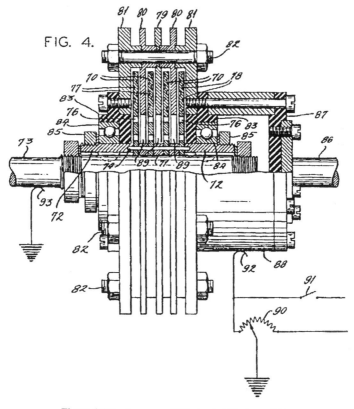

Figure 9.2.3 A proposed ER clutch (Winslow, 1959).

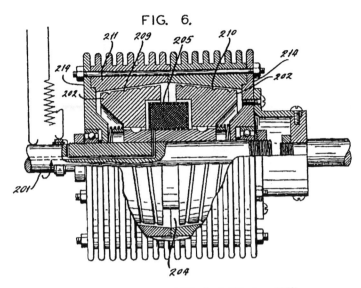

Figure 9.2.4 A proposed MR clutch (Winslow, 1959).

In this patent Winslow suggests that the coupling force for ER fluids is basically proportional to the field strength squared. A possible explanation of this is that the particles, assumed to have zero intrinsic polarity, have an induced polarity proportional to the field. The coupling strength is then proportional to the field times the particle polarisation. The coupling strength was found to be fairly independent of the relative velocity. His comments on the necessary character of ER and MR fluids are of some interest:

> When properly formulated fluids of the foregoing type are subjected to an applied field, a slight migration of the particles tends to occur. The particles seek regions of high field intensity, and in so doing tend to associate with other particles in the form of chains or fibres. This migration to form a fibrous mass is aided by forces which move the particles mechanically as when the fluid is subjected to shear. In dilute mixes, the formation of these fibres in the direction of the field is readily observed under the microscope. In high concentration, the individual fibres are not discernable . . . . In general, in order to achieve an enhanced force transmitting effect, it has been found that the volume concentration of electrically or magnetically chargeable particles should exceed about 38 % . . . . In using these field responsive fluids for various applications, certain practical difficulties have been encountered, particularly in applications involving the transmission of large values of mechanical power. In general it is found that if, under the action of shear, the fluid film be of sufficiently low viscosity to allow easy slippage when slippage is desired, i.e. with little or no applied field, then the particles may be partially separated from the oil vehicle by centrifugal action, resulting in an oil slip layer when slippage is not desired.
>
>   In order to preclude a gravity settling of particles when the coupling is idle for long periods, the fluid is compounded to have a definite thixotropy sufficient to hold the particles immobile under kinetic agitation of the molecules, but to allow for Brownian movement of the particles in the presence of mechanical shear . . . . Suitable dimensions which have been found for the spacing range from 0.002 inches (0.05 mm) for couplings used for very low slip speeds to 0.060 inches (1.5 mm) for couplings used for very high slip speeds.
>
>   By the term 'field responsive fluid', as used in the present disclosure and claims, is meant any fluid of the type composed of a suitable liquid and suspended, substantially solid, particles which are attractable in the presence of an electric or magnetic field to impart shear resistance to the fluids. Fluids suitable for the coupling described may be made from a variety of materials. The following formulas and procedures should be considered exemplary rather than limitative of fluids contemplated by the present invention.

For the electric field responsive fluids, I add 100 parts by weight of dry micronised silica gel powder of desiccant grade to a solution containing about 40 parts by weight of an electrically stable dielectric oily vehicle of from 2 to 20 centipoise viscosity at 25°C, about 10 parts by weight of an oil soluble dispersing agent; such as sorbitol sesqui-oleate, sold as Arlacel C, ferrous oleate, lead naphthenate, etc., about 10 parts by weight of a water soluble dispersing agent, such as sodium oleate, sodium naphthenate, or polyoxyalkalene derivative of sorbitol oleate, sold as Tween 80, etc, and then about 15 parts by weight of water.

This mix, which is initially in a somewhat pasty condition, is circulated through a pump, such as shown and described in my prior application, Serial No. 716,626, filed December 16, 1946, now abandoned. The pumping is continued until the mix becomes a readily flowing thixotropic syrup. During the latter part of the pumping, the fluid is exposed to drying conditions to remove about half of the original 15 parts of water to bring the resistivity of the fluid into the range between 10^8 and 10^{10} ohm per cm." [*sic*, units really ohm cm]

The term 'thixotropic' is used in this specification in its usual sense; this is, to mean that the fluid quickly sets to a gel when no longer molested by mechanical forces and just as quickly reverts to a liquid syrup when again subjected to mechanical forces; this phenomenon repeating itself each time the coupling passes through a cycle of slipping and non-slipping.

As an alternative, I may substitute for the 100 parts by weight of silica gel of the above formula about 60 parts by weight of dry micronised synthetic resins of the exchange type as now commonly used in adsorption processes. Ion exchange resins are found to have an adsorption capacity for the foregoing soaps or dispersing agents which is comparable to silica gel. The fluidising of the resin type mixes may be similarly accomplished by a pumping operation.

For the magnetic field responsive fluids, I may add 100 parts by weight of reduced iron oxide powder sold as 'Iron by Hydrogen Merck' to a solution containing 10 parts by weight of a lubricant grade oily liquid of from 2 to 20 centipoise at 25°C, and 2 parts by weight of ferrous oleate or ferrous naphthenate as dispersant. The somewhat pasty mix is pumped, as before, until fluidised. Toward the end of the pumping operation, I prefer to add about 1 part by weight of an alkaline soap, such as lithium stearate or sodium stearate, to impart thixotropic body to the fluid.

In general, the thixotropic body of either the electric or magnetic field responsive fluid may be increased by substituting for part of the oleates or naphthenates a corresponding part of laurate, palmitate, or stearate.

An alternative magnetic field responsive fluid may be made by substituting for the reduced iron oxide an extremely fine grade of iron powder made by retorting iron carbonyl in a manner well known to the art.

The purpose of the prolonged pumping operation in the above procedures is two-fold: Agglomerates of the primary particles are broken up so that the largest particles present are micron size; and the surfaces of the primary particles are conditioned or smoothed in a manner not fully understood. In this way, the normal viscosity of the fluid is reduced and the field-induced viscosity is increased.

The function of the soaps or soap-like additives in the above formulas is three-fold: (1) they enable very concentrated yet workable fluids, such that on working in a pump the agglomerates are broken up by large shearing stresses; (2) they provide a particle coating with low sliding coefficient of friction; (3) they serve to render the fluid thixotropic with the advantages already described. Other functions of the soaps or dispersing agents involve the electric double layer which determines whether the particles will have repulsion or attraction apart from action of the applied field, and in the case of electric field responsive fluids, the dielectric strength of the films on the particles is also involved.

Since 1990, the operating principle has been developed and some practical problems overcome to create practical engineering devices, including telescopic dampers. For example, with iron particles there is obviously a possible problem of abrasive wear of seals. The expense of fine iron particles produced by the iron pentacarbonyl route continues to be a limitation, but the usability of iron in oil has now reached the stage of practical application.

9.3 ER Materials

An ER material is one for which any of the rheological properties depend upon the electric field as volts/metre. In the context of dampers, this means, for example, a material with a shear stress or viscosity that depends on the electric field.

A polar molecule is one with a dipole moment. It is electrically neutral overall, but the charges are displaced giving a dipole moment measured in basic SI units of C m (coulomb metres). Any liquid with a polar molecule would be expected to show some effect of an electric field, because the molecules would tend to align with the field. Hence any alcohol for example (methanol, ethanol, etc, having an – OH alcohol tail) would qualify. Such effects are, however, too weak for useful application. In general, many molecules have some dipole moment, but in most bulk materials the molecules are arranged in a random way, so the total effect is insignificant. However, in a crystal the molecules are aligned with one another, so the effect can be cumulative. This occurs, for example, in quartz (silicon dioxide) crystals, which results in piezoelectric effects and in the possibility of using mechanical vibrations of a quartz crystal to give a high-speed electronic clock, as used in computers and for frequency control of some radios. Also, a conducting or semiconducting particle in a nonconducting liquid when in a field will have an induced charge distribution whilst remaining electrically neutral overall, so having an induced polarity. Hence, many solid powders exhibit an ER effect, as described in Winslow's original patent.

Practical ER damper liquids, then, have a large mass fraction of a suitable solid powder blended into a low viscosity oil carrier liquid. Various solids may be used to achieve the electrorheological effect, but a common one is so-called aluminium silicate, because it is effective, chemically inert, cheap and easily available.

In fact, aluminium silicate is misleadingly named, it is not a true single-molecule compound of aluminium, silicon and oxygen. Ideally it is a combination of three molecules:

Molecule	Formula	Structure
Aluminium oxide	Al_2O_3	O=Al–O–Al=O
Silicon dioxide	SiO_2	O=Si=O
Water	H_2O	H–O–H

In an ideal aluminium silicate mixture, there are two silicon dioxide molecules for each aluminium oxide molecule, hydrated by two H_2O molecules into a repeated sequence with secondary bonding into an alternating two-layer crystal, one being a tetrahedral layer of silica, the other an octahedral layer of alumina. The empirical formula of the complete unit is $Al_2Si_2O_9H_4$. Hence the ideal mass ratio, deduced from the relative atomic masses (atomic weights), is 54.1 % silica, 45.9 % alumina. When this is in hydrated form, the ratios are silica 46.5 %, alumina 39.5 %, water 14.0 %. Aluminium and silicon can also form other oxides (e.g. silicon monoxide) and may each be hydrated to various degrees (different numbers of H_2O molecules per base molecule). Also, the ratio of silicon oxide molecules to aluminium oxide molecules may vary, giving different forms of rock (andalusite, etc.).

Commercial aluminium silicate is mined diatomaceous earth. This comprises deposits of aluminium silicate shells from dead diatoms, microscopic unicellular algae called *bacillarophyta*, the organic parts of which contribute to crude oil after millions of years of heat and pressure underground. Hence commercial aluminium silicate is impure, varying in the ratio of aluminium to silicon, and typically containing small amounts of other metal oxides, e.g. magnesium oxide, titanium oxide, and iron oxide. It is also known as China Clay, Kaolin and Kieselguhr. It is noncombustible, already being fully oxidised in pure form, and nontoxic, and is in fact used as a medicine for stomach disturbances.

Pure aluminium oxide has a density of 3684 kg/m^3, quartz (silicon dioxide) 2650 kg/m^3, and silicon monoxide 2130 kg/m^3. Common fused silica is 2070–2210 kg/m^3. Alumina monohydrate is 3014, alumina trihydrate is 2420–2530 kg/m^3. The resulting density of kaolin is about 2600 kg/m^3, evidently varying somewhat with the exact constitution, crystal state and degree of hydration. The colour is white in pure form, but typically a shade of pink or red due to the presence of iron oxide. The specific thermal capacity is 1050 J/kg K, and in fired (pottery) form the modulus is 55 MPa and the linear thermal expansion coefficient is 29 ppm/K. The electrical resistivity varies considerably with temperature over the wide possible operating range in general, but at damper temperatures it is very high. At room temperature the resistivity exceeds $10^{12}\ \Omega$ m, and at 100 °C is about $5 \times 10^9\ \Omega$ m.

Dampers are required to operate over a wide temperature range, typically specified as −40°C to +130°C. This poses a problem for electrorheological fluids in maintaining reasonably unchanged properties. The variation seems to arise because of the variation of the oil volume due to thermal expansion, which affects the volumetric fraction of solids, which is critical. Over this temperature rise the volume of the oil increases by about 15–20 %. Also the properties of ER fluids are sensitive to contamination, particularly by water.

With ER dampers, it is not easy to achieve a large force, and high operating voltages are needed with a thin gap to give a strong field. There is some electrophoretic effect, with charged solid particles drifting towards the charged plates, so there is a leakage current much in excess of the current expected from the conductivity of the solid and liquid phases separately. This must be limited. The required properties of the fluid cannot be determined in the absence of a specific design for such a damper, which is not described until Section 9.4. However, it could have an annular flow design with an effective plate area of 1.0 dm^2 and with a gap (annular width) of 0.5 mm. This is then a viscous type of damper with effectively adjustable viscosity or shear strength, the properties are not governed by the flow through the usual damper valves which are designed to avoid sensitivity to liquid properties other than density.

Representative target ER fluid properties would then be (Petek, 1992):

Viscosity	< 50 mPas at 25 °C, zero voltage
Operating voltage E_{max}	= 4 kV
Liquid thickness t	= 0.5 mm
Max field strength E_{max}/t	= 8 kV/mm = 8 MV/m.
Yield stress at E_{max}	> 4 kPa
Current density at E_{max}	< 100 mA/m^2

The leakage current would then be 1 mA at peak voltage, which is 4 W electrical control power per damper.

The oil is basically an insulator, so the leakage current is mainly due to the particles, which may form long conducting or semiconducting fibrils. Also, charge transfer may occur when particles are in contact. After separation, such individual particles are left electrically charged, and so will drift through the oil towards the appropriate terminal. This is called electrophoresis.

A low viscosity at zero operating voltage is desirable to obtain fast switching response and the maximum effect in active-damping ride improvement. However, for simple switching between low and high damping modes it may be useful to have some damping at zero voltage, and this also offers some fail-safe capacity. Unfortunately, any damping effect provided in this way is very temperature sensitive.

ER dampers can be operated as direct-voltage or alternating-voltage devices. In the latter case, convenient for the high-voltage electrical power supply, the electrical capacitance of the damper may be significant. However, this also applies to the direct-voltage device because the time constant of operation has been reported to depend on limitations of the supply current in charging up the damper-capacitor. The electrical capacitance may apparently be higher than would be expected from the basic plate size, spacing and ER liquid properties because of the dipole moment of the ER liquid which gives the fluid a high dielectric constant. The device time constant may therefore be determined by the capacitance and the characteristics of the power supply rather than by the material alignment response time. AC operation reduces the problems of electrophoretic drift, but adds to the current requirement because of the damper electrical capacity. A disadvantage of AC operation is that for the same mean force as with DC a higher peak control voltage is needed. Nevertheless, low-frequency AC may be the best method.

Mechanism of ER Viscosity Change.

Considering a polar molecule or particle with a dipole moment, with an electric field, Figure 9.3.1, in a static or very low speed fluid flow the particle will align with the field. In a slowly shearing flow, this particle will then add to the effective viscosity by drag effect on the fluid due to the different fluid velocities relative to the particle at the two ends, the particle drifting along at a mean speed. As the shear rate increases, at constant voltage, the particle will become partially aligned with the flow, having somewhat less effect on the effective viscosity, i.e. there would be a nonlinear resistance to flow. Finally, at a high shear rate, of value depending on the field strength, the particle will become aligned and there will be little extra viscosity effect.

Considering a constant velocity shear rate with varying field strength, Figure 9.3.2, as the voltage increases the particle will be pulled more out of alignment with the flow and will give a greater effective viscosity.

ER materials may also exhibit a reduction of marginal viscosity with field strength. There are at least two possible mechanisms for this. The field may cause clumping of the particles into large groups. For a given mass of solids, the viscosity increase compared with the host oil depends on the size distribution of the particles. Any bunching would make them less effective. Also, with the field on, many of the particles will be in the form of 'fibrils' (strings) which will allow the oil to flow more freely in between. Therefore, the effect of the field is to create a yield stress, but the subsequent (yielded) marginal viscosity may be greater or less than that of the host oil.

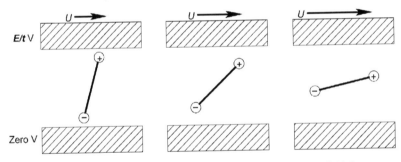

Figure 9.3.1 Polar particle subject to a constant electric field in varying fluid shear rates.

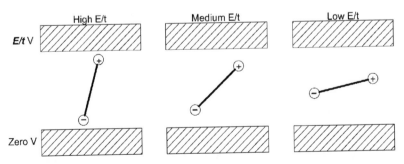

Figure 9.3.2 Polar particle subject to a varying electric field in constant fluid shear rate.

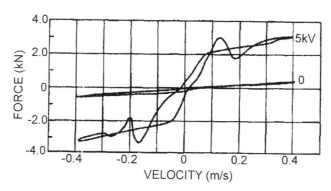

Figure 9.3.3 Stick-slip phenomenon of ER liquid (Petek *et al.*, 1995).

Mechanism of ER Shear Strength Change

Dipole molecules can reduce their energy by clustering together. In a field they form chains, not unlike a weak polymer, known as 'strings' or fibrils ('a minute thread-like structure'). Also, these can become entangled, giving the former liquid some of the properties of a solid, in particular in this case some shear yield strength. A practically achievable yield stress is 4 kPa, which is very low by ordinary material standards (structural steel is 250 MPa, solid polymers may be 10–50 MPa), but sufficient to be of some practical use. The yield stress has been observed to be subject to a stick–slip phenomenon, with a force overshoot as the flow changes direction, Figure 9.3.3.

The response time for ER fluid is short, possibly 1 ms, being the time required for two responses to occur: (1) the individual particles may align with the field simply by rotation; (2) the particles must form any fibrils or bunches. These responses are slowed by the viscosity of the host oil, which should therefore be kept quite low. The response time is temperature sensitive, becoming rather slow at low temperature, probably simply due to the high oil viscosity. Leakage current due to electrical conductivity limits the upper temperature, requiring high current and power from the power supply to create sufficient working voltage.

ER materials have a pre-yield region at low stress with some strain, and appear to be conventional viscoelastic materials in this range.

To quote from Gamota & Filisko (1991):

> The generally accepted model for an ER material is that when the material is subjected to an electric field, the particles align themselves and form a fibrous microstructure. Furthermore, it has been suggested that the yield phenomenon may be related to the breaking of the fibres or chains forming the microstructure. The yield behaviour cannot be explained by random breakage of individual fibres as has been suggested by others. It is suggested [here] that a gross planar movement or flow occurs within the microstructure which is initiated once a critical activation stress is overcome. The movement is analogous to a long range coordinated shearing of planes as is seen in the deformation of a ductile material. However, unlike ductile materials, the ER material is able to re-establish its microstructure.

The ER material, then, has an appealing biological type of self-healing quality

Figure 9.3.4 shows a proposed mechanical analogue model for ER material. The U component is a Coulomb friction element. The other parts are standard linear elements. In the pre-yield state the U component is fixed in position, so the complete Zener unit on the right is effective (a spring in series with the parallel spring/damper Voigt–Maxwell unit) giving a standard viscoelastic material. With field on, at low stress an ER material approaches zero strain rate, so it is effectively then a solid, not a viscous liquid. The Coulomb friction term U is strongly field dependent. The viscous element C2

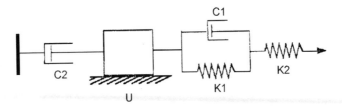

Figure 9.3.4 A mechanical analogue for an ER material. Reproduced from Gamota and Filisko (1991) Dynamic mechanical studies of electrorheological materials: Moderate frequencies, *Journal of Rheology*, 35, 3.

accounts for the viscosity when field-on and yielding. This component is also field sensitive, but the variational coefficient may be positive or negative, and the actual effective damping coefficient may become negative here. This model is adequate for qualitative understanding of ER materials, and for quantitative modelling for normal applications.

9.4 ER Dampers

Electrorheological dampers are of distinct internal design. At zero field, the ER liquid has negligible yield stress, and behaves as a Newtonian fluid with simple viscosity μ. To add the ER effect, it is necessary to produce a strong electric field over a significant area. The practical approach is to use an annular flow design, as seen in Figure 9.4.1. This particular example design uses a free piston to accommodate oil expansion and the rod insertion volume. A conventional 'double-tube' configuration could be used, although there would then be a total of three concentric tubes. In this example design, it is assumed that the force will be controlled entirely by the electrical field, i.e. there are no conventional valves in the piston. The inner tube must be located accurately within the outer tube, maintaining a fairly uniform clearance of less than 1 mm. This would require some insulating spacers which would partially obstruct the fluid flow in the annulus — the effect could be small and is neglected in this basic analysis.

To understand the resulting damper characteristics, it is convenient to analyse the viscous pressure drop and the ER shear drop separately. The damper force is then represented by

$$F_D = C_D V_D + C_Q V_D^2 + C_E E$$

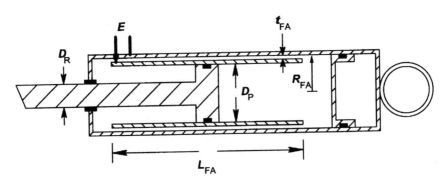

Figure 9.4.1 Basic design of an ER (electrorheological) damper, showing the fluid flow annulus of length L_{FA}.

This is compared with a true Bingham solution later. The prospective ER liquid characteristics are:

$$\rho = 1500 \, \text{kg/m}^3$$
$$\mu = 40 \, \text{mPa} \cdot \text{s}$$
$$C_{\tau E} = 0.4 \, \text{Pa} \cdot \text{mm/V}$$
$$\tau_{Y\max} = 4 \, \text{kPa}$$

The maximum shear stress is limited by the field (V/m) that it is practical to apply. Practical geometrical design values are:

$$D_R = 12 \, \text{mm}$$
$$D_P = 30 \, \text{mm}$$
$$R_{FA} = 18 \, \text{mm}$$
$$t_{FA} = 0.5 \, \text{mm}$$
$$L_{FA} = 200 \, \text{mm}$$

The fluid annulus is the thin space of mean radius R_{FA} and thickness t_{FA} between the outer body and the inner tube. Displacement of the piston by X causes a flow $A_{PA}X$ through the fluid annulus, where A_{PA} is the annular area of the piston, depending on the piston and rod diameters:

$$A_{PA} = \frac{\pi}{4}(D_P^2 - D_R^2)$$

The fluid annulus cross-sectional area, using the annulus central radius R_{FA}, is

$$A_{FA} = 2\pi R_{FA} t_{FA}$$

The area factor is then

$$f_A = \frac{A_{PA}}{A_{FA}}$$

The fluid annulus plays the part of the valve with high velocity compared with the piston. Practical values (see above) are

$$A_{PA} = 593.8 \, \text{mm}^2$$
$$A_{FA} = 59.5 \, \text{mm}^2$$
$$f_A = 10.5$$

The velocities in the annulus are therefore about 10 times the piston velocity, so dynamic pressure losses are quite small (under 50 kPa), much less than for a conventional damper design. The quadratic damper force coefficient C_Q is given approximately by

$$C_Q = \tfrac{1}{2}\rho \alpha f_A^2 A_{PA}$$

depending on details of the design, where α is the kinetic energy correction factor. In this example case, $C_Q \sim 100 \, \text{N s}^2/\text{m}^2$, so it is not an important term, but worth including if high speeds are to be analysed.

The viscous pressure loss in the fluid annulus may be obtained by applying equations from Chapter 5, Fluid Dynamics. Assume an oil viscosity of 40 mPa s. At a design maximum piston velocity of 2 m/s the mean velocity in the fluid annulus is 21 m/s, and the Reynolds number based on hydraulic diameter ($2t$) is 1600, which means that the flow will remain laminar. The volumetric flow rate is

$$Q = A_{PA} V_D$$

where V_D is the damper velocity (extension positive). The viscous pressure drop is

$$P_V = \frac{6\mu L_{FA} Q}{\pi R_{FA} t_{FA}^3}$$

The pressure drop P_V acts on the piston annular area, so the viscous damper force is

$$F_{D,V} = A_{PA} P_V$$

In terms of the damper velocity, the viscous damper force is therefore

$$F_{D,V} = \frac{6\mu L_{FA} A_{PA}^2}{\pi R_{FA} t_{FA}^3} V_D$$

giving the field-off damper coefficient $C_D = F_{DV}/V_D$. This may also be used (approximately) to calculate the viscous contribution to the damper force when the field is applied provided that the appropriate (marginal) viscosity is used.

To obtain the ER effect, an applied electric potential difference E (volts) between the inner and outer tubes gives a radial electric field strength E/t_{FA} (V/m). This gives the ER liquid a yield shear stress τ_{ER} according to

$$\tau_{ER} = C_{\tau E} \frac{E}{t_{FA}}$$

where $C_{\tau E}$ is a coefficient depending only on the properties of the ER fluid, e.g. the concentration of particles, but not on the damper geometry. The ER shear stress acts over the two cylindrical surfaces of the fluid in the annulus. To initiate any movement of the ER fluid, a force must be applied axially to the fluid in the annulus, overcoming the total shear resistance. This ER fluid shear force on the fluid in the annulus (not a damper force), is

$$F_{FA,ER} = 2(2\pi R_{FA}) L_{FA} \tau_{ER}$$

This force is produced by a pressure drop acting on the fluid annulus cross-sectional area. The effective ER shear pressure drop is therefore

$$P_{ER} = \frac{F_{FA,ER}}{A_{FA}} = 2\frac{L_{FA}}{t_{FA}} \tau_{ER} = 2\frac{L_{FA} C_{\tau E} E}{t_{FA}^2}$$

This resistance pressure acts on the piston annulus area, so the resulting ER shear damper force is

$$F_{D,ER} = P_{ER} A_{PA} = \frac{2 L_{FA} A_{PA} C_{\tau E} E}{t_{FA}^2}$$

which is proportional to the applied electric potential E. Any force less than this will not move the damper.

In summary, then, the damper extension force can be calculated approximately from

$$F_D = C_D V_D + C_Q V_D^2 + C_E \text{abs}(E) \qquad V_D > 0$$

$$F_D = C_D V_D - C_Q V_D^2 - C_E \text{abs}(E) \qquad V_D < 0$$

$$F_D \text{ indeterminate} \qquad V_D = 0$$

with

$$C_D = \frac{6\mu L_{FA} A_{PA}^2}{\pi R_{FA} t_{FA}^3}$$

$$C_Q \approx \tfrac{1}{2} \rho \alpha f_A^2 A_{PA}$$

$$C_E = \frac{2 L_{FA} A_{PA} C_{\tau E}}{t_{FA}^2}$$

If the damper is not moving then the applied potential does not control the actual force, only the limit force. The force may be anywhere in the range plus or minus the limit force. For a practical ER damper, at low speeds the applied potential is the main factor, but at high speeds the viscous effect is likely to be large, and temperature dependent.

Some example values, using the data given earlier, and a moderate damper speed, are

$$
\begin{aligned}
V_D &= 0.500 \text{ m/s} \\
Q &= 0.297 \text{ litre/s} \\
V_{FA} &= 5.250 \text{ m/s} \\
Re &= 39.4 \\
P_V &= 2.016 \text{ MPa} \\
F_{D,V} &= 1.197 \text{ kN} \\
C_D &= 2.394 \text{ kNs/m} \\
E &= 5000 \text{ V} \\
E/t &= 10.00 \text{ MV/m} \\
\tau_{ER} &= 4.00 \text{ kPa} \\
P_{ER} &= 3.200 \text{ MPa} \\
F_{DER} &= 1.900 \text{ kN} \\
C_E &= 0.380 \text{ N/V}
\end{aligned}
$$

The force ratio at this speed is

$$F_{D,V}/F_{D,ER} = 0.630$$

so the controllability at low speeds is potentially quite good.

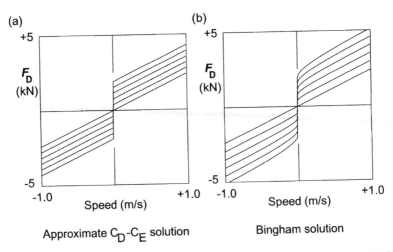

Figure 9.4.2 Theoretical characteristics of an example design of ER damper: (a) approximate CD–CE solution; (b) Bingham flow solution.

A more accurate calculation requires consideration of the Bingham flow pattern in the annulus. Figure 9.4.2(a) shows the theoretical characteristics of an example ER damper design, assuming no variation of viscosity. This is similar to the case of a damper with an adjustable mechanical valve in which the preload force is varied (and would be even more so with a parallel leak hole). Figure 9.4.2(b) shows the result of a true Bingham flow solution, which gives somewhat higher forces. This is because the shearing action is concentrated near to the wall.

The ratio of viscous force to ER force, preferably small for good controllability, is

$$f_F = \frac{F_{D,V}}{F_{D,ER}} = 6 \frac{A_{PA}}{A_{FA}} \frac{\mu V_D}{C_{\tau E} E}$$

This shows that the small fluid annulus thickness required to obtain a strong field and large ER effect tends to increase the viscous effects even more. This makes very thin annuli unsuitable, and limits the operation of ER dampers to high voltages to give an adequate field strength (V/m) for current ER liquids. Note that the viscosity is that with the oil heavily loaded with powder, even when field off.

The fluid annulus mean radius is a little more than the piston radius. Making these equal as an approximation gives

$$f_F \approx \frac{3}{2} \frac{D_P}{t_{FA}} \frac{\mu V_D}{C_{\tau E} E}$$

Again, this shows that a small fluid annulus thickness relative to the piston diameter, desirable to obtain a strong ER effect, makes the viscous effects relatively larger.

The capacitance of an actual ER damper may be measured easily, or calculated approximately for a proposed design. The damper annulus is equivalent to two flat plates of area equal to the annulus surface area and spacing equal to the annular clearance. The capacitance of two such plates is

$$C = \frac{\varepsilon A}{t}$$

where A is the plate area (one surface), t is the spacing, and ε is the permittivity of the intervening material. The electrical permittivity of free space (vacuum), and almost exactly that of air, is

$$\varepsilon_0 = \frac{1}{\mu_{M0}c^2} = 8.654 \times 10^{-12} \text{F/m}$$

where μ_{M0} is the magnetic permeability of free space and c is the speed of light. The permeability of free space is given a value by definition ($4\pi \times 10^{-7}$H/m), so the permittivity is a derived quantity. The relative permittivity ($\varepsilon/\varepsilon_0$) of damper oil and ER materials would probably be less than 10. Even allowing this value for ε, with practical geometrical dimensions the capacity is

$$C = \frac{2\pi \varepsilon R_{FA} L_{FA}}{t_{FA}} \approx 4\text{nF}$$

This is a small value, and although it should be considered in design of the control system it seems unlikely to be of great importance at practical ER damper operating frequencies. As mentioned in Section 9.3, the damper electrical capacity has, however, been reported to be a significant factor in the damper time constant.

A demonstration electrorheological damper is reported by Petek (1992a,b,c, 1995) with laboratory and vehicle tests. This damper included a conventional passive fluid valve in the piston allowing additional flow through the piston during damper compression, to give asymmetrical forces (more extension force than compression force) automatically, i.e. without explicit electrical control of this aspect. Figure 9.4.3 shows some laboratory results.

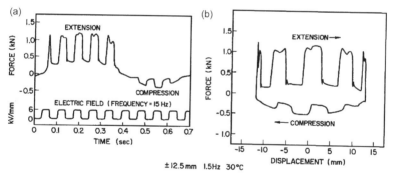

Figure 9.4.3 Laboratory tests of an ER damper with square-wave control voltage applied: (a) force vs time; (b) force vs position (Petek, 1992a,b,c).

9.5 ER Controlled Valve

A possible improvement to the basic ER damper concept, overcoming some limitations and difficulties, is to use the ER effect only to control a valve rather than to provide the whole resistance. The action is simply that the ER resistance increase when field-on causes an initial pressure drop which makes the main valve close, or at least move and change its resistance, e.g. as a needle valve, so most of the pressure drop is by ordinary fluid dynamics rather than by ER effect. This has been explored by Choi (2003), who built a demonstrator damper and tested it on a vehicle successfully. Some problems were encountered with valve stability, but the concept seems to have possibilities. Figures 9.5.1 and 9.5.2 show the approach used. This idea could also be applied to MR dampers.

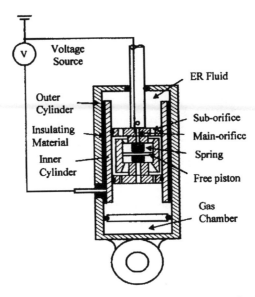

Figure 9.5.1 Configuration of a damper with an ER-controlled valve (Choi, 2003).

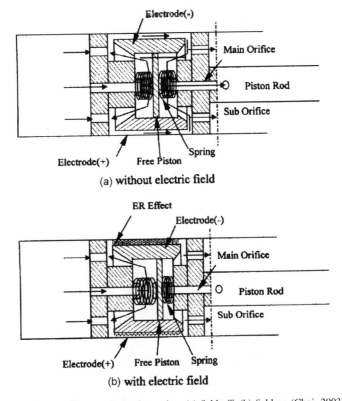

Figure 9.5.2 ER-controlled valve action: (a) field off; (b) field on (Choi, 2003).

9.6 MR Materials

A magnetorheological (MR) material is one for which the rheological properties, such as yield stress and viscosity, depend upon the magnetic field. MR liquids are sometimes described as 'low voltage' in contrast to ER liquids, but this is misleading. They are not subject to any voltage, only to a magnetic field, but the field is generated by a current of order one amp at low voltage in an field coil external to the liquid.

As with an ER liquid, the MR liquid is formed by suspending numerous small solid particles, typically a few micrometres in diameter, in a low-viscosity mineral or silicone carrier oil. The average diameter is about 8 µm with a normal range of 3–10 µm. The solid particles are ferromagnetic, basically just soft iron. Fibrous carbon may be added, and also a surfactant to minimise settling out. The result is a very dense 'dirty' grey to black oil. The shear strength achievable with magnetic field on is typically 50–100 kPa at fields of 150–250 kA/m. The magnetic activation is not sensitive to electrical conductivity, so temperature has less effect than for ER devices.

Small iron particles are used in many applications, including in the manufacture of tapes for tape recording, magnetic computer discs, and so on. The desired size is small by mechanical standards, much smaller than iron filings, so the iron powder is prepared chemically. Iron oxide may be reduced with hydrogen to form the powder directly. Alternatively, the carbonyl route may be used. The chemical name carbonyl means a molecular 'tail' with structure $-C=O$. The action of carbon monoxide on coarse iron particles at high temperature forms iron pentacarbonyl, $Fe(CO)_5$, which is a highly toxic pyrophoric pale yellow liquid with melting point $-20°C$, boiling point $103°C$ and molecular weight 195.85. This has a ring structure including one iron atom and five carbons, each carbon also having one oxygen atom. Iron tetracarbonyl $(Fe(CO)_4)_3$, a green crystalline solid, and (di-) iron nonacarbonyl $Fe_2(CO)_9$, an orange crystalline solid, also form. Sunlight is sufficient to decompose iron pentacarbonyl into iron nonacarbonyl and CO. Heating iron nonacarbonyl decomposes it to iron pentacarbonyl, carbon monoxide and iron. Heating iron pentacarbonyl gas causes decomposition back to iron and carbon monoxide. Under the right conditions, the resulting iron vapour condenses into the desired very fine spherical particles, mostly 1–45 µm in diameter, of high purity, 97% and better, with some traces of metal oxides, carbon and air, and good electromagnetic properties. The good sphericity makes for a powder with free flowing characteristics. The carbonyl process can be controlled to produce powder with a fairly tight size distribution, having a particle mean diameter of 8 µm and standard deviation of about 2 µm. The individual particles have a layered 'onion' structure and are hard on the surface, which makes them less oxidisable (more corrosion resistant) than would otherwise be the case.

The almost pure iron particles that result are sometimes called carbonyl iron, which is not a chemical term, but a very misleading name, really meaning iron-from-the-carbonyl. Other names include CIP (carbonyl iron powder) and ferronyl iron. The particles used in MR dampers are just small iron spheres, not iron pentacarbonyl.

Typical general MR liquid properties are:

Mass solids	(%)	75.0	80.0	85.0
Volume solids	(%)	23.4	28.9	36.5
Density	(kg/m^3)	2452	2844	3384
Volumet. expansion	(ppm/K)	770	720	650
Sp. thermal capacity	(J/kg K)	920	800	700

Application of the magnetic field H causes a flux density B, with an initially linear increase of Bingham solid/liquid yield stress according to the gradient

$$C_{\tau H} = \frac{d\tau_Y}{dH} \approx 0.5 \, \text{Pa} \cdot \text{m/A}$$

or, related instead to the associated flux density,

$$C_{\tau B} = \frac{d\tau_Y}{dB} \approx 80 \, \text{kPa/T}$$

(kPa per tesla). This saturates to a maximum strength in a typical asymptotic way. A low-solids MR liquid requires a greater field to bring it up to saturation, and has a lower final strength. The parameter H_1 in Table 9.6.1 is the ratio of maximum shear strength to the gradient, and so is the scale factor in an exponential equation for the strength versus field.

Table 9.6.1 MR Fluid Properties v. Solids Content

Mass solids	(%)	75.0	80.0	85.0
Magnetic properties:				
$d\tau_Y/dH$	(Pa m/A)	0.22	0.34	0.50
τ_Y max	(kPa)	40	50	60
H_1	(kA/m)	220	170	120
Fluid shear yield stress and viscosity at zero applied field:				
τ_{Y0}	(kPa)	0.015	0.020	0.025
μ_0	(mPa s)	80	100	300

At zero applied field, the particles may have some residual magnetism and so the liquid will typically have some residual yield strength, but this is very small compared with the operating strength. At zero field, the subsequent viscosity, for a given host oil, is very sensitive to the solids content.

There is a limiting saturation field and flux density, as usual with magnetic materials, typically at a flux of 0.7 T giving a yield shear stress of about 60 kPa as seen in the table. These figures are very dependent on the particular MR liquid constitution.

MR liquids may be modelled approximately as a Bingham plastic with field-dependent parameters. However, more elaborate models have also been proposed, including the Spencer model (modified Bouc–Wen) which includes some hysteretic effects (Savaresi et al., 2004). The model proposed by Gamota and Filisko (1991) for ER materials, Figure 9.3.4, may be suitable for many applications.

By way of illustration of the effect of the magnetic field, Figure 9.6.1 shows the force observed on an MR damper as a function of extension speed with field coil current as parameter. Notable here is that the current clearly creates an effective yield stress in the liquid and reduces the subsequent fluid marginal viscosity (gradient of the curves), which even goes negative.

Despite the encouraging properties of MR liquids, the introduction of MR dampers has been delayed by various difficulties, including settlement (sedimentation) and thickening. Settlement is quite slow

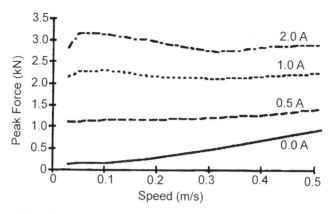

Figure 9.6.1 Controllable MR damper force versus extension speed, illustrating the effect of field on the effective Bingham parameters yield stress and marginal viscosity (Weiss *et al.*, 1993).

because of the smallness of the particles, and the sediment is soft, so remixing occurs rapidly. The settlement velocity may be estimated by Stokes' equation, which gives

$$V = \frac{(\rho_S - \rho_L)gD^2}{18\mu}$$

For 8-µm-diameter iron particles in a representative oil, this is about 6 µm/s, or 20 mm/h. With the use of appropriate additives of the kind used in engine oils, as explained by Winslow, settlement is reduced and remixing even after a long period easily occurs in a few damper strokes.

In-Use Thickening (IUT) is a gradual increase of viscosity when field-off over a long period of use. This is a problem because a low field-off damper force may be important in obtaining the full benefits of fast adaptive damping. Even a doubling of the base force may be serious, and early MR liquids showed a tripling of the force well within a normal operating life. Subsequent investigation has shown that the problem arises because of spalling of fine material from the friable surface of the iron balls when they are scraped together by the action of the field in pressing them together and of the fluid motion in driving them laterally. This surface layer is rich in iron oxides, carbides and nitrides, and so is rather brittle. The amount of material lost is very small, but this very fine nanometre-sized abraded dust is disproportionately effective in influencing the base fluid viscosity. Replacing 1% of the mass of 8 µm iron by nanometre-sized particles has been shown to triple the MR fluid viscosity. This problem has now been overcome by proprietary methods. Speculatively, softening of the iron balls, eliminating brittle failure, may help, as may coating them. The manufacturers of the 'ferronyl iron' do offer various metallurgical states and coatings. Also, it would be possible to produce iron alloys in microsphere form because other metal carbonyls are certainly available, e.g. nickel and chromium, and, presumably manganese.

It has been found that MR liquid life can be modelled well by the lifetime dissipated energy per unit volume (LDE). A normal car damper has a typical average power dissipation of 5–10 W on normal roads. A distance of 160 000 km at a mean speed of 20 m/s is 8 Ms. Using 10 W, the lifetime energy dissipation per damper is then 80 MJ. The liquid volume is typically 100 cm^3. A typical total in service LDE value for a car damper is therefore 80 MJ / 100 cm^3, 0.8 MJ/cm^3, approaching 1 TJ/m^3. Obviously this will vary considerably with the particular case, but this is within the practical life of current MR fluids, which have been demonstrated to be good up to more than 2 TJ/m^3.

MR fluids have their hazards. The iron penatcarbonyl is pyrophoric (igniting on contact with air) and very toxic, but this would only be of concern to the original material manufacturer. The iron powder product is nontoxic, and even used for inclusion in multi-mineral pills for human and animal consumption. The dry iron powder is combustible (oxidisable) with an ignition temperature around 420°C, and like any finely divided powder is a combustion hazard. This is because of the very large surface area which allows a rapid reaction if ignition occurs.

One iron sphere of diameter 8 μm has a volume of 268×10^{-18} m^3 and a surface area of 201×10^{-12} m^2. One cubic centimetre of material contains 4×10^9 balls with surface area 8042 cm^2. Fortunately, once the iron is mixed with an oil it is no longer freely exposed to air or oxygen, so the combustion hazard is effectively absent in service conditions.

9.7 MR Dampers

An MR (magnetorheological) material is fundamentally different in practice from an ER material, because the realistically achievable shear stress is much higher for MR, e.g. 50 kPa, more than one order of magnitude better than ER. As a result, the design of an MR damper can be more conventional, with the valve in the piston, although the piston size is increased somewhat from a normal damper. Figure 9.7.1 shows a basic design for a single-tube variant; double tube designs are also possible.

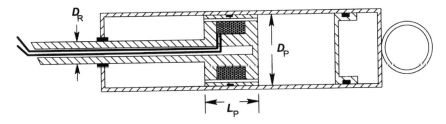

Figure 9.7.1 A basic design for a single-tube MR damper.

The magnetic field is generated by the axial coil, for which connecting leads are conveniently brought out through the rod.

The product of the current I and the number of turns N_T (all of the same handedness) is $M = N_T I$, the magnetomotive 'force' or magnetic potential. The field strength that this will produce in the liquid depends on the entire magnetic circuit. This should have a low reluctance, so soft iron or steel is used for the piston. For best results a low-carbon steel with high permeability and high flux saturation level is desirable. For an air-gap electromagnet, the gap has high resistance relative to the iron (the material permeabilities being in a ratio of 1000 or more), so a gap of any significance dominates the circuit resistance. For an MR liquid, the permeability may be quite high (it may be 80% iron), and a careful magnetic circuit design is necessary. At high flux density, the iron may saturate, and be a limiting factor, so the cross-section of the iron must be adequate all around the magnetic circuit. The total flux in the circuit is the same at all sections around the circuit, so the critical point of the iron is the section of least cross-sectional area.

The magnetic circuit of the example design is axisymmetric, Figure 9.7.2. The flux passes through the coil axially, expands radially outward through the disc at one end, through one MR fluid gap, back along the iron sleeve, radially inward through the disc at the other end and back into the core completing the circuit. Design of a magnetic circuit is more complex than for an electric current circuit because of the nonlinear behaviour of the materials, and nowadays is likely to be done using a suitable software package. However, a linear model analysis, as will be done here, is useful for preliminary design and gives useful understanding and an appreciation of the units and numerical values.

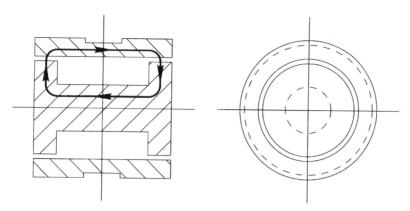

Figure 9.7.2 Magnetic circuit in an MR damper piston.

A magnetic circuit is analogous to an electric circuit. In the latter case, an electric potential around the circuit (from, e.g., a battery) measured in volts (V) produces a current in amps (A). According to the local cross-section of the wire, there is a local current density (A/m²). Locally, the current density is related to the electric field strength in V/m (volts per metre) by the electrical conductivity (A/V m). In most circuits the analysis is greatly simplified by using lumped parameters such as a resistor with a specified resistance, voltage and current, rather than dealing with current densities and fields.

The magnetic circuit has a magnetic potential (e.g. from a coil with a current) measured in amp-turns (units just A). The result is a magnetic flux around the circuit measured in webers (Wb). According to the local cross-sectional area of the circuit there is a local magnetic flux density measured in tesla (T, which is just Wb/m²). The local flux density is related to the local magnetic field strength (A/m) by the permeability (Wb/A m, the same as henry per metre, H/m, 1 H = 1 Wb/A). The magnetic circuit is most easily understood in terms of its physical parts, each of which has a magnetic reluctance, analogous to electrical resistance. Reluctances add in series and parallel in the same way as do resistances.

The permeability of vacuum is, by definition,

$$\mu_{M0} = 4\pi \times 10^{-7} \text{ H/m}$$

For air, the value is effectively the same. Ferromagnetic materials have a high value, possibly several thousand times as large. The ratio of the absolute permeability to the permeability of a vacuum is called the relative permeability. The relative permeability of MR fluids is typically around 5 or 6. The high iron content can be imagined to almost short out the reluctance, which depends mainly on the oil gaps between the iron spheres, about 1/6 of the total distance.

Figure 9.7.3(a) shows a curve of flux density against magnetic field strength $B(H)$ for a steel such as might be used in an MR damper piston. Here, the considerable nonlinearity is apparent. Also, there is saturation. There is a flux density limit, so a field beyond 2 kA/m has negligible effect. This fairly realistic curve has been generated by the expression

$$B = B_1\{1 - e^{-H/H_1}\}^p$$

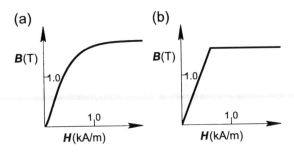

Figure 9.7.3 Magnetic response of a damper piston steel: (a) real; (b) linear model with saturation.

where B_1 is the asymptotic value, H_1 scales the magnetic field axis, and the index p shapes the part near to the origin, with $B_1 = 1.7$ T, $H_1 = 300$ A/m and $p = 1.7$. This equation inverts to

$$H = -H_1 \log_e \left\{ 1 - \left(\frac{B}{B_1}\right)^{1/p} \right\}$$

Figure 9.7.2(b) shows a simple linear model, in which the consequent flux is proportional to the field up to a sudden saturation. Before saturation, $B = \mu_M H$ with constant μ_M, the material absolute permeability, approximated as linear.

The MR fluid itself is characterised magnetically by a fluid shear yield stress related to the magnetic field or flux density. This is modelled as a coefficient, such that

$$\tau_Y = C_{\tau H} H$$
$$\tau_Y = C_{\tau B} B$$

Either may be used, but obviously not the two cumulatively. Within the linear model, $B = \mu_M H$, so the coefficients are related by

$$C_{\tau H} = \mu_M C_{\tau B}$$

The coefficients have a value of around 0.6 Pa m/A and 80 kPa/T respectively. The liquid has a magnetic flux response curve $B(H)$ somewhat similar to that of iron or steel, but with a less defined saturation point. It is modelled here as linear, with a constant permeability ($B(H)$ gradient) having a relative permeability of 6, and a saturation flux density of 0.7 T.

Given the piston geometry and the above linear parameters for the materials, below saturation the reluctance of the various parts around the magnetic circuit can be calculated. This then gives the total reluctance. Assuming a coil current, the coil amp-turns divided by the total magnetic circuit reluctance then gives the circuit flux (Wb). The flux densities can then be calculated. The flux density in the MR fluid then gives the MR yield stress, the consequent damper force, and the damper MR coefficient $C_M = F_D/I$. By linearity, the limit (saturation) current may also be determined, giving the consequent saturation limit shear stress and limit MR yield stress damper force.

Table 9.7.1 gives a computer printout for a basic analysis of an MR damper, to illustrate the calculations and the numerical values. The calculation procedure is as follows. The basic geometry is analysed to obtain the basic lengths and areas. The viscous annulus is assumed to be the full length of the piston, with an effective magnetic annulus only at the radial magnetic gaps. If the coil

Table 9.7.1 MR damper analysis

Specification		
Piston geometry		
Rod diameter	=	12.000 mm
Piston diameter	=	36.000 mm
Fluid annulus R	=	11.500 mm
Fluid annulus t	=	1.000 mm
Viscous annulus L	=	40.000 mm
Magnetic annulus L	=	20.000 mm
Piston iron/steel		
Relative permeability	=	2000.000
Flux den saturation	=	1.700 T
MR fluid		
Density	=	2000.000 kg/m^3
Viscosity	=	200.000 mPa s
MR coeff $C_{\tau B}$	=	80.000 kPa/T
MR coeff $C_{\tau H}$	=	0.603 Pa m/A
Rel. permeability	=	6.000
Flux den saturation	=	0.700 T
Coil		
Number of turns	=	250.000
Operating conditions		
Damper velocity	=	0.500 m/s
Coil current	=	0.500 A

Analysis		
Piston radii		
Coil core	=	8.000 mm
Fluid annulus inner	=	11.000 mm
Fluid annulus outer	=	12.000 mm
Sleeve OD at seal	=	16.000 mm
Sleeve OD main	—	18.000 mm
Areas		
Piston annulus	=	904.779 mm^2
Fluid annulus	=	72.257 mm^2
Area ratio	=	12.522 —

Newtonian viscous analysis		
Volumetric flowrate	=	0.452 L/s
Fluid velocity	=	6.261 m/s
Hydraulic diameter	=	2.000 mm
Reynolds number	=	0.125 k
Viscous P drop	=	0.601 MPa
Damper force	=	0.544 kN
Damper viscous CD	=	1.088 kNs/m

Magnetorheological analysis		
Magnetic circuit		
A1 (core)	=	2.011 cm^2
A2 (inner radial)	=	5.027 cm^2
A3 (outer radial)	=	6.912 cm^2
A4 (sleeve at seal)	=	3.519 cm^2

Table 9.7.1 *(Continued.)*

A5 (sleeve main)	=	5.655 cm²
Iron A minimum	=	2.011 cm²
A fluid (each end)	=	7.226 cm²
L/A iron	=	223.048 m⁻¹
L/A fluid	=	2.768 m⁻¹
Reluctance iron	=	0.089 MA/Wb
Reluctance fluid	=	0.367 MA/Wb
Reluctance total	=	0.456 MA/Wb
Coil		
Number of turns	=	250.000
Coil C-S area	=	0.600 cm²
Wire length in coil	=	14.923 m
Wire diameter	=	0.490 mm
Resistance	=	1.346 Ω
Inductance	=	0.137 H
Satn flux den iron	=	1.700 T
Satn flux den fluid	=	0.700 T
Magnetic flux	=	0.274 mWb
Flux den iron max	=	1.364 T
Flux den fluid	=	0.379 T
MR yield stress	=	0.030 MPa
Fluid annulus force	=	87.747 N
MR pressure drop	=	1.214 MPa
MR damper force	=	1.099 kN
Damper coeff CM	=	2.198 kN/A
Iron flux saturation is limiting factor.		
Limiting current	=	0.623 A
Damper force (Isat)	=	1.370 kN

can be kept to a slightly reduced outer radius then over this section the fluid annulus thickness will be greater and the viscous pressure drop would be reduced advantageously. A Newtonian viscous analysis is next performed to obtain a viscous damper force and a damper current-off viscous damping coefficient. This is a similar calculation to that for an ER damper, and reduces to a single expression for C_D.

The magnetic circuit must then be analysed to relate the applied current to a magnetic field across or flux density in the MR fluid. The reluctances around the circuit must be added (or integrated). The reluctance of a simple constant-section component (technically, literally a cylinder, circular or otherwise) is just

$$R_M = \frac{L}{\mu_M A}$$

Within the linear model, for a given material the permeability is constant, so the L/A values may be added first. Table 9.7.2 shows approximate values for the iron circuit. Some judgement is required here about the shape of the flux lines, but this is not too critical in a preliminary design.

For the fluid, there are two sections in series. The length of each is the radial gap of the annulus. The area is the $2\pi RL$ value appropriate to the annulus, with a length of one disc.

Table 9.7.2 Reluctance of the iron magnetic circuit L/A values

Section	L (mm)	A (cm^2)	L/A (m^{-1})
1 core	20	2.0	100
2 radial	5	5.0	10
3 sleeve 1	20	5.6	36
4 at seal	10	3.5	29
5 as 3	20	5.6	36
6 as 2	5	5.0	10
Total			221

The material permeabilities are the relative permeabilities times the permeability of free space (vacuum), $\mu_{M0} = 4\pi \times 10^{-7}$ H/m. The reluctances are the $L/\mu_M A$ values. This gives the total magnetic reluctance R_M of the circuit. The units of reluctance are A/Wb (ampères per weber), that is the current-turns required per weber of magnetic flux produced.

An approximate analysis of the coil may also be made. It has a specified number of turns. The coil sectional space available follows from the geometry — the coil competes for space with the magnetic core and the length of the end discs. The coil space divided by the number of turns gives the maximum wire sectional area and an estimate of possible wire diameter. This is desirably large for low resistance and high currents, but competes with the number of turns. The number of turns times the mean radius gives the wire length in the coil.

With the wire resistivity of about 1.7×10^{-8} Ω m this gives a coil resistance, in this case about 1.4 Ω. The coil magnetic inductance H_M is given by

$$H_M = \frac{N_T^2}{R_M}$$

where N_T is the number of turns, with a value of about 150 mH (millihenry, or mWb/A). To produce a steady current of 0.5 A requires only about 1 V, but rapid changes of current, as required for fast control, may demand tens of volts to overcome the inductance.

The total magnetic flux is known, as are the various circuit sectional areas, so the flux densities at all sections may now be calculated. The circuit is limited by the first point of saturation, which could occur at the MR liquid or in the iron. The critical section of the piston iron is naturally at the minimum cross-sectional area, in this case the coil core. From the flux density in the fluid the MR yield stress may be calculated, from which follow the various pressures and the MR damper force and the MR damper coefficient

$$C_M = \frac{F_{D,MR}}{I}$$

in N/A (newtons per ampère).

Evaluation of the damper MR coefficient C_M may be condensed down into a simple expression, given that the geometry remains appropriate and the total magnetic reluctance has been established:

$$C_M = \frac{4\pi A_{PA} R_{FA} L_{MFA} C_{\tau B} N_T}{A_{FA} A_{Mf} R_M}$$

where A_{Mf} is the area of magnetic flux through the fluid (at each end), 7.226 cm² in this example. This expression may be further simplified to

$$C_M = 4\frac{A_{PA}}{A_{FA}}\frac{N_T C_{\tau B}}{R_M}$$

The main complication, then, in a linear analysis, is in the evaluation of the magnetic reluctance and the limiting coil current for saturation at some section of the circuit.

Simplistically, then, having established C_D and C_M, the total damper force is given by

$$F_D = C_D V_D + C_M I$$

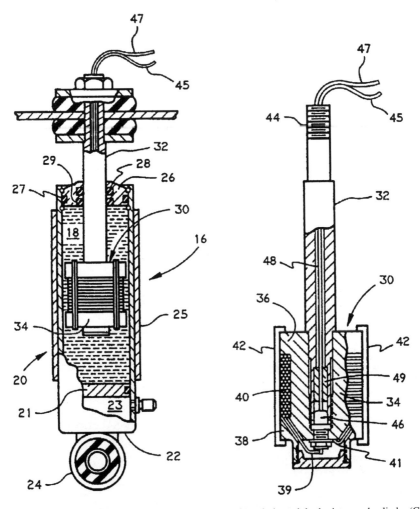

Figure 9.7.4 Proposed design of MR damper with a magnetic path through both piston and cylinder (Carlson and Chrzan; 1994a). J.D. Carlson and M.J. Chrzan (1994) US Patent 5,277,281.

but with further complications in practice. Really the flow in the magnetic part of the annulus should be solved as a Bingham flow as for the ER damper. Also, there is the variation in marginal viscosity with magnetic field to consider.

Comparing the greatest flux density obtained with the saturation flux density, the current to produce the saturation limit is easily deduced (with linearity). The MR damper force at saturation then also follows easily. This can, of course, be attacked directly.

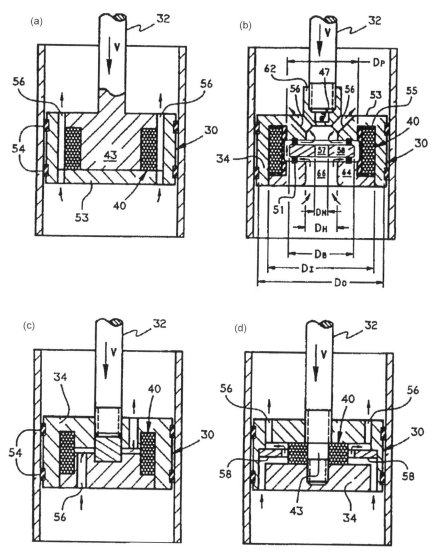

Figure 9.7.5 Proposed designs of MR damper piston with various fluid paths and magnetic configurations (Carlson and Chrzan, 1994a).

For high speeds, the fluid exit energy becomes significant. Then an extra term in V^2 should be included, using the quadratic damping coefficient

$$C_Q = \tfrac{1}{2}\rho \alpha f_A^2 A_{PA}$$

which is about 280 N s^2/m^2 in this example. This could be reduced by good detail design.

It will be appreciated that the design is an example only. Geometric optimisation of the piston is a tricky problem, even within a linear model. To obtain a good MR damper performance, the piston needs to be longer and of larger diameter than a conventional damper piston to provide the coil space and magnetic path sections, and poses an interesting nonlinear packaging problem with many constraints. Many other configurations are possible. For example, Figure 9.7.4 shows a design in which the magnetic path passes through the piston and cylinder, the piston having nonmagnetic guide strips to locate it. Figure 9.7.5 shows various proposed designs of piston.

Figure 9.7.6 shows a model developed to represent complete MR dampers with fair accuracy. The top spring is for the gas pressure variation with rod insertion volume and is not ER/MR related. This may be compared with the ER/MR material model of Figure 9.3.4.

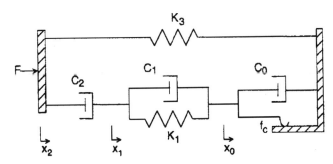

Figure 9.7.6 Model of complete MR damper. Reproduced with permission from Maranville and Ginder (2003) *Proc. of SPIE*, Vol. 5056, pp.524–533.

10
Specifying a Damper

10.1 Introduction

The full specification of a damper can be immensely complex, covering all the dimensional data, plus solid material specifications, manufacturing methods, liquid specifications, gas pressurisation, and performance specifications with tolerances. However, for a normal damper many of these are fairly standard and may be taken for granted. Essentially, the damper must be connected to the vehicle and exert the desired forces. Hence the primary specification features may be considered to be:

(1) end fitting design;
(2) length range;
(3) $F(V)$ curve.

This is a functional specification. That specification may be achieved in a variety of ways, but to guarantee the performance over a range of conditions, the method of achieving it may also be specified. Hence the specification may well include:

(4) configuration;
(5) diameter;
(6) oil properties.

The question of the life of a damper, that is wear rate and maintenance of an acceptable $F(V)$ over a long period of use, is a difficult one. Durability tests, both bench and field, will normally be required. To help to achieve durability, the specification may include information on:

(7) seals;
(8) surface finish;
(9) corrosion resistance.

According to the particular case, any other technical details may be added, as required. Finally, last and by no means least, there is

(10) cost.

The Shock Absorber Handbook/Second Edition   John C. Dixon
© 2007 John Wiley & Sons, Ltd

10.2 End Fittings

The end fittings are dictated by the vehicle. Usually the lower end of the damper will use a transverse eye with rubber bush fixed onto a stud protruding from a suspension arm. The actual specification will therefore be the type of fitting and the dimensions, i.e. tube material, inner diameter to accept the bush, wall thickness as length (usually equal to the bush length). In more detail there will be a tolerance on the accuracy of the end tube position, plus a method of attachment, e.g. welding, and a minimum strength. The bush itself needs to be specified too, in particular the dimensions, but also the stiffness of the rubber. In practice, this will frequently be done indirectly by indicating a standard part number.

If the upper end is connected directly into the top of the wheel arch, then the protruding rod diameter, length and thread must be stated, plus the size and hardness of the rubber pads and the supporting metal plates.

10.3 Length Range

The range of suspension motion is normally limited by separate bump stops and droop stops, so the maximum spacing of the damper mounting points is known. It is essential that the damper should be able to span this entire range, in order to prevent damage to the damper or unpredictable handling because of improper restriction of suspension motion. Hence the damper has:

(1) a minimum maximum length, i.e. a minimum fully extended length;
(2) a maximum minimum length, i.e. a maximum fully compressed length.

The actual points between which the lengths are measured must, of course, be clear. In the case of transverse eyes then the eye centres will probably be used; this must not be confused with the overall length. With an axial end fixing rod, the measuring point must also be defined.

Obviously the damper stroke must normally exceed the full range of relative motion of the connection points. However, this alone is not sufficient. The exception to the above is when the damper is intended to act as the bump or droop stop, and designed appropriately to do so, often with the incorporation of bump and droop rubbers.

10.4 $F(V)$ Curve

The $F(V)$ curve, over the appropriate range of compression and extension velocities, is the essence of the damper specification, usually expressed as forces at discrete velocities. In practice there must be some tolerance to allow for manufacturing variation, the width of which will depend upon the quality of the damper. The tolerance is not well specified by only a force or only a percentage tolerance, so a combination of these may be given, e.g. ± 20 N $\pm 10\%$ of nominal force. Alternatively, maximum and minimum acceptable forces may be presented in graphical form. Tolerances will also be placed on the gas force, and on the effect of temperature

As described earlier, the actual desired force curve for the damper must take into account the desired damping for the vehicle and the effect of the installation motion ratio.

10.5 Configuration

According to the application, some particular configuration of damper will generally be preferred, and required. Hence the damper will be required to be a single-tube or double-tube type, possibly with a floating gas-separator piston, possibly with a remote reservoir and so on.

10.6 Diameter

The required forces may be achievable by a high pressure on a small piston. However such a design will be very sensitive to leaks, and hence to wear. Also a small diameter will have poorer cooling. Therefore a minimum or actual nominal diameter for the piston may be specified, and also perhaps the rod diameter and, for a double-tube type, the diameter of the outer tube.

10.7 Oil Properties

According to the particular application, the oil type, and viscosity and density at standard temperature, may be specified. Usually a standard damper mineral oil will simply be defined by a manufacturer's reference number. For more difficult cases, a low-viscosity-index synthetic oil may be required, but again indicated by a manufacturer and oil type number.

10.8 Life

The useful life of a damper is difficult to predict because of variations in conditions of use, i.e. of driving styles and local road roughness, and is difficult to test because the life is usually measured in tens of thousands of kilometres on normal roads. The useful life is normally limited by leakage due to rod seal wear or piston seal wear. Hence the life is enhanced by careful choice of seal design and materials with a very good finish on the hard rubbing surfaces of rod and cylinder, all of which need to be specified. A large piston diameter, giving larger liquid displacement volumes and lower operating pressures, improves the tolerance to leakage.

10.9 Cost

The importance of the manufacturing cost of a damper may seem to be too obvious to require discussion. However, the significance of the price will vary considerably with the application. For a high-grade racing car or rally car the accuracy and predictable of the $F(V)$ curve, and the life reliability, may be so important that a high price is not problematic. On the other hand, for an economy passenger car, which is price sensitive, less critical on exact damping level and produced in large quantities, the dampers will have to be a much lower price.

Of course, the cost is not so much a part of the technical specification as the result of it. The technical specification must not be higher than is appropriate to the vehicle, or the price will be adversely influenced.

11
Testing

> "... passing from wild eagerness to stony despair..."
> John Galsworthy, *The Forsyte Saga*.

11.1 Introduction

Testing of dampers may be categorised under three main headings:

(1) rig testing of part or whole of the damper;
(2) road testing of the damper on the vehicle;
(3) vehicle annual safety certification.

Rig testing of complete dampers or their separate parts may be placed under three further headings:

(1) to measure performance;
(2) to check durability;
(3) to test theoretical models.

Testing of theory is required to validate methods of analysis and to give confidence in theory for design work. This is likely to involve testing of individual parts, of complete valves in a steady-flow circuit (described in Chapter 6), or testing of complete dampers to relate damper characteristics to valve characteristics, to investigate piston or rod seal friction effects, etc.

Performance testing is required to check that prototypes or samples of production dampers meet their specifications within tolerance, and are adequately consistent one to another. In competition, performance testing is required to check that a given valve set-up gives the expected behaviour and, again, that dampers are consistent and in matched pairs. Consistency tests and matching tests are frequently disappointing because of the sensitivity of the dampers to small dimensional discrepancies in the valves and to small leakage paths. Adjustables are frequently inconsistent one to another in their response to the adjustment setting. Testing may therefore be used to select matched pairs or to refine manufacture and assembly to the necessary level.

When left standing, a normal double-tube damper will accumulate air in the extension chamber. The damper can be restored to its correctly charged state by stroking it several times—a process called purging. This forms a basic test: if it takes more than three strokes to purge, this is a definite sign of a fault. Incomplete purging is indicated by a considerable reduction in extension force near to full stroke.

The Shock Absorber Handbook/Second Edition   John C. Dixon
© 2007 John Wiley & Sons, Ltd

Durability testing is sometimes performed by rig testing, and this can be useful for initial testing of new materials or production methods, but the primary durability testing is by road testing.

Road testing may be divided into four main categories:

(1) long-distance testing of durability on public roads;
(2) short-distance durability testing on severe test roads;
(3) ride and handling testing on public roads;
(4) ride and handling testing on special test roads.

Long-distance road testing of dampers alone would generally be uneconomic, but is undertaken in conjunction with reliability testing of all the other parts of complete vehicles. Short-distance severe testing of complete vehicles is sometimes used, driving over pavé type surfaces or similar. Because of the large amplitudes of suspension motion, this type of test is very severe on the dampers which may fail by low cycle fatigue fracture of the mountings, overheating of the seals, oil vaporisation etc. with consequent loss of damping effect.

Testing of handling is mainly undertaken on special circuits; for safety reasons, public roads are not generally suitable for extreme cornering testing. Ride testing is of course viable on public roads, but special roads with particular surface conditions obviously offer some advantages. Testing of the complete vehicle may be intended to assess the suitability of proposed dampers for a particular vehicle, or to relate actual vehicle behaviour in ride and handling to theoretical predictions in order to validate vehicle dynamics theory for design purposes.

Regular safety certification of vehicles is required in many countries. Damper tests tend to be quite cursory, but are, of course, much better than no test at all. Typically the vehicle is simply depressed firmly by hand on each corner in turn; after release the subsequent motion should be free of multiple oscillations. This is essentially a visual test of the natural vibration amplitude ratio which is closely related to the vehicle damping ratio. The test may also be applied by suddenly stopping a descending four-post hoist. Dampers and struts are also inspected for signs of fluid leakage, which is predictive of damper failure due to extensive fluid loss in due course.

It seems that the best that a damper can achieve is to be within specification and not faulty. The possibilities for being at fault seem endless. Often, the result of testing even reputable dampers is badly shaped curves and failures of consistency. The lesson, though, is clear enough. Dampers must be carefully designed by a knowledgeable engineer and must be carefully manufactured if performance, consistency and life are to be satisfactory. The process of testing ruthlessly reveals any shortcomings, but, painful as it may be, it is much better to discover these problems before the dampers go on the vehicle.

11.2 Transient Testing

Laboratory testing of dampers has a long history. The earliest tests were of transient free vibration. In 1929 Weaver used a full-scale physical heave-only quarter-car model with a massive concrete block on a suspension spring. By operating this in free vibration he was able to subject a damper to a realistic cycle of transient operation. The instrumentation was a mechanical X–Y plotter that gave $F(X)$ loops (called card diagrams in these days), the shape of which give a good indication of damping ratio, and also reveal, to the experienced eye at least, certain damper faults (see for example James and Ullery, 1932).

The advantage of such free vibration tests is that they are relatively cheap and easy to perform, and do not require equipment to continuously cycle the damper. Also, the single-shot test causes virtually no heating-up of the damper. The disadvantages are that the basic results are less clear cut than for a repetitive sinusoidal motion test, and that the velocity is limited because the maximum initial deflection is restricted by the damper stroke. For extension, of course, this velocity limit is likely to correspond quite well with real maximum velocities experienced in service.

For such a test, the undamped peak velocity V_M is related to the initial displacement X_0 from equilibrium by

$$\tfrac{1}{2}kX_0^2 = \tfrac{1}{2}mV_M^2$$

$$V_M = X_0\sqrt{\frac{k}{m}} = \omega_N X_0 = 2\pi f_N X_0$$

For example, at an initial deflection of 100 mm from equilibrium, the undamped peak velocity V_M will be about 0.6 m/s.

It is, however, not possible by this method to apply bump velocities that might be met in practice from driving off kerb edges or striking kerbs.

With modern data acquisition and analysis methods, the transient test data could be used to produce an $F(V)$ plot rather than just the basic raw $F(X)$ plot, although with a fairly high damping ratio it would be necessary to perform two tests, one for compression and one for extension.

Figures 11.2.1–11.2.8 show transient motion graphs of $X(t)$, $V(t)$, $F_D(X)$ and $F_D(V)$ for a variety of damper and frictional forces. These were produced by time-stepping simulation. The graph of damper force against position is not in general a causal relationship of course, it is just the relationship that occurs in the particular motion. The $F_D(V)$ curves are limited to the values actually occurring in the simulated transient test.

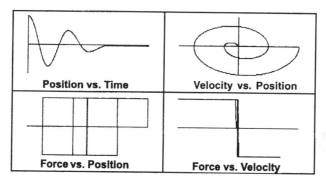

Figure 11.2.1 Natural vibration of spring + mass + damper from $X = X_0, V = 0$. Friction type is Coulomb.

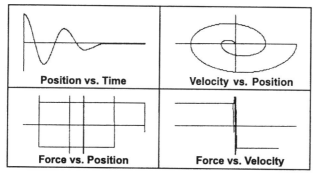

Figure 11.2.2 Natural vibration of spring + mass + damper from $X = X_0, V = 0$. Friction type is static + dynamic.

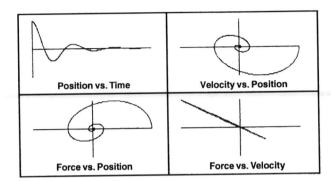

Figure 11.2.3 Natural vibration of spring + mass + damper from $X = X_0, V = 0$. Friction type is linear symmetrical.

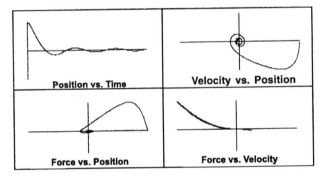

Figure 11.2.4 Natural vibration of spring + mass + damper from $X = X_0, V = 0$. Friction type is quadratic symmetrical.

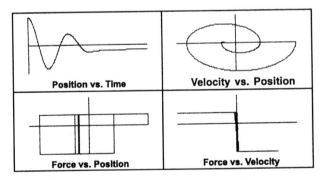

Figure 11.2.5 Natural vibration of spring + mass + damper from $X = X_0, V = 0$. Friction type is asymmetrical Coulomb (fluid blow-off).

Testing

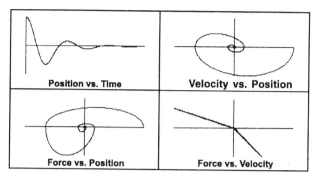

Figure 11.2.6 Natural vibration of spring + mass + damper from $X = X_0, V = 0$. Friction type is asymmetrical linear (bilinear).

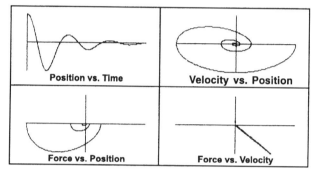

Figure 11.2.7 Natural vibration of spring + mass + damper from $X = X_0, V = 0$. Friction type is unidirectional linear.

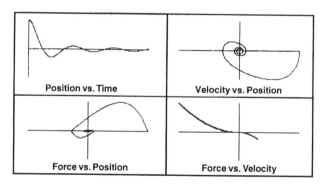

Figure 11.2.8 Natural vibration of spring + mass + damper from $X = X_0, V = 0$. Friction type is asymmetrical quadratic.

The $F(X)$ curves in these diagrams are curves that are obtained in transient motion. Obviously these are not as easily interpreted as the $F(X)$ curves occurring in sinusoidal testing (Chapter 7, Figure 7.12.1 et seq.). The transient $F(X)$ curve is most easily understood where the damping ratio ζ of the system is low, i.e. where the motion is nearest to sinusoidal. To achieve this, the test system requires increased mass and stiffness. Noting that

$$\omega_N = \sqrt{\frac{k}{m}}$$

$$\zeta = \frac{1}{2\sqrt{km}}$$

by maintaining the ratio k/m constant, then the damping ratio will be reduced whilst the natural frequency, and the peak velocities reached for a given initial displacement, change relatively little. Increasing stiffness k alone will also increase the speeds achieved. Hence the results of the test may be broadened by moving away from the mass and stiffness values taken directly from the vehicle.

Transient testing performed in this way was of value in the early days of motoring, but has nowadays been largely superseded because equipment is now readily available, at some cost, to perform continuous cyclical testing. However transient testing is still relevant and may become common again, albeit with modern techniques and equipment.

11.3 Electromechanical Testers

Measurements under steady-state conditions are generally easier to make than those in transient conditions. However the complete damper cannot really be subject to a true steady-state test. The nearest that is achievable is a triangular waveform, in which the damper is moved at constant speed for a limited period, which must result in a displacement not exceeding the maximum stroke of the damper, followed by constant velocity back again.

However a triangular waveform is not easy to apply, requiring a hydraulic actuator and suitable control equipment. Early cyclic tests were achieved by reciprocating the damper in a roughly sinusoidal manner, with a slider-crank mechanism using a connecting rod, Figures 11.3.1 and 11.3.2. The force was measured by elastic deflection of a transverse beam holding the other end of

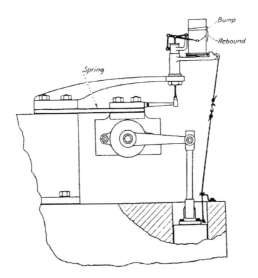

Figure 11.3.1 Early tester shown with a rotary vane damper (James and Ullery, 1932).

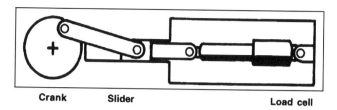

Figure 11.3.2 Reciprocating tester.

the damper. By including a substantial flywheel, the electric motor used to drive such apparatus need produce only the mean power requirement, although there will be some crank speed variation.

The inclination of the connecting rod introduces a substantial harmonic into the damper motion, which is therefore quite significantly nonsinusoidal for practical connecting rod lengths. This can be eliminated by using a Scotch Yoke mechanism which gives a true sinusoid, Figure 11.3.3.

In a cyclical test, the frequency or amplitude is varied to give a desired peak velocity, using

$$X = X_0 \sin \omega t$$
$$V = \omega X_0 \cos \omega t$$
$$A = -\omega^2 X_0 \sin \omega t$$

with the velocity amplitude (i.e. amplitude of the sinusoidal velocity graph)

$$V_0 = \omega X_0 = 2\pi f X_0$$

With the electromechanical drives described it is usual to adjust the frequency either by use of a variable speed DC motor or by a variable ratio gearbox. Variation of stroke may be possible by disassembly of the apparatus, so the stroke is set to give the desired maximum speed, within the limits of the damper and test apparatus.

With electrically driven testers there will usually be some variation of the crank angular velocity, because it is impractical to provide a very large flywheel.

For a linear damping coefficient C_D, at a stroke S and amplitude $X_0 = S/2$, and frequency f, the peak sinusoidal speed is the velocity amplitude

$$V_0 = \omega X_0 = 2\pi f X_0$$

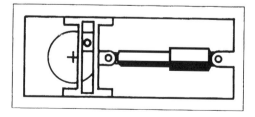

Figure 11.3.3 Scotch Yoke drive.

The peak power dissipation is $C_D V_0^2$. For sinusoidal motion, the mean power dissipation is one-half of that. The energy dissipation E_C in one cycle is

$$E_C = \frac{C_D V_0^2}{2f} = 2\pi^2 C_D f X_0^2$$

The maximum energy deviation from the mean power dissipation is $\frac{1}{4}E_C$. With a constant power input, allowing a 2% cyclical speed variation may require a rotational inertia of 0.05 kg m^2.

Alternatively, or in addition, it is possible to use a motor with suitable electronic control to maintain correct speed during the cycle, provided that the motor is capable of delivering the peak power requirement.

Because of the limitations described above, electromechanical testers are usually limited to small low-powered units. These are suitable for limited testing and low-speed comparative work, including matching at low speeds. For larger testers it is usually preferred to use hydraulic drive.

11.4 Hydraulic Testers

Where high power inputs and flexible control is required, hydraulically driven testers are favoured. The hydraulic ram is double acting, typically operating at a pressure around 1 MPa with a force capability of 10 kN. Very-high-quality valves are required for the ram, to regulate the oil supply accurately with a sophisticated control system and a large pump, making for an expensive system.

The ram position is controlled by a voltage input. There will be provision for an adjustable constant voltage which will set a steady position anywhere in the range of the ram. In addition, there is a signal input allowing variation of the position relative to the mean. This variable signal may take any form required. In practice the system control unit will provide some or all of the following:

(1) sinusoidal wave;
(2) triangular wave;
(3) square wave;
(4) random motion;
(5) external input.

For the cyclical motions, the frequency and amplitude can be set as required. The random signal is white noise over a limited frequency range. The upper frequency limit of response is effectively set by the frequency response of the valves and ram. Because of the flow rate limitations of the valves, the frequency response will depend upon the amplitude.

Hydraulic testers can have their own problems, of course, including signal processing limitations and imperfect motion control.

Such a system can be used for testing dampers or for testing the complete suspension for a quarter-car including wheel and tyre, with the ram acting as the road profile, determined by a known road profile or by random generation with an appropriate frequency distribution.

Four such systems can be used to support a complete car through the four tyres, and give a complete ride stimulus with the semi-random motions of the four rams given values correctly related to each other, requiring appropriate frequency-dependent correlation between left and right tracks and the rear stimulus being the same as the front, but with a time delay given by the vehicle wheelbase divided by the notional vehicle speed.

The ram may alternatively be controlled to give a specified force rather than position, and again this has constant and variable elements.

In practice, the sinusoidal activation is the one mainly used for damper testing. It is also possible to use an hydraulic tester to perform transient tests or single-cycle tests instead, to reduce damper heating.

11.5 Instrumentation

The basic parameters to be measured may include instantaneous values of:

(1) position X;
(2) velocity V;
(3) acceleration A;
(4) force F;
(5) pressure P;
(6) temperature T.

These need sensors, plus data processing and suitable display. In addition the data stream will be processed to give items such as cyclic extremes of position and force.

The raw data, except for temperature, change through the cycle, and therefore cannot be displayed effectively by analogue or digital meters. Therefore this form of data is usually presented on an oscilloscope, showing variation with time, or, more usefully, as a loop such as force against position, $F(X)$. A storage type of oscilloscope is much superior in showing the cycle shape. With digital data acquisition, it can of course be displayed on the computer VDU.

The derived values are approximately steady state and can be displayed on digital meters (although, even at constant speed and amplitude, the values creep as the temperature changes).

The position sensor will be a potentiometer slider or, better, an LVDT (linear variable differential transformer) built into the ram. Either of these will give a voltage output directly and linearly related to the position, with a position signal voltage coefficient k_{vx} of perhaps 100 V/m (0.1 V/mm).

The velocity may be derived from a velocity sensor, but more likely will be obtained by electronic differentiation of the position signal. Acceleration can be obtained by differentiation of the velocity signal, but two stages of differentiation will greatly exaggerate any noise in the original position signal. Therefore, an accelerometer sensor is likely to be used, which is basically just a load cell with a known mass, with acceleration derived from $A = F/m$. Also, velocity could be integrated from the acceleration.

The damper force is measured by some form of load cell. For an electromechanical tester this may simply be a slightly flexible beam supporting the static end of the damper, giving a small deflection which can be measured to give an instantaneous force signal. In the early testers the motion was amplified mechanically to produce the 'card diagram'. On a hydraulic tester, a more elaborate and expensive load cell will be used, but this is no different in principle. The force creates a small deflection in the sensing element, the deflection being detected by, for example, a strain gauge, amplified to produce a force signal voltage coefficient k_{Vf} of perhaps 1 mV/N.

Measurements of damper liquid pressure require the installation of suitable pressure sensors into the damper body. Normally this will require the welding or brazing of a tapped boss to accept a standard sensor. These are preferably positioned at the extreme ends of the damper so that the piston seals are not damaged by passing over the hole and so that welding distortion of the working tube does not cause leakage. It may be wise to use several pressure tappings to guard against problems due to the dynamic pressure of high speed oil jets emerging from the valves. The operating frequency of a damper is low by most standards, and especially by the standards of electronic equipment, so the frequency response of sensors is unlikely to be a problem.

The temperatures of the damper oil and body change fairly slowly, being the result of cumulative energy dissipation throughout a test, combined with limited cooling. The temperature certainly affects the performance, tending to reduce the damping forces at a given speed. However the gas pressurisation force and gas spring stiffness increase with temperature. Some monitoring of temperature is therefore desirable. This is easily achieved by the use of standard thermocouple sensors with associated digital processors and displays which are available quite cheaply. The most suitable sensors are thin flexible insulated wires, and should be rated for a temperature up to 200°C or better. These are easily taped to the body of the damper. One at mid-stroke is sufficient for basic monitoring.

A thermocouple sensor is actually a pair of wires, electrically connected, preferably by a small weld, at the sensing end only, and which must otherwise be electrically isolated from each other. The two wires are of different materials. A temperature gradient in a wire produces a voltage gradient, so the total voltage between the ends of one wire depends on the material and the end temperatures. By joining the sensing ends of wires of different materials, a small output voltage is available, typically 20 µV/°C. This is actually slightly nonlinear, and also the approximately ambient temperature at the non-sensing end needs to be allowed for, but this is all catered for by the standard display units.

From the point of view of the user, the thermocouple indicates the temperature of the electrical connection point of the wires. To ensure that this is close to the temperature that it is desired to measure, the thermocouple must be in good thermal contact with the damper, and not excessively cooled. This may be ensured by taping the wire on with an insulating cover of soft sponge of about 10–20 mm over the tip, with some run of wire, e.g. 50 mm, kept against the damper to reduce conductive cooling of the tip.

An oscilloscope can be used to display various combinations of parameters, with the trace photographed to provide a permanent record. Typical plots of interest are:

(1) position vs time;
(2) velocity vs time;
(3) acceleration vs time;
(4) force vs time;
(5) force vs position;
(6) force vs velocity.

The kinematic parameters $X(t)$, $V(t)$ and $A(t)$ are mainly of use as a visual confirmation that the cycle shape is as desired. For a hydraulic-ram-driven tester a closed loop control system regulates the valves to achieve the specified input $X(t)$ function, but this may be imperfect, especially if there are sudden and large changes of force. An $X(t)$ display also gives a ready confirmation of the stroke being used, which is surprisingly difficult to check with any accuracy at all by simple direct visual observation of the moving ram and damper.

Full facility testers will have velocity and acceleration signals available for display. For a sinusoidal test, these should also be sinusoidal, and will show any deviations of $X(t)$ because they are mathematical derivations which enhance any higher harmonics.

Display of the force–position loop is the most frequent use of the oscilloscope. Interpretation of this figure is explained in Section 11.7.

Interestingly, it is possible to display $F(V)$ characteristic on the oscilloscope during a sinusoidal test because the velocity sweeps through a range of speeds $-\omega X_0$ to $+\omega X_0$. In basic $F(V)$ form this is useful enough, but is improved by displaying $F(\text{abs}(V))$ to bring it into the more familiar graphical form and to permit some enlargement of scale. The next section describes how to obtain $\text{abs}(V)$ electronically. An $F(V)$ curve obtained in a single sinusoidal motion is not quite the same as one obtained by using a variety of amplitudes or frequencies and picking the force values off at maximum speed. In the former case, the low speeds occur at large excursions.

11.6 Data Processing

The raw data of instantaneous values, primarily position X and force F, pressures and so on to be processed to give other items such as:

(1) velocity V;
(2) speed, $\text{abs}(V)$;

(3) extreme positions;
(4) extreme velocities;
(5) extreme forces;
(6) forces at mid-stroke;
(7) extreme pressures;
(8) rate of temperature change.

A testing system will normally incorporate some of these. In particular, the extremes of position and of force, or force at mid-stroke, are required as a minimum to establish the damper peak cyclic velocity and force to permit plotting of a damper $F(V)$ curve.

If a velocity signal is not available it may be obtained by the circuit of Figure 11.6.1. It is assumed that the source of V_{E1} is low impedance, and the output load on V_{E2} is high impedance. The time

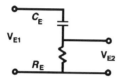

Figure 11.6.1 Differentiation circuit.

constant of this circuit is $t_N = RC$. An RC circuit is nonoscillatory, but the time constant can be considered to have an associated frequency f_N, with

$$2\pi f_N = \omega_N = \frac{1}{t_N} = \frac{1}{R_E C_E}$$

For oscillatory variations of V_{E1} which are slow relative to this natural frequency, the capacitance is a high impedance relative to the resistance, so the current is then

$$I_E = C_E \frac{dV_{E1}}{dt}$$

Hence, the output voltage is

$$V_{E2} = I_E R_E = R_E C_E \frac{dV_{E1}}{dt}$$

Therefore, the output is the time derivative of the input with the factor $R_E C_E$. Supplying a position signal

$$V_{E1} = k_{VX} X$$

to the input therefore gives

$$V_{E2} = R_E C_E k_{VX} \frac{dX}{dt}$$

or

$$V_{E2} = k_{VV} V$$

with

$$k_{VV} = R_E C_E k_{VX}$$

The resistance and capacitance values must be chosen to give a frequency f_N safely above the highest test frequency. For simple sinusoidal testing up to 3.2 Hz, then $C = 1$ μF and $R = 10$ kΩ give $\omega_N = 100$ rad/s and $f_N = 16$ Hz. A position signal with $k_{VX} = 100$ V/m will then give a velocity signal with $k_{VV} = 1$ V/(m/s). If a greater frequency range is to be covered, smaller R_E and C_E values must be used, with a smaller output coefficient, but this can be amplified if required.

To display $F(abs(V))$ on an oscilloscope requires production of the signal $abs(V)$, i.e. a fully rectified velocity signal. However this cannot be done by the use of simple diode rectification because of the forward voltage drop of the diodes. There are several possible solutions, including ones with op-amps to correct the forward voltage drop.

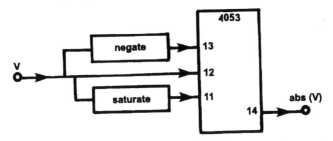

Figure 11.6.2 Velocity signal rectification using a multiplexer chip.

An alternative method favoured by the author is shown in Figure 11.6.2, using an analogue switch chip such as the 4053, plus an inverter and saturating high-gain amplifier. The 4053 multiplexer chip is a three-pole two-way solid-state switch of which only one pole is needed. The output on pin 14 takes its value from the input on pin 12 or pin 13 depending upon whether the control pin 11 is high or low. Therefore connect the velocity signal to pin 12 and an inverted velocity signal to pin 13. Use a high gain saturating op-amp to obtain a two state voltage from the signal, and apply to the control pin 11. Pin 14 then gives a rectified (absolute) velocity signal with very little forward voltage drop. This is ideal for driving the oscilloscope for an $F(abs(V))$ display.

The extreme values are obtained by standard peak-holding circuitry. Mid-stroke values are obtained by sampling the variable when the position value equals the static value, i.e. when the oscillatory part of the sinusoidal signal undergoes a sign change, briefly connecting the variable signal to a storage circuit.

On the whole, development of good quality instrumentation tends to be very time consuming, and it is more practical to purchase standard equipment where possible. For those interested in enhancing their instrumentation, Horowitz and Hill (1989) is useful reading.

Modern equipment will have data logging in digital form, which is a great convenience. Data processing is then achieved by computer, with suitable software, and can be displayed in any favoured form.

11.7 Sinusoidal Test Theory

Considering a simple linear damper with damping coefficient C_D (N s/m), actuated sinusoidally at stroke S, amplitude $X = S/2$, at frequency f Hz,

$$\omega = 2\pi f$$
$$X = X_0 \sin \omega t$$
$$V = \omega X_0 \cos \omega t$$

resulting in a damping force

$$F_D = \omega C_D X_0 \cos \omega t$$

The energy dissipated per cycle, E_C, is

$$E_C = \oint F_D dX = \oint F_D V dt$$
$$= \oint (C_D X_0^2 \omega^2 \cos^2 \omega t) dt$$
$$= C_D X_0^2 \omega \oint (\cos^2 \omega t) d\omega t$$

Integrating around the whole cycle,

$$E_C = \pi C_D X_0^2 \omega = 2\pi^2 C_D X_0^2 f$$

and the mean power dissipation P_m is

$$P_m = 2\pi^2 C_D X_0^2 f^2 = \tfrac{1}{2} C_D V_0^2$$

A typical passenger car damper may have a mean effective damping coefficient of about 2 kN s/m, and would be tested up to an absolute maximum speed of 2 m/s; for example 64 mm stroke (32 mm amplitude) at 10 Hz. This extreme case gives a maximum force of 4 kN, an energy dissipation of 404 J/cycle, and a mean power dissipation of 4.0 kW.

This has a number of implications for damper testing. First, the total thermal capacity of an average car damper is about 400 J/K, so at 4 kW the temperature rise is 10 K/s, and worse for the working fluid only. This is far above what would normally be sustained in use on the road; in normal running the damper dissipation is about 10 W, and the full stroke could only occur in an isolated way with a temperature rise of 2 K. For sustained testing at such high damper speeds, water cooling is required, and indeed is used for durability testing, and even on the car for off-road racing. From the practical standpoint of performance testing, it means that readings must be taken very quickly at such high speeds, and for comparison between dampers they must be taken in a consistent manner to avoid discrepancies from temperature effects. Where the facility is available, the use of single-shot cycles is a possible solution, or preferably a few successive cycles, but this is liable to give false results if certain effects are present, e.g. emulsification or cavitation. After a run-up to such high speeds, a damper will be very hot, and care must be taken not to overheat the seals or oil.

The second implication of the figures is that a fairly substantial device is required to drive a damper for a vigorous test: a minimum of a 4 kW drive, with the possibility of delivering substantially more instantaneous power, possibly by a flywheel for an electromechanical system.

Of course the above figures are for worst-case extreme testing. At a normal peak speed of 0.3 m/s the mean power dissipation becomes less than 100 W, Figure 11.7.1.

Considering a vertical damper driven at the bottom by a hydraulic tester, fixed at the top, the force required to be exerted at the top is

$$F_{DI} = F_D(X) + F_D(V) + F_G + F_F + m_1 g$$

where $F_D(V)$ is the velocity-dependent force of the damper, and $F_D(X)$ is the position-dependent force, e.g. due to gas compression.

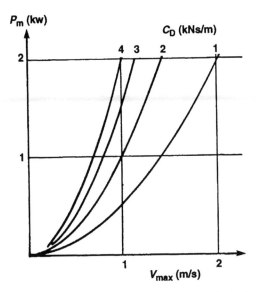

Figure 11.7.1 Damper mean power dissipation versus velocity amplitude, for sinusoidal motion.

The mass m_1 is that of the upper part of the damper plus the load cell and the end fittings which must also be supported. The mass m_1 is normally only a fraction of kilogram, and the weight $W_1 = m_1 g$ is only a few newtons. The force F_G is the static pressure force on the upper part of the damper arising from the damper internal gas pressure value exerted on the rod area. F_F is the static friction force on the upper part of the damper due to piston friction, bearing friction and seal friction within the damper.

The force F_{D2} at the lower end is

$$F_{D2} = F_D(X) + F_D(V) + F_G + F_F + m_2 g + m_2 A$$

where m_2 is the mass of the lower part of the damper, plus connecting elements and part of the lower load cell, and A is the acceleration of the lower end mass. Normally only a single load cell will be used. From these two equations, it is apparent that the difference between the two end forces is

$$F_{D2} - F_{D1} = (m_2 - m_1)g + m_2 A$$

For a linear damper with coefficient C_D, with sinusoidal displacement as before, the acceleration is

$$A = -X_0 \omega^2 \sin \omega t$$

and the lower force is

$$F_{D2} = F_D(X) + C_D X_0 \omega \cos \omega t + F_G + F_F + m_2 g + m_2 X_0 \omega^2 \sin \omega t$$

The inertia (acceleration) term is associated with the sine term, whereas the damping resistance force is associated with the cosine term. This illustrates that the acceleration force is out of phase with the velocity dependent damping force, so the mass of a load cell and part of the damper at the reciprocating end has little effect on the peak or mid-stroke forces used for plotting the damper $F(V)$ curve.

On the other hand, if the $F(V)$ curve is to be shown on an oscilloscope, or otherwise derived from a single cycle through the speed range, the low speeds occur when the acceleration force is significant, so the resulting curve will be affected in this area. Obviously the size of this effect depends upon the total end mass. For example at 3.2 Hz and stroke 100 mm the peak speed V_0 and peak acceleration A_0 are

$$V_0 = \omega X_0 = 1.0 \text{ m/s}$$
$$A_0 = \omega^2 X_0 = 10 \text{ m/s}^2$$

The total difference of F_{D2} values between the two extreme zero velocity positions is $2\,m_2 A_0$. For $m_2 = 0.5$ kg this is 10 N difference, which is significant at low speeds.

Also, for pressurised dampers the $F(X)$ variation due to gas pressure must be considered. The effective stiffness may be around 100 N/m which gives a total force difference of 10 N over the 0.1 m stroke, which in this case balances the acceleration force.

In general, the stiffness plus acceleration forces F_{KA} are

$$F_{KA} = kX + mA = (k - m\omega^2)X_0 \sin \omega t$$

Therefore, complete cancellation occurs throughout the stroke for the resonant condition

$$k = m\omega^2$$

Hence, for an unpressurised damper it may be better to use a load cell at the fixed end, but for a pressurised damper with a suitable choice of lower mass a load cell at the reciprocating end may be advantageous.

A complete solution is to add to the force signal a small extra signal proportional to displacement, positive or negative as required according to k, m and ω, to eliminate the F_{KA} term.

Figure 11.7.2 shows the $F(X)$ and $F(V)$ curves for a bilinear damper in a simulated sinusoidal test with some unbalanced stiffness or acceleration force, illustrating how the force becomes two-valued at zero speed, creating a loop instead of a line. Coulomb friction also offsets the force, but not just at zero speed, rather it does so uniformly throughout the motion. Also, the Coulomb friction is normally small, so in practice it is possible to mix out the positional component (stiffness and acceleration forces) if these are measured and calculated suitably, to give a good image of the $F(V)$ curve on an oscilloscope. Such curves will frequently still exhibit residual hysteresis in the $F(V)$ curve, especially for higher stroking frequencies. This arises from any form of compliance in the system, and in particular from compressibility effects due to emulsification of the oil, or, of course, from rubber bush mountings.

Figures 7.12.1–7.12.13 show the $F(X)$, $F(V)$ and $F(\text{abs}(V))$ curves for a range of force types in sinusoidal testing. Study of these may facilitate the interpretation of experimental damper graphs in terms of the contributory force types.

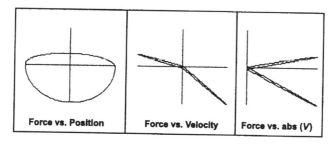

Figure 11.7.2 $F(X)$, $F(V)$ and $F(\text{abs}(V))$ for a sinusoidal test. Force type: bilinear damper plus small stiffness.

11.8 Test Procedure

This section describes the procedure typically used for measuring the $F(V)$ curve of a damper by sinusoidal testing. This is an example only, of course, and procedures vary from one site to another, with different equipment or for other purposes. In practice, any test will usually be one of a batch of tests on similar dampers, or a series of tests on an adjustable damper.

First, of course, some details of the test must be planned. The basic decisions are:

(1) the maximum speed to be tested;
(2) the number of points and the lesser speeds.

The maximum speed is governed by the type of damper and the speeds expected in service for that type, and consideration of whether the full range is actually of immediate interest. The choice of upper speed may also be influenced by consideration of heating effects, especially if further runs may be required, e.g. after adjustment. There is normally relatively little cooling during a rig test, so the temperature rise of the damper is closely related to the total energy dissipated. The mean power dissipation is proportional to the square of the peak speed, so the last couple of velocity values are decisive, and a small reduction of speed range tested will substantially reduce the heating problems.

Sufficient lesser speeds are then needed to give an adequately defined curve, possibly with emphasis in certain speed ranges of particular interest. Usually the test is of a type tested frequently before, so the choice of speeds is easily made.

Appendix D shows two test sheets, one of limited speed used for Formula 1 and similar racing dampers, the other suitable for passenger and rally car dampers with much greater speed range. Of course the full range, and all intermediate points, are not always used.

The variation of peak velocity may be achieved by varying either frequency or amplitude, or indeed, both, but fixed-frequency variable-amplitude is preferred. The frequency is chosen to allow the required maximum speed with an amplitude within the damper capability (i.e. less than half of the maximum stroke), also considering the natural heave frequency in practice for the type of vehicle. This may be around 1 Hz for a passenger car, but around 5 Hz for a ground-effect racing car. Finally, the

Table 11.8.1 Procedure for sinusoidal test

(1)	Mount damper
(2)	Affix thermocouple
(3)	Set and check adjustable damper settings
(4)	Measure damper-limited extreme positions
(5)	Calculate stroke and mid-stroke position
(6)	Warm up damper
(7)	Measure static forces F_{SC} and F_{SE} at mid-stroke
(8)	Calculate F_G and F_F
(9)	Measure F_{SC} at spaced points
(10)	Calculate stiffness K_D
(11)	Set position statically at mid-stroke
(12)	Set and check test frequency
(13)	Record starting temperature
(14)	Measure F_C and F_E for various speeds
(15)	Record final temperature
(16)	Remeasure F_G and F_F whilst hot
(17)	Demount damper

frequency actually used will usually be 0.8, 1.6, 3.2 Hz or, possibly, 6.4 Hz, because these figures give convenient round numbers for the radian frequency (5, 10, 20, 40 rad/s) giving simple values relating amplitudes to particular peak velocities. Alternatively, tests may be performed at basic sprung and unsprung mass natural frequencies, with varied amplitude.

The test then proceeds as follows:

(1) Mount damper. The upper end is mounted first to support the damper weight. The ram must then be adjusted accurately in position to align with the lower mounting to facilitate bolting up.
(2) Affix thermocouple. A complete wrap around of tape with a small foam block is used.
(3) Set and check adjustable settings. In haste, it is easy to test an adjustable damper on the wrong settings. Record the settings.
(4) Measure extreme positions as limited by the damper. The digital indicators are set one to instantaneous force, one to instantaneous position. By creeping the damper up to the limits under manual control and observing the sudden force change, the normal motion limits Z_{max} and Z_{min} can be read and recorded. If, as on some racing dampers, there is a bump-stop rubber fitted over the rod, and this has not been removed for some reason, then the most extreme compression position may be observed visually.
(5) The stroke may now be calculated as

$$S = Z_{max} - Z_{min}$$

and the mid-stroke position Z_{mid} as

$$Z_{mid} = \tfrac{1}{2}(Z_{max} + Z_{min})$$

(6) Warm up. Set the static position to mid-stroke and operate the damper at a moderate stroke for perhaps one or two minutes, according to the type of damper. Preliminary observation of the forces and the $F(X)$ loop are made in case any abnormalities are detected, requiring special investigation or termination of the test.
(7) Measure static forces at mid-stroke. The damper is crept very slowly through the mid-stroke position, by manual control, in each direction, and the compression and static forces F_{SCin} and F_{SCout} are recorded.
(8) Calculate the gas pressure force F_G and the Coulomb friction force F_F, using

$$F_G = \tfrac{1}{2}(F_{SCin} + F_{SCout})$$
$$F_F = \tfrac{1}{2}(F_{SCin} - F_{SCout})$$

The gas pressure force result acts as a check that the damper is correctly pressurised, where appropriate, before proceeding. The actual pressure can be calculated using the rod area, but usually the expected value of the gas force F_G is known in advance. The Coulomb friction force is a check on correct design, machining and assembly of the seals and should be small (e.g. 10 N or less).

(9) Measure F_{SC} at spaced points. Choose two positions spaced at z, a substantial fraction of the stroke. Measure the creep compression force F_{SC} at these two positions.
(10) Calculate the damper stiffness:

$$K_D = \frac{F_{SC1} - F_{SC2}}{z_1 - z_2}$$

A gas-pressurised damper is really nonlinear, but a linear mean stiffness remains a useful approximation in most cases.

(11) Set the damper statically at the mid-stroke position, in preparation for velocity tests.
(12) Set/check test frequency. As selected for the particular damper and vehicle type. Lock the frequency.
(13) Record temperature for start of velocity tests.
(14) Measure F_{DC} and F_{DE} for various speeds. Adjust the amplitude to the successive values chosen (see the example test sheets). At this stage the digital displays are set to display the extreme forces F_{DC} and F_{DE} (or mid-stroke forces).

There is no display of actual amplitude, the control of which must therefore be set with extra care. The force values are recorded. The oscilloscope will normally be set to display the $F(X)$ loop which can be observed for abnormalities. These, plus any other non-routine behaviour, such as excess noisiness, are also recorded. The highest speeds are usually swept through quickly to avoid excessive temperature rise, and the amplitude is returned to zero immediately after the last reading is taken.

(15) Record final temperature.
(16) Remeasure hot values for F_G and F_F. The rise of temperature will generally cause an increase in F_G, but there should be little change in F_F.
(17) Demount the damper. If the test is complete, set the digital meters to indicate instantaneous force and position. Manually adjust the position to the original position when mounting. Trim to zero force to facilitate bolt removal.

In the test as described, both control of the test and recording of data is performed manually. Data is then typed into a computer as the test proceeds.

Modern computer control and automatic data logging have many advantages. A pre-programmed test proceeds expeditiously, and readings are generally taken more quickly or more consistently, possibly with less heating at high speeds. Against that, manual control gives flexibility of response, and any unusual behaviour is easily investigated at the time. Automatic data logging is certainly a convenience, but in practice not much of an advantage; this is because the quantity of data is relatively limited, so it is easy to type it in. Therefore, although automatic control and data logging is very valuable for some tests, it is of little help on the case of the basic sinusoidal test where the quantity of data is small. It shows to great advantage in more elaborate tests, e.g. quarter-car testing with random road inputs, where it becomes essential.

Whatever method of data acquisition and display is used, the principal result is the $F(V)$ curve for the damper for compression and extension through the speed range of interest.

11.9 Triangular Test

With hydraulic testing apparatus, it may be preferred to use a triangular displacement waveform instead of the sinusoidal one. A symmetrical triangular displacement waveform gives nominally constant velocity over the whole stroke, equal in the two directions, i.e. a square velocity wave, Figure 11.9.1. With the force read at mid-stroke this is held to be nearer to steady state than is the sinusoidal actuation, which may be an advantage. One the other hand, the ends of the stroke involve highly transient conditions, with nominally infinite acceleration, rates of change of internal pressures, forces and so on, which may be unsettling for the damper, especially at high speed. The sinusoidal motion may be somewhat more realistic in that the real vehicle motions are nearer to sinusoidal. Certainly, a triangular waveform is harder on the equipment, including the end fittings, than is a sinusoidal waveform for the same peak speed, because of the slamming at the ends of the stroke. This is much less of a problem for a damper with rubber bush mountings than one with rigid rose joints (such as racing dampers).

For a stroke S, the amplitude is $X_0 = S/2$. The frequency is f and the cyclical period T_P is just

$$T_P = \frac{1}{f}$$

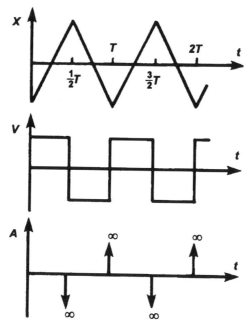

Figure 11.9.1 Kinematics of ideal triangular waveform displacement test.

The time for one stroke is of course half of this, giving a triangular test speed

$$V = \pm 2fS = \pm 4fX_0$$

Figure 11.9.2 shows the idealised $F(X)$ and $F(V)$ curves for a bilinear damper, the latter simply being two points. In practice the highly transient action at the ends of the stroke cause distortion of the $F(X)$ loop, but for normal speeds the force settles and the mid-stroke reading of force is a good one. Because the triangular displacement test contains only one velocity, there is no possibility of displaying an instantaneous $F(V)$ curve on a oscilloscope.

For a linear damper of coefficient C_D, the energy dissipated per cycle E_C in a triangular-waveform test is

$$E_C = 2SF = 2SC_DV = 4C_DfS^2 = 16C_DfX_0^2$$

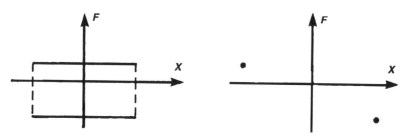

Figure 11.9.2 $F(X)$ and $F(V)$ curves for an ideal bilinear damper subject to idealised triangular waveform displacement.

The mean power dissipation P_m is

$$P_m = 4C_D f^2 S^2 = 16 C_D f^2 X_0^2$$

Because the speed is constant, this may also be expressed simply as

$$P_m = C_D V^2$$

The velocity achieved is $2/\pi$ (equalling 64%) of that in the sinusoidal test for a given stroke and frequency limit, and for a given test speed the mean power dissipation is twice that of the sinusoidal test; this alone may be a decisive factor in favour of a sinusoid for testing at high speeds.

Sinusoidal and triangular testing have both been used successfully for damper testing. The one to be preferred in any particular case depends upon the damper, its end fittings, the apparatus available, the speeds to be tested and the precise purpose of the test.

11.10 Other Laboratory Tests

Other forms of damper testing are certainly possible. Actuating the damper with a nominally white noise (random) displacement has been advocated and demonstrated. In the case of a simple damper this does not seem to offer any decisive advantage, although this type of test is certainly of value for some more complex systems incorporating a damper, such as a physical quarter-car model or a full vehicle test.

Where the prime purpose is to obtain an $F(V)$ curve, a possible approach might be to vary the velocity smoothly from zero to some upper value, i.e. to use a constant acceleration test. For an allowable damper stroke S, in time t of acceleration A going from zero speed to peak speed V,

$$V = At \qquad S = \tfrac{1}{2} A t^2 \qquad V^2 = 2AS$$

For example, at $A = 2.5$ m/s^2 in a time 0.4 s the speed reached is 1 m/s with displacement of 200 mm. At 10 m/s^2 the same stroke is reached in 0.2 s with a terminal speed of 2 m/s. Where facilities permit such a pre-programmed single shot test this is worth considering. With two strokes of the damper, after subtracting the gas pressure and friction forces the full compression and extension characteristic could be obtained with very little temperature rise. This type of test does not seem to have been used thus far, and its range of validity is not yet established. Corrections for stiffness and inertia should be made. At high accelerations, compressibility would become a problem.

It is, however, established practice to deduce the $F(V)$ curve from a single sinusoidal cycle, or a few cycles. This has the advantage over peak-velocity testing that there is less heating, and the whole test is faster. However, corrections for stiffness and inertia should be made, and it is only suitable for dampers which are not position-dependent. This is sometimes known as a CVP test (Continuous Velocity Pick Off) in contrast to a PVP test (Peak Velocity Pick Off) in the case of a series of sinusoidal runs at varying amplitudes or frequency.

Apart from the basic performance tests, it may be desirable to measure $F(V)$ for reasons other than checking the behaviour of a given design of damper. Because dampers depend for their characteristics on fluid flow in very small passages, they are sensitive to production variation. In critical cases, mainly in the province of motor sport, it may be desirable to perform an acceptance test of individual dampers to confirm that they have been correctly assembled with the correct parts. The author's experience of this has been salutary. Dampers have frequently been found to be seriously deficient in performance, and on stripping been found to have an amazing array of part and assembly errors, including interchanged compression and extension shims, adjusters which have no effect, adjusters which work opposite to the declared way, and adjusters which are effective but erratic in relation to the position scale, e.g. increasing force up to the penultimate setting but the final setting causing a considerable reduction. One damper exhibiting very erratic behaviour at low speed was found to contain some fibres from a paper towel, which evidently were being trapped in the valve. Any of these kinds of faults could cause obscure problems in operation of the vehicle and waste a great deal of track testing time.

Even when correctly assembled and free of such gross errors, dampers of the same specification are frequently found to have forces differing by 20% or more. This can be due to minor machining differences such as different deburring of valve holes, small burrs preventing correct seating and so on. Perhaps the main lesson here is that it is actually quite difficult to build good and consistent dampers. Hence, apart from the basic acceptance testing, some racing and rally teams find it useful to select matched pairs of dampers from a batch of nominally the same specification.

Post-use testing is always of interest to check for deterioration of performance. This should not be a problem in racing, but rally dampers have a very hard life, and the dampers of road vehicles are, through sheer long use, always at risk from wear of the seals.

Durability testing has been performed in the laboratory, but this is relatively unusual. For the most part, damper wear is fairly well understood and mainly dependent on the rod seal and piston seal materials, on the seal operating temperature and the finish of the surface of the facing material, and finally on the oil used. Durability is normally so good as to make laboratory testing of this aspect difficult, but special machines have been built, including water cooling, and are in regular use. The laboratory contribution to durability testing is, however, more commonly to make accurate measurements of the performance change resulting from extended use.

Occasionally other somewhat unusual tests may be undertaken where the laboratory offers controlled conditions not readily available for the complete vehicle. This may include, for example, testing the damper when it is deliberately cooled to a very low temperature, such as may be experienced in harsh winter conditions. This places a limit on the damper oil, since the low temperatures cause great increase of viscosity and the damper may become unacceptably harsh.

11.11 On-Road Testing

The ultimate test of a damper is always on the vehicle. A perfectly good damper may fail this test if it is not suitable for the vehicle, although, of course, it may be the ideal damper for some other vehicle. Road testing may be considered in principle in two categories:

(1) public road testing;
(2) special track testing.

although the distinction may sometimes be blurred. Tests on public roads are performed over long mileages in conjunction with other general testing. The purposes of such testing are:

(1) quality of ride;
(2) quality of handling;
(3) durability.

The vehicle ride and handling may be tested over a wide range of ordinary roads. With the considerable body of existing experience of damper selection built up by manufacturers, the suspension engineer can specify in advance for most vehicles a damper which will be quite close to optimum, and indeed possibly within the range of variation of individual opinion. Road testing would then be in essence an acceptance test, possibly with some fine tuning. Sometimes, however, a great deal of experimental on-car tuning will be required before results are considered right, and the damper will finish with settings significantly removed from the original specification.

The test driver will be conscious of the ride motion of the vehicle over the various surfaces, and will note any peculiarities which occur on any particular surface or under any particular conditions such as very high or very low ambient temperature. He will also be aware of the handling behaviour of the vehicle, and of the peculiarities or unsatisfactory behaviour which may occur due to the dampers.

The effect of the dampers on handling may be investigated on special handling test tracks comprising numerous curves of various radii, usually with a good-quality surface. Testing on special rough tracks is usually intended to pose a severe challenge to vehicle reliability or ride and handling performance. Because of the extreme suspension motion that occurs, the dampers may have real problems coping with sustained fast driving on rough roads, due to the resultant high temperatures and possibility of aeration of the damper oil. It is for this reason that rally cars require special dampers, nearly always having separated gas or being designed to operate in the emulsified state, and being of generous size to give good cooling.

The track testing of racing car dampers is a rule unto itself, being exclusively concerned with obtaining the minimum lap time. The emphasis is very much on handling rather than ride, although the latter may in reality be of some importance, especially in long races, because of the adverse effect of a bad ride on driver fatigue and performance. In particular the dampers affect traction and tire grip, and influence the fore–aft distribution of lateral load transfer during body roll at corner entry and exit, which has a significant effect on transient handling behaviour. This is especially important for racing cars with extreme aerodynamics, which are very sensitive to ride height.

Although computer simulation of ride and handling qualities continues to develop apace, it seems certain that damper testing, both in the laboratory and on the road, will continue to be essential on a more-or-less permanent basis.

Testing

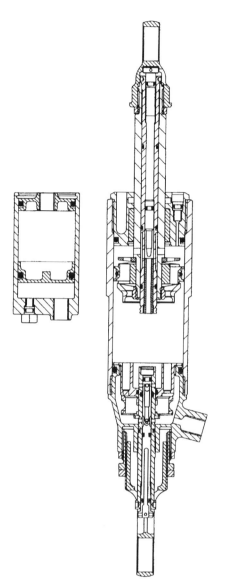

Figure 11.11.1 This cross-section of a Penske racing damper shows the complexity of the modern hydraulic damper, with associated cost for precision manufacture. Features visible here include the remote reservoir on the right, with free-floating piston, adjustable spindle within the rod, and O-ring sealed assembly to permit easy rebuilding to achieve altered characteristics (by permission of Penske racing Shocks).

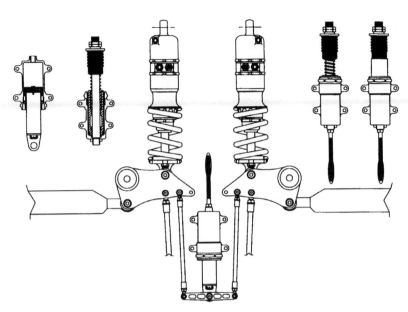

Figure 11.11.2 This layout is indicative of racing suspension practice, with the pushrods, driven by the outer end of the lower wishbones or upright, actuating the rockers, which in turn drive the various spring/damper units. The centre unit is effective in heave only, not in roll. This permits heave springing and damping to be adjusted without altering the roll and load-transfer characteristics. This facilitates the set-up of vehicles with strong aerodynamics, which are sensitive to ride height, as found in, for example, Indy, Formula One, and similar events (Penske Racing Shocks).

Appendix A
Nomenclature

Chapter 1 Introduction

A_x	m/s²	longitudinal acceleration
A_y	m/s²	lateral acceleration
D	m	diameter
F_D	N	damper force
F_F	N	friction force
F_N	N	normal force
f_{NH}	Hz	natural frequency of body in heave
f_{NP}	Hz	natural frequency of body in pitch
f_{NR}	Hz	natural frequency of body in roll
F_{PC}	N	friction pack compression force
g	m/s²	gravitational field strength
h_D	m	height of drop test
H_G	m	height of centre of mass
I_{PB}	kg m²	pitch inertia of body (sprung mass)
$k_{\phi S}$	rad/m s²	suspension roll angle gradient
K_P	N m/rad	pitch stiffness of suspension
K_W	N/m	suspension wheel rate (vertical stiffness)
L	m	wheelbase
L_A	m	arm length
M	N m	moment
m	kg	mass
M_F	N m	friction moment
M_P	N m	pitch moment
N_F		number of friction surfaces
P	Pa (N/m²)	pressure
Q	m³/s	volumetric flow rate
T	m	vehicle track (tread)
V_D	m/s	damper velocity
V_I	m/s	impact velocity (of wheel)
V_{SB}	m/s	suspension bump velocity
V_{SD}	m/s	suspension droop velocity ($-V_{SB}$)

The Shock Absorber Handbook/Second Edition John C. Dixon
© 2007 John Wiley & Sons, Ltd

Z_H	m	amplitude of heave oscillation
z_{SB}	m	bump displacement of suspension ($\equiv z_S$)

Greek

θ	rad	pitch angle
θ_A	rad	arm angle
μ	Pa s	fluid viscosity (N s/m^2)
μ_F		coefficient of limiting friction (Coulomb)
μ_{FD}		dynamic coefficient of limiting friction
μ_{FS}		static coefficient of limiting friction
ϕ_S	rad	suspension roll angle
ω_{NH}	rad/s	natural frequency of body in heave (rad/s)
ω_{NP}	rad/s	natural frequency of body in pitch (rad/s)
ω_{NR}	rad/s	natural frequency of body in roll (rad/s)

Chapter 2 Vibration Theory

a	m	front axle to centre of mass
b	m	rear axle to centre of mass
C	N s/m	damping coefficient (N/m s^{-1})
C_{DC}	N s/m	damping coefficient in compression
C_{DE}	N s/m	damping coefficient in extension
C_{D0}	N s/m	zeroth moment vehicle ride damping coefficient
C_{D1}	N s/rad	first moment vehicle ride damping coefficient
C_{D2}	N m s/rad	second moment vehicle ride damping coefficient
C_{K0}	N/m	zeroth moment vehicle ride stiffness coefficient
C_{K1}	N/rad	first moment vehicle ride stiffness coefficient
C_{K2}	N m/rad	second moment vehicle ride stiffness coefficient
C_{K3}	s^{-2}	vehicle ride stiffness coefficient
C_{K4}	s^{-4}	vehicle ride stiffness coefficient
C_M	N m s/rad	pitch modal damping coefficient
C_Q	N s^2/m^2	quadratic damping coefficient (N/(m^2 s^{-2}))
C_R	N m s/rad	roll damping coefficient
c_{SD}	N s/m kg	specific damping coefficient
D	s^{-1}	Heaviside time derivative operator
e		exponential base (2.71828...)
$F_{D,Q}$	N	quadratic damping force
f_N	Hz	natural frequency
F	N	force
F_{CF}	N	Coulomb friction force
F_F	N	friction force
F_{Fmax}	N	maximum friction force
F_N	N	normal force
f_R	Hz	effective frequency of road profile
I_M	kg m^2	modal inertia
I_P	kg m^2	pitch inertia
i_P		pitch dynamic index
I_R	kg m^2	roll inertia

K	N/m	stiffness, suspension wheel rate
K_H	N/m	heave stiffness
K_M	N m/rad	pitch modal stiffness
K_P	N m/rad	pitch stiffness
K_R	N m/rad	roll stiffness
k_{ss}	N/m kg	specific stiffness by mass
L	m	wheelbase
m	kg	mass
m_B	kg	body mass
m_f	kg	front mass
m_r	kg	rear mass
m_S	kg	sprung mass
M	N m	moment
M_P	N m	pitch moment
r		forcing frequency ratio (f_R/f_N)
R_F		rear/front frequency ratio
R_K		rear/front stiffness ratio
R_M		rear/front mass ratio
S	m	pitch mode shape
T		transmissibility
t	s	time
V	m/s	velocity
V_{DC}	m/s	damper compression velocity ($-V_{DE}$)
V_{DE}	m/s	damper extension velocity ($-V_{DC}$)
z	m	displacement
Z_B	m	body amplitude
z_B	m	body displacement
Z_{CF}	m	Coulomb friction displacement
Z_R	m	effective amplitude of road profile
z_R	m	local road profile elevation
z_S	m	suspension deflection
$\dot{z}$	m/s	velocity (dz/dt)
$\ddot{z}$	m/s^2	acceleration (d^2z/dt^2)

Greek

α	s^{-1}	damping factor
ζ		damping ratio
ζ_C		Coulomb damping ratio
ζ_M		modal damping ratio
ζ_Q		quadratic damping ratio
θ	rad	pitch angle, mode angle
Θ	rad	mode angle amplitude
μ_F		Coulomb coefficient of limiting friction
μ_{FD}		dynamic μ_F (with relative motion)
μ_{FS}		static μ_F (no relative motion)
ϕ	rad	phase angle of vibration
ω_D	rad/s	damped natural frequency
ω_M	rad/s	modal undamped natural frequency
ω_N	rad/s	undamped natural frequency

ω_{ND}	rad/s	damped natural frequency
ω_R	rad/s	effective frequency of road profile

Chapter 3 Ride and Handling

a	m	centre of mass to front axle
b	m	centre of mass to rear axle
C_C	N s/m	damping coefficient of seat cushion
C_{D0}	N s/m	vehicle ride damping zeroeth moment coefficient
C_{D1}	N s	vehicle ride damping first moment coefficient
C_{D2}	N s m	vehicle ride damping second moment coefficient
C_{K0}	N/m	vehicle ride stiffness zeroth moment coefficient
C_{K1}	N/rad	vehicle ride stiffness first moment coefficient
C_{K2}	N m/rad	vehicle ride stiffness second moment coefficient
C_T	N s/m	damping coefficient of tire
C_W	N s/m	damping coefficient of suspension at the wheel
D	s^{-1}	Heaviside time derivative operator
e_D		damper force transfer factor (asymmetry)
f_{ARB}		anti-roll bar roll stiffness factor
f_{WH}	Hz	wheel hop natural frequency
F_C	N	damper compression force
F_E	N	damper extension force
F_f	N	front suspension force on body
F_m	N	damper mean force
F_r	N	rear suspension force on body
f_R	Hz	road fluctuation effective temporal frequency
I_P	kg m^2	pitch inertia (second moment of mass)
i_P		pitch dynamic index
I_R	kg m^2	roll inertia
i_R		roll dynamic index
K_C	N/m	stiffness of seat cushion
K_f	N/m	individual suspension stiffness at front
K_H	N/m	heave stiffness
K_P	N m/rad	pitch stiffness
k_P	m	pitch radius of gyration
k_R	m	roll radius of gyration
K_R	N m/rad	roll stiffness
K_R	N/m	ride stiffness
K_r	N/m	individual suspension stiffness at rear
K_T	N/m	vertical stiffness of tire
K_W	N/m	suspension stiffness (wheel rate)
L	m	wheelbase
m_B	kg	mass of body (sprung mass)
m_f	kg	front end mass
M_P	Nm	pitch moment
m_P	kg	mass of a passenger
m_r	kg	rear end mass
n_{SR}	cycles/m	road fluctuation spatial frequency
Q_H		handling quality parameter

Q_R		ride quality parameter
R		road roughness grading factor
R_f		natural frequency ratio rear/front
R_K		suspension stiffness ratio rear/front
S	m/rad	mode shape
T	m	vehicle track (tread)
T_R	s	period of a road fluctuation
z	m	vertical displacement
Z_B	m	vertical position of vehicle body
z_B	m	ride displacement of body
z_f	m	vertical displacement of body at front axle
Z_P	m	vertical position of passenger
z_P	m	ride displacement of passenger
Z_R	m	vertical position of road profile
z_R	m	ride displacement of road profile
z_r	m	vertical displacement of body at rear axle
Z_W	m	vertical position of wheel
z_W	m	ride displacement of wheel

Greek

ζ_H		damping ratio for body heave
ζ_{WH}		damping ratio for wheel hop
θ	rad	pitch angle
λ_R	m	road wavelength
ω_{NH}	rad/s	natural frequency in heave
ω_{Nf}	rad/s	natural heave frequency of front end
ω_{NR}	rad/s	roll natural frequency
ω_{Nr}	rad/s	natural heave frequency of rear end
ω_{NP}	rad/s	natural frequency in pitch
ω_R	rad/s	road fluctuation effective temporal frequency
ω_{SR}	rad/m	road fluctuation spatial frequency

Chapter 4 Installation

C_1	N s/m	damper coefficient
e	m	offset
F_D	N	damper force
f_R		rising rate factor
l	m	rocker arm length
l_{WP}	m	length from wheel centre to pivot axis (in plan)
n		damper characteristic exponent
$R_{A/B}$		motion ratio of item A relative to item B
R_{APH}		motion ratio for pitch in heave
R_D		damper motion ratio dz_D/dz_S
R_{DC}		damper coefficient ratio
$R_{D\phi}$	m/rad	damper velocity ratio in roll
R_R		rocker motion ratio
R_{RL}		rocker arm length motion ratio

$R_{R\psi}$		rod angle motion ratio
$R_{R\psi 0}$		rod angle motion ratio at $\theta = 0$
S_F	m	damper free stroke
S_I	m	damper installed stroke
t	s	time
U_{DS}		damper stroke utilisation
V	m/s	velocity
V_D	m/s	damper velocity
V_S	m/s	suspension bump velocity
z_B	m	body displacement
z_D	m	damper compression
z_S	m	suspension displacement in bump
z_W	m	wheel displacement

Greek

α_D	rad	damper out-of-plane angle
γ	rad	wheel camber angle
ε_{APH}	rad/m	axle pitch/heave coefficient
ε_{BC}	rad/m	bump camber coefficient
θ_R	rad	rocker position angle
θ_{RD}	rad	rocker deviation angle
ϕ_R	rad	rocker included angle
ψ	rad	rocker rod offset angle from tangent
ω	rad/s	angular speed
ω_R	rad/s	rocker angular velocity

Subscripts

1	input
2	output
z	static (zero) ride height

Chapter 5 Fluid Mechanics

A	m²	area, cross-sectional area
A_E	m²	effective orifice area
A_O	m²	total orifice area
A_P	m²	piston area
A_R	m²	rod area
C	m	wetted circumference of cross section
C_d		orifice discharge coefficient
C_{GA}	kg/m³ Pa	gas/liquid absorption coefficient
C_{GLV}		gas absorption volume coefficient
C_P	J/K	thermal capacity (at constant P)
c_P	J/kg K	specific thermal capacity
C_V		velocity coefficient
d	m	orifice diameter
d		relative density

Nomenclature

D	m	diameter
D_H	m	hydraulic diameter
E	J/kmol	characteristic energy
e	m	surface roughness parameter
F_D	N	drag force
F_G	N	gravity force
F_{ST}	N	surface tension force
f		pipe friction factor
f_{GV}		fraction of gas by volume
f_{Gm}		fraction of gas by mass
g	m/s^2	gravity
K		loss coefficient
K	Pa	bulk modulus
k	W/m K	thermal conductivity
K_V	m^2/s	vortex strength
k_μ		variation coefficient of viscosity with gas volume fraction
$k_{\mu T}$	°C^{-1}, K^{-1}	viscosity–temperature coefficient
L	m	length
m	kg	mass
m_E	kg	mass of emulsion
m_G	kg	mass of gas
m_{GA}	kg	mass of gas absorbed by the liquid
m_{GF}	kg	mass of gas free (not absorbed)
m_L	kg	mass of liquid
N		number, e.g. parallel orifices, molecules
N_{EX}		bubble expansion number
n		velocity profile reciprocal index
P	Pa	static pressure, pressure drop across orifice
P_{ST}	Pa	surface tension pressure in bubble
P_{St}	Pa	stagnation pressure
Q	m^3/s	volumetric flow rate
Q	s/Pa	ERMR fluid quality factor
q	Pa	dynamic pressure
r	m	entry edge radius
R	m	radial position
R_A	J/kg K	specific gas constant of air
Re		Reynolds number
R_G	J/kg K	gas constant
R_U	J/kmol K	universal (molar) gas constant
R_V		velocity ratio and effective area ratio
s	m	pipe radius, or half of plate spacing
T	°C, K	temperature
T_K	K	absolute temperature
u	m/s	fluid velocity
U_T	m/s	ideal (Bernoulli) orifice exit velocity
U_B	m/s	bubble velocity
V	m^3	volume
V_R	m/s	radial velocity
V_T	m/s	tangential velocity
X	m	mean spacing between particles

| y | m | position in velocity profile |

Greek

α	K^{-1}	coefficient of volumetric thermal expansion
α		profile kinetic energy correction factor
β		profile momentum correction factor
β_E	Pa^{-1}	emulsion compressibility
β_L	Pa^{-1}	liquid compressibility
β_G	Pa^{-1}	gas compressibility
ε		diffuser loss factor
η_D		diffuser recovery coefficient
η_R		static pressure recovery coefficient
θ		diffuser single-side angle
μ	Pa s	dynamic viscosity
ρ	kg/m^3	density
σ_S	N/m	surface tension

Subscripts

E	emulsion
F	foam
G	gas
L	liquid
M	mixture
S	solid
1,2	station numbers

Chapter 6 Valve Design

A	m^2	area
A_A	m^2	actual orifice area
A_E	m^2	effective orifice area
A_L	m^2	leak area
A_M	m^2	maximum area
A_P	m^2	area of parallel orifice
A_{PA}	m2	piston annulus area
A_S	m^2	area of series orifice
A_V	m^2	valve area
C_d		discharge coefficient
C_Q	N s^2/m^2	quadratic damping coefficient
D_P	m	piston diameter
D_R	m	rod diameter
D_T	m	tube diameter
F	N	force
f_A		area ratio A_{PA}/A_T

f_N	Hz	natural frequency
k	N/m	stiffness
K_B	N/m	bellows equivalent stiffness
k_{TP}		transition pressure ratio
k_{TV}		transition volumetric flow ratio
L_T	m	tube length
m	kg	mass
P	Pa	pressure or pressure drop
P_V	Pa	viscous pressure drop
P_{vfc}	Pa	pressure for valve just fully closed
P_{vfo}	Pa	pressure for valve just fully open
Q	m³/s	volumetric flow rate
Q_M	m³/s	flow rate through variable area when fully open
Q_V	m³/s	flow rate through variable area
Q_{vfc}	m³/s	flow rate at variable valve just fully closed
Q_{vfo}	m³/s	flow rate at variable valve just fully open
Re		Reynolds number
Re_{max}		Maximum laminar $Re = 2000$
R_N	m	vortex nozzle diameter
R_O	m	vortex chamber outlet hole radius
U_T	m/s	ideal (theoretical) flow velocity
V_D	m/s	damper velocity
V_R	m/s	radial velocity
V_T	m/s	flow speed in tube, tangential velocity
$\dot{W}$	W	energy dissipation rate
x	m	displacement

Greek

α		energy correction factor
ρ	kg/m	density
μ	Pa s	dynamic viscosity

Chapter 7 Damper Characteristics

A_{PA}	m²	piston annulus area
A_{PF}	m²	piston effective friction area
A_P	m²	piston area
A_R	m²	rod cross-sectional area
C_D	Ns/m	damper coefficient
C_{DC}	Ns/m	damper coefficient in compression
C_{DE}	Ns/m	damper coefficient in extension
C_{DJ}	N/m Hz	damper jacking coefficient
C_{DQ}	N s²/m²	damper quadratic coefficient
C_{DT}	Ns/m	total vehicle damping coefficient
D_P	m	piston diameter
D_R	m	rod diameter
e_D		damper force transfer factor (asymmetry)
f	Hz	frequency

Symbol	Unit	Description
F	N	force
$F_{A,PCC}$	N	force of P_{CC} on piston annulus area
F_D	N	externally applied damper force
F_{DC}	N	damper compression force
F_{DJ}	N	damper jacking force
F_{DS}	N	damper static force (gas plus friction)
F_{DSC}	N	damper static force (compression +ve)
F_{DE}	N	damper extension force
F_F	N	Coulomb friction force
F_{FP}	N	friction force on piston
F_{FR}	N	friction force on rod
F_G	N	damper gas compression force
F_{NS}	N	piston seal normal force
$F_{P,PCC}$	N	force of compression chamber pressure P_{CC} on piston
$F_{P,PEC}$	N	force of expansion chamber pressure on piston
$F_{R,PCC}$	N	force of P_{CC} on rod area
k_{FC}	Pa s/m³	resistance coefficient of foot compression valve
k_{FE}	Pa s/m³	resistance coefficient of foot extension valve
k_{FPA}		piston friction area factor A_{PF}/A_P
k_{PC}	Pa s/m³	resistance coefficient of piston compression valve
k_{PE}	Pa s/m³	resistance coefficient of piston extension valve
K_D	N/m	damper stiffness (gas pressure)
K_{DJ}	N/m	damper jacking stiffness
K_{ST}	N/m	vehicle total suspension stiffness
L	m	damper length between fixtures
L_S	m	axial length of piston seal
m	kg	moving damper mass
m_S	kg	vehicle sprung mass
P	Pa	pressure or pressure difference
P_{CC}	Pa	compression chamber pressure
P_{EC}	Pa	extension chamber pressure
P_{FC}	Pa	pressure drop across foot compression valve
P_{FE}	Pa	pressure drop across foot extension valve
P_G	Pa	reservoir gas pressure
P_{PC}	Pa	pressure drop across piston compression valve
P_{PE}	Pa	pressure drop across piston extension valve
P_R	Pa	reservoir liquid pressure
Q	m³/s	volumetric flow rate
Q_{FC}	m³/s	volumetric flow rate in foot compression valve
Q_{FE}	m³/s	volumetric flow rate in foot extension valve
Q_{PC}	m³/s	volumetric flow rate in piston compression valve
Q_{PE}	m³/s	volumetric flow rate in piston extension valve
R_{DJ}		damper jacking ratio
R_{DJN}		damper jacking ratio at the undamped natural heave frequency
t	s	time
V_D	m/s	damper velocity
V_{DC}	m/s	damper compression velocity
V_{DCmax}	m/s	maximum compression speed without cavitation
V_{DE}	m/s	damper extension velocity
V_{DEmax}	m/s	maximum extension speed without cavitation

w	m	cylinder wall thickness
X	m	damper displacement
X_0	m	motion displacement amplitude
Z	m	heave motion amplitude
Z_{DJ}	m	damper jacking height

Greek

α		damping factor
ε_A		axial strain
ε_H		hoop strain
λ		progressivity factor
Λ	m^3	volume
μ	Pa s	dynamic viscosity
ν		Poisson's ratio
ρ	kg/m^3	density
σ_A	Pa	axial stress
σ_H	Pa	hoop stress
ω	rad/s	motion frequency in rad/s

Chapter 8 Adjustables

A_M	m2	valve maximum area
A_P	m^2	valve parallel hole area
A_S	m^2	valve series hole area
F	N	force
f_E		equalisation factor
f_H		handling optimisation factor
F_V	N	tyre vertical force
F_{Vmean}	N	mean tyre vertical force
k_A	m^2/Pa	valve area coefficient
P	Pa	pressure difference across valve
P_{vfc}	Pa	valve fully closed pressure (just opening)
P_{vfo}	Pa	valve fully open pressure
Q	m^3/s	valve flow rate
V	m/s	velocity
z_B	m	body (sprung mass) ride displacement
z_S	m	suspension bump deflection
z_W	m	wheel ride displacement

Chapter 9 ER and MR Dampers

A_{FA}	m^2	area of fluid annulus (sectional)
A_{PA}	m^2	area of piston annulus
B	T	magnetic flux density (T = tesla = Wb/m^2)
C_D	N s/m	damper viscous damping coefficient
C_E	N/V	ER effect damper coefficient
C_M	N/A	MR effect damper coefficient

C_Q	N s²/m²	damper quadratic damping coefficient
$C_{\tau E}$	Pa m/V	ER yield stress coefficient on electric field strength
$C_{\tau H}$	Pa m/A	MR yield stress coefficient on field strength
$C_{\tau B}$	Pa/T	MR yield stress coefficient on flux density
D_P	m	piston diameter
D_R	m	rod diameter
E	V	ER applied electric potential
f_A		area ratio A_{PA}/A_{FA}
f_F		force ratio $F_{D,V}/F_{D,E}$, etc.
F_D	N	damper force
$F_{D,V}$	N	damper linear force (viscous)
$F_{D,ER}$	N	damper ER force
$F_{D,MR}$	N	damper MR force
$F_{D,Q}$	N	damper quadratic force (dynamic)
F_{FA}	N	axial force on fluid in annulus
H	A/m	magnetic field strength
I	A	applied current in coil (MR)
L_{FA}	m	length of fluid annulus
L_{MFA}	m	length of magnetic fluid annulus
M	A	magnetomotive force (A-turns)
N_T		number of turns in coil (MR)
p		an index
P_{ER}	Pa	ER effect pressure drop
P_{MR}	Pa	MR effect pressure drop
P_V	Pa	viscous pressure drop
Q	m³/s	volumetric flow rate
R	m	radius
R_{FA}	m	radius of fluid annulus (mean)
R_M	A/Wb	magnetic reluctance
t_{FA}	m	thickness of fluid annulus
V_D	m/s	damper velocity (extension postive)

Greek

ε_0	V/m	permittivity of free space
ε_{ER}	V/m	permittivity of ER liquid
μ	Pa s	fluid dynamic viscosity
μ_{M0}	H/m	magnetic permeability of free space
μ_{MI}	H/m	magnetic permeability of MR piston iron
μ_{MR}	H/m	magnetic permeability of MR liquid
ρ	kg/m³	density
τ	Pa	shear stress
τ_Y	Pa	shear yield stress
τ_{ER}	Pa	shear yield stress due to ER effect
τ_{MR}	Pa	shear yield stress due to MR effect
Θ	Wb	MR magnetic flux (Wb = webers)

Chapter 10 Specifying a Damper

No nomenclature for Chapter 10.

Chapter 11 Testing

A	m/s^2	acceleration
C_D	N s/m	damping coefficient (N/m s^{-1})
C_E	C	electrical capacitance
E_C	J	energy dissipation per cycle
E_F	J	flywheel energy
F	N	force
F_A	N	damper acceleration force
F_D	N	damper force
F_{DC}	N	damper compression force
F_{DE}	N	damper extension force
F_{D1}	N	damper force at top end
F_{D2}	N	damper force at bottom end
F_E	N	damper extension force
F_F	N	damper force from Coulomb friction
F_G	N	damper force from gas pressure
F_K	N	damper stiffness force
F_{KA}	N	damper stiffness plus acceleration force
F_{SC}	N	damper static force in compression
F_{SE}	N	damper static force in extension
f	Hz	frequency in Hz
f_N	Hz	natural frequency in Hz
G		gear ratio
I_E	A	electrical current
k	N/m	stiffness
k_D	N/m	damper effective stiffness
k_{VF}	V/N	voltage coefficient of force signal
k_{VV}	V s/m	voltage coefficient of velocity signal (V/(m/s))
k_{VX}	V/m	voltage coefficient of position signal
m	kg	mass
m_1	kg	mass of upper (fixed) part of damper
m_2	kg	mass of lower (reciprocating) part of damper
P	Pa	pressure
P_m	W	mean power dissipation
R_E	Ω	electrical resistance
S	m	stroke
T	°C	temperature
T_P	s	cyclical period
t	s	time
V	m/s	velocity
V_0	m/s	velocity amplitude
V_E	V	electrical voltage
V_M	m/s	maximum speed
X	m	position (displacement)
X_0	m	initial displacement or amplitude
Z	m	test ram position
Z_{max}	m	measured maximum extension position of ram
Z_{mid}	m	ram position at damper mid stroke
Z_{min}	m	measured maximum compression position of ram

Greek

Δ_f	Hz	test speed variation
ζ		damping ratio
ϕ	rad	phase angle
ω	rad/s	radian frequency
ω_N	rad/s	natural frequency in rad/s

Appendix B
Properties of Air

B.1 Standard Properties

The properties of air are of interest in the context of damper cooling, and for its behaviour internally under pressure, and when forming an emulsion. Table B.1 gives the basic values for standard conditions.

The effective critical point for air (not a pure substance) is:

$$P_C = 3.72 \, \text{MPa}$$
$$T_C = -140.7\,°\text{C} \, (132.5\,\text{K})$$

Table B.1 Standard properties of dry air at sea-level, 15°C

Constituents by mass		Nitrogen	(N_2)	0.7553
		Oxygen	(O_2)	0.2314
		Argon	(Ar)	0.0128
		Carbon dioxide	(CO_2)	0.0005
Temperature		T_C	15	°C
		T_K	288.15	K
Pressure (absolute)		P	101325	Pa
Density		ρ	1.2256	kg/m^3
Dynamic viscosity		μ	17.83×10^{-6}	N s/m^2
Kinematic viscosity		ν	14.55×10^{-6}	m^2/s
Molar mass		m_m	28.965	kg/kmol
Specific gas constant		R_A	287.05	J/kg K
Specific heats		c_P	1005	J/kg K
		c_V	718	J/kg K
Ratio of specific heats		γ	1.400	–
Thermal conductivity		k	0.02534	W/m K
Speed of sound		V_S	340.6	m/s
Prandtl number		Pr	0.710	

The Shock Absorber Handbook/Second Edition John C. Dixon
© 2007 John Wiley & Sons, Ltd

Avogadro's number is 6.0225×10^{26} molecules/kmol, so the mass of an average air molecule is 48.1×10^{-27} kg. At standard temperature and pressure (15°C, 101325 Pa) the molecular density is 25.5×10^{24} molecules/m³.

B.2 Effect of Temperature

For cooling analysis the properties of air are required from low ambient, e.g. minus 40°C, up to maximum damper temperatures of 130°C.

Air can be treated for most purposes as an ideal gas. The following equations are all of good engineering accuracy over the relevant range.

The relative molecular mass (molecular weight) of dry air is

$$M_A = 28.965$$

with a corresponding molar mass

$$m_A = 28.965 \text{ kg/kmol}$$

The specific gas constant is

$$R_A = 287.05 \text{ J/kg K}$$

The absolute (kelvin) temperature T_K in terms of the Celsius temperature T_C is

$$T_K = 273.15 + T_C$$

and in terms of the Fahrenheit temperature is

$$T_K = 273.15 + (T_F - 32)/1.8$$

At absolute pressure $P(\text{N/m}^2 = \text{Pa (pascal)})$ the density ρ is

$$\rho = \frac{P}{R_A T_K}$$

or, by comparison with a reference condition P_0 and T_{K0}

$$\left(\frac{\rho}{\rho_0}\right) = \left(\frac{P}{P_0}\right)\left(\frac{T_{K0}}{T_K}\right)$$

The specific thermal capacity at constant pressure c_P is given by the empirical expression

$$c_P = 1002.5 + 275 \times 10^{-6}(T_K - 200)^2 \text{ J/kg K}$$

which, by comparison with tables, is within 0.1% from 200 to 450 K (−70 to 180°C).

The specific heat at constant volume c_V is then

$$c_V = c_P - R_A$$

Properties of Air

A direct empirical expression for specific thermal capacity at constant volume is

$$c_V = 717.8 + 0.07075(T_K - 300) + 0.26125 \times 10^{-3}(T_K - 300)^2$$

which is within 0.2% from 0 to 400°C and within 1% from −100 to 500°C.

The ratio of specific thermal capacities γ is

$$\gamma = \frac{c_P}{c_V}$$

The thermal conductivity k is given by

$$k = \frac{0.02646\, T_K^{1.5}}{T_K + 245.4 \times 10^{-12/T_K}} \text{ W/m K}$$

This (unlikely looking) equation has been adapted by the author from an imperial units equation used for the production of reference tables of range 100–1000 K. A simpler expression adequate for cooling calculations is

$$k = 0.02624 \left(\frac{T_K}{300}\right)^{0.8646} \text{ W/m K}$$

which is within 1% for −30 to 230°C and within 10% for −100 to 700°C.

The dynamic viscosity μ is given by

$$\mu = \frac{1.458 \times 10^{-6} T_K^{1.5}}{T_K + 110.4} \text{ Pa s } (\text{N s/m}^2)$$

This expression is used for the production of reference tables (100–800 K) so, presumably, is more than sufficiently accurate for engineering purposes.

The kinematic viscosity ν (SI units m²/s) is, by definition

$$\nu = \frac{\mu}{\rho}$$

The Prandtl number is, by definition,

$$Pr = \frac{c_P \mu}{k}$$

For consistency this may be found by substitution. A direct empirical expression in the case of air is

$$Pr = 0.680 + 4.69 \times 10^{-7}(T_K - 540)^2$$

In practice, for normal air cooling

$$Pr \approx 0.70$$

The volumetric (cubical) thermal expansion coefficient of any permanent gas (at constant pressure) is given by

$$\beta = \frac{1}{T_K}$$

The Grashof number (used for convection cooling) is

$$Gr = \frac{\beta g \rho^2 X^3 (T_S - T_A)}{\mu^2}$$

where X is a length dimension, T_S is the surface temperature and T_A is the ambient air temperature. This can be expressed as

$$Gr = C_{Gr} X^3 (T_S - T_A)$$

with a Grashof coefficient

$$C_{Gr} = \frac{\beta g \rho^2}{\mu^2} = \frac{\beta g}{\nu^2}$$

Using $\beta = 1/T_K$, this becomes

$$C_{Gr} = \frac{g \rho^2}{T_K \mu^2}$$

Appendix C
Properties of Water

Water is occasionally used as a damper fluid in special applications, but is mainly of interest as a cooling medium for severe duty applications or testing.

In liquid or gaseous form (steam), the molecular formula is H_2O, with a relative molecular mass ('molecular weight') of 18.015 kg/kmol.

Avogadro's number is 6.0225×10^{26} molecules/kmol, so the mass of one water molecule is 29.9×10^{-27} kg.

At a reference temperature of 15°C, the following values are applicable:

$$\rho = 999.1 \text{ kg/m}^3$$
$$\mu = 1.139 \times 10^{-3} \text{ N s/m}^2$$
$$\nu = 1.140 \times 10^{-6} \text{ m}^2/\text{s}$$
$$k = 0.596 \text{ W/m K}$$
$$c_P = 4186 \text{ J/kg K}$$
$$Pr = 7.82$$
$$B = 2.15 \text{ GPa (bulk modulus)}$$
$$\sigma = 73.5 \text{ mN/m}$$

At an average cooling temperature of 50°C, the values are:

$$\rho = 988.0 \text{ kg m}^{-3}$$
$$\mu = 0.547 \times 10^{-3} \text{ N s/m}^2$$
$$\nu = 0.553 \times 10^{-6} \text{ m}^2/\text{s}$$
$$k = 0.644 \text{ W/m K}$$
$$c_P = 4186 \text{ J/kg K}$$
$$Pr = 3.56$$
$$B = 2.29 \text{ GPa}$$
$$\sigma = 67.9 \text{ mN/m}$$

The commonly quoted density of water of 1000 kg m^{-3} is the value at 4°C only, reducing by about 4% at 100°C to 996 kg m^{-3}, and to 864 kg m^{-3} at 200°C (under pressure), in a highly nonlinear

---

The Shock Absorber Handbook/Second Edition   John C. Dixon
© 2007 John Wiley & Sons, Ltd

manner, the value at 4°C being a maximum. Hence a simple constant thermal expansion coefficient is not applicable over any significant temperature range. For cooling purposes an average value of

$$\rho = 998 \, \text{kg/m}$$

can be used (2% accuracy for 0–100°C). Where a more accurate value is desired

$$\rho = 1001.3 - 0.155 \, T_C - 2.658 \times 10^{-3} T_C^2$$

where T_C is celsius (centigrade), has 0.2% accuracy from 0 to 200°C. Above 80°C of course, pressurisation of a water cooling circuit becomes necessary to prevent evaporation or boiling, but this has little effect on the density.

The dynamic viscosity μ reduces considerably with temperature according to:

$$\log_{10} \mu = -2.750 - 0.0141 \, T_C + 91.9 \times 10^{-6} T_C^2 - 311 \times 10^{-9} T_C^3$$

for μ in Pa s (N s/m^2), which is 0.5% accurate for 3–100°C.

The specific thermal capacity c_P varies slightly from 0 to 100°C, but for cooling may generally be considered constant at

$$c_P = 4200 \, \text{J/kgK}$$

with accuracy 0.05% for 0–100°C but 7% error at 200°C. For better accuracy above 100°C,

$$c_P = 4209 - 1.31 \, T_C + 0.014 \, T_C^2 \; \text{J kg}^{-1} \, \text{K}^{-1}$$

which is within 0.2% for 3–200°C.

The thermal conductivity k of water varies significantly, with

$$k = 0.5706 + 1.756 \times 10^{-3} T_C - 6.46 \times 10^{-6} T_C^2$$

which is within 0.3% from 1 to 200°C.

The Prandtl number may best be found from its definition:

$$Pr = \frac{c_P \mu}{k}$$

It varies considerably, reducing sharply with temperature.

The bulk modulus B at 15°C is 2.15 GPa. This constant value is within 8% from 0 to 100°C. The variation with temperature is given by

$$B = 2.29 \times 10^9 (1 - 48 \times 10^{-6} (T_C - 53)^2)$$

which is within 1% for 0–100°C.

The surface tension σ of water against air is 73.5 mN/m at 15°C. A constant value of 68 mN/m is within 12% from 0 to 100°C. The expression

$$\sigma = 0.0760 - 1.677 \times 10^{-4} T_C$$

is within 0.6% from 0 to 100°C.

Appendix D
Test Sheets

The following pages show the test result sheets which have been used for some years in the author's laboratory. For many purposes these have now been superseded by automatic control and data acquisition. However, they illustrate the principles of basic sinusoidal testing well, and also show the other parameters which can usefully be measured or calculated.

Z_{max}	mm	ram position for fully extended damper
Z_{min}	mm	ram position for fully compressed damper
S	mm	stroke, $Z_{max} - Z_{min}$
Z_{mid}	mm	$\frac{1}{2}(Z_{max} + Z_{min})$ is mid stroke
Z_0	mm	convenient rounded value of Z_{mid}
X_{max}	mm	maximum allowable amplitude
F_{SCi}	N	'static' (creep in) central compression force
F_{SCo}	N	'static' (creep out) central compression force
F_G	N	gas force $\frac{1}{2}(F_{SCi} + F_{SCo})$
F_C	N	Coulomb friction force $\frac{1}{2}(F_{SCi} - F_{SCo})$
A_R	m²	rod cross-sectional area
P_G	Pa	gas pressure (F_G / A_R)
Z_1	mm	first position for stiffness measurement
Z_2	mm	second position
F_1	N	first force (static)
F_2	N	second force
K	N/mm	stiffness $(F_1 - F_2)/(Z_1 - Z_2)$

The Shock Absorber Handbook/Second Edition John C. Dixon
© 2007 John Wiley & Sons, Ltd

DAMPER TEST TABLE F1 RACE OU/JCD

DAMPER	DATE:
	RUN:

SPEC:

$Z_{max} =$ $F_{SCi} =$ $Z_1 =$
$Z_{min} =$ $F_{SCo} =$ $Z_2 =$
Stroke $=$ $F_G =$ $F_1 =$
$Z_{mid} =$ $F_C =$ $F_2 =$
$Z_0 =$ $A_R =$ $K =$
$X_{max} =$ $P_G =$

SPEED m/s	AMPL mm	AMPL setting	Rbd	Bmp	Rbd	Bmp	Rbd	Bmp
ADJ:								
Zero	0.00	0						
0.02	1	0.5						
0.04	2	1.0						
0.06	3	1.5						
0.08	4	2.0						
0.12	6	3.0						
0.16	8	4.0						
0.20	10	5.0						
0.24	12	6.0						
0.32	16	8.0						
0.40	20	10.0						

Sinusoidal test, amplitudes shown for frequency 3.2 Hz.

Test Sheets

DAMPER TEST TABLE RALLY 1 OU/JCD

DAMPER		DATE:
		RUN:
SPEC:		

Z_{max} = F_{SCi} = Z_1 =

Z_{min} = F_{SCo} = Z_2 =

Stroke = F_G = F_1 =

Z_{mid} = F_C = F_2 =

Z_0 = A_R = K =

X_{max} = P_G =

SPEED m/s	AMPL mm	AMPL setting	Rbd	Bmp	Rbd	Bmp	Rbd	Bmp
ADJ:								
Zero	0.00	0						
0.04	2	1.0						
0.08	4	2.0						
0.16	8	4.0						
0.24	12	6.0						
0.32	16	8.0						
0.40	20	10.0						
0.52	26	13.0						
0.60	30	15.0						
0.80	40	20.0						
1.00	50	25.0						
1.20	60	30.0						
1.40	70	35.0						

Sinusoidal test, amplitudes shown for frequency 3.2 Hz.

Appendix E
Solution of Algebraic Equations

E.0 Introduction

The need for solution of low-order algebraic polynomial equations with real coefficients occurs in vehicle ride analysis, and also in the solution of some analytical models of valves. Some computer packages offer iterative numerical solutions. Terse sets of equations are available in Spiegel (1968, Mathematical Handbook), etc, but the solution methods required to implement a good computer program do not seem to be readily available, so they are summarised here. Even the simple quadratic equation produces some numerical difficulties, and division by zero or attempting to square root a negative number can easily occur. Fortran provides a complex number type.

E.1 The Linear Equation

The linear equation is

$$ax + b = 0$$

and could hardly be expected to need a subroutine for its solution. Even here, however, there is the danger of dividing by zero when evaluating $x = -b/a$. Even worse, perhaps a and b are both zero. Such occasional problems, not pressing at the algebraic solution stage, are handled easily enough by a human, but need careful consideration in a computer program.

E.2 The Quadratic Equation

The general form of the quadratic is

$$ax^2 + bx + c = 0$$

The general solution in the usual standard form is

$$x = \frac{-b \pm \sqrt{b^2 - 4ac}}{2a}$$

The other standard form is

$$x = \frac{2c}{-b \pm \sqrt{b^2 - 4ac}}$$

The discriminant is d, where

$$d = b^2 - 4ac$$

For a positive discriminant, there are two real solutions. For $d = 0$ there are two coincident real solutions. For $d < 0$ the roots are complex (or simply imaginary for $b = 0$).

Special cases require careful consideration. When $a = 0$, to a mathematician there is an infinite solution. To an engineer, this is probably a case of calling a subroutine with an incorrect a value, and an error should be reported by the subroutine.

The case of $b = 0$ gives $x = \pm\sqrt{-c/a}$.

The special case of $c = 0$ gives $x_1 = 0$ and $x_2 = -b/a$.

The standard general solution is not the best method for numerical evaluation, because of inaccuracies if a or c is small, causing subtraction of similar values. The solution is to obtain the more accurate root first, by avoiding the subtraction, and then using the general property that the product of the roots equals c/a. This can be expressed conveniently as

$$t = -\tfrac{1}{2}(b + \text{sgn}(b)\sqrt{d})$$
$$x_1 = \frac{t}{a} \qquad x_2 = \frac{c}{t}$$

When the quadratic equation is used in iterative solutions of other equations, it is often desired to find only the quadratic root of smaller magnitude, this giving the nearer of two prospective roots to an existing root estimate. This smaller magnitude root alone may be obtained directly by

$$x = \frac{-2c}{b + \text{sgn}(b)\sqrt{b^2 - 4ac}}$$

E.3 The Cubic Equation

Since the introduction of computers, most commercial programs have solved the cubic by taking advantage of the fact that a cubic always has at least one real root, iterating on a real root from a user-supplied root estimate, and then dividing out that root to give an easily-solved remnant quadratic.

In contrast, analytical solutions of the cubic have a long and fascinating history, and form the basis of a useful alternative computer solution not requiring a root estimate. Around 1500 BC the Babylonians were regularly solving such equations by numerical methods in their sexagesimal system (base 60) (Neugebauer, 1969). Omar Khayham wrote about geometric cubic solutions using conics circa year 1100 (also giving quadratic solutions by geometric and algebraic methods). The cubic was apparently first solved partially algebraically by Scipione del Ferro, around 1500, but not published. In 1535, Niccolo 'Tartaglia' Fontana achieved an algebraic solution, and revealed it in confidence to gambling analyst Girolamo Cardano. The latter published it in 1545, without permission, but with acknowledgement of its source, possibly also solved independently, in his book *Ars Magna*, all part of the excitement of the Italian Renaissance. This solution is usually known as Cardano's method, following the usual academic tradition of recognising first publication. This solution failed in the case of three real roots. This case was solved algebraically in 1615 by Francois Viète (or Vieta), an amateur mathematician who had a colourful career, including breaking the Spanish code in the French–Spanish war of the period. Viète had obtained a simple expression for cos 3θ in terms of cos θ and $\cos^3\theta$, and recognised that this could be used to solve the cubic. This is known as Viète's method, or trigonometric analogy.

The algebraic equations for the solution may be found in mathematical handbooks, but a good computer implementation is not trivial and requires some effort, again with careful attention to special cases.

The general form of the cubic is

$$ax^3 + bx^2 + cx + d = 0$$

The reduced form with no term in x^2 may always be obtained by substituting

$$x = y - \frac{b}{3a}$$

giving

$$y^3 + py + q = 0$$

with

$$p = \left(\frac{c}{a}\right) - \frac{1}{3}\left(\frac{b}{a}\right)^2$$

$$q = \frac{2}{27}\left(\frac{b}{a}\right)^3 - \frac{1}{3}\left(\frac{b}{a}\right)\left(\frac{c}{a}\right) + \left(\frac{d}{a}\right)$$

The cubic discriminant is then

$$u = q^2 + p^3$$

For $u > 0$ there is one real solution and a pair of complex conjugate ones. For $u = 0$ (requiring $p < 0$ unless $p = q = 0$) there are generally two distinct real solutions x_1 and $x_2 = x_3$, unless $s = 0$ (below) in which case all three roots are equal. For $u < 0$ there are three distinct real solutions. The method of solution varies with u.

For $u > 0$ (one real solution)

$$s = \sqrt[3]{q + \sqrt{u}}$$

$$t = \sqrt[3]{q - \sqrt{u}}$$

The real solution is

$$x_1 = s + t - \frac{1}{3}\left(\frac{b}{a}\right)$$

The two complex solutions are

$$z = x \pm iy$$

with

$$x = -\frac{1}{2}(s+t) - \frac{1}{3}\left(\frac{b}{a}\right)$$

$$y = \frac{\sqrt{3}}{2}(s-t)$$

For $u = 0$, the above applies, giving

$$s = t = \sqrt[3]{q}$$

$$x_1 = 2s - \frac{1}{3}\left(\frac{b}{a}\right)$$

$$x_2 = -s - \frac{1}{3}\left(\frac{b}{a}\right)$$

$$x_3 = x_2$$

For $u < 0$ (three real solutions, requiring $p < 0$), after Viète, considering the reduced form, substitute

$$y = \lambda \cos\theta$$

giving

$$\lambda^3 \cos^3\theta + p\lambda\cos\theta + q = 0$$

so

$$4\cos^3\theta + \frac{4p}{\lambda^2}\cos\theta + \frac{4q}{\lambda^3} = 0$$

Compare this with the standard trigonometric identity (due to Viète)

$$\cos 3\theta = 4\cos^3\theta - 3\cos\theta$$

which gives

$$4\cos^3\theta - 3\cos\theta - \cos 3\theta = 0$$

For equivalence,

$$-3 = \frac{4p}{\lambda^2}$$

and

$$-\cos 3\theta = \frac{4q}{\lambda^3}$$

Solution of Algebraic Equations

Hence, where $p < 0$ is assured for this case,

$$\lambda = \sqrt{-\frac{4p}{3}}$$

$$\theta = \frac{1}{3}\arccos\left(-\frac{4q}{\lambda^3}\right)$$

$$= \frac{1}{3}\arccos\left(-\sqrt{\frac{27q^2}{4p^3}}\right)$$

The result of the arccos may be incremented by 2π and 4π radians from the primary value. Hence three possible results are obtained:

$$y_1 = \cos\theta$$
$$y_2 = \lambda\cos\left(\theta + \tfrac{2}{3}\pi\right)$$
$$y_3 = \lambda\cos\left(\theta + \tfrac{4}{3}\pi\right)$$

and in each case

$$x = y - \tfrac{1}{3}\left(\frac{b}{a}\right)$$

giving the three real solutions.

The other method of cubic solution is to find one real root x_1, by any means available, e.g. iteration, and then to obtain the associated quadratic factor:

$$(ax^2 + mx + n)(x - x_1) = ax^3 + bx^2 + cx + d$$

where

$$m = b + ax_1$$
$$n = c + mx_1$$

This is effective if the first (real) root x_1 is easily found, but not better in general. Numerical methods will reliably find one real root of a cubic equation, and will do so quite efficiently if an estimate of the root is available. Iterative solutions are widely used to find a real root of the cubic, particularly when a reasonable root estimate is available.

The writing of a program to solve the cubic analytically is a worthwhile exercise. Most of those who do so find that during development the program exhibits some unusual characteristics, contrasting with most programs which produce wrong results at the least opportunity.

E.4 Quartic Equation

Several methods exist for solution of the general quartic equation (with real coefficients)

$$ax^4 + bx^3 + cx^2 + dx + e = 0$$

The first solution was possibly due to Ludovico Ferrari, also published by Cardano in 1545. Mathematician Valmes was burned at the stake by the Spanish Inquisition for saying that he had a solution. All analytical methods amount to much the same thing, obtaining a resolvent cubic the solution of which allows the quartic to be factorised into two quadratics with real coefficients. It was known that this could always be done by a cubic resolvent before it was known how to solve the cubic analytically.

One common method is based on reducing the above quartic to eliminate the cubic term, leading to the resolvent cubic equation.

Alternatively, and more concisely, the quartic may be solved by Brown's method, again factoring into quadratics by a cubic solution, but attacking the general form directly, expressing it as the difference of two squares. Considering

$$x^4 + \frac{b}{a}x^3 + \frac{c}{a}x^2 + \frac{d}{a}x + \frac{e}{a} = 0$$

as

$$\{x^2 + (A+C)x + (B+D)\}\{x^2 + (A-C)x + (B-D)\} = 0$$
$$(x^2 + Ax + B)^2 - (Cx + D)^2 = 0$$

gives

$$A = \frac{b}{2a}$$
$$A^2 + 2B - C^2 = \frac{c}{a}$$
$$AB - CD = \frac{d}{2a}$$
$$B^2 - D^2 = \frac{e}{a}$$

Solving the above equations simultaneously gives the resolvent cubic equation

$$8B^3 + \left(-\frac{4c}{a}\right)B^2 + \left(-\frac{8e}{a} + \frac{2bd}{a^2}\right)B + \left(4\frac{ec}{a^2} - \frac{eb^2}{a^3} - \frac{d^2}{a^2}\right) = 0$$

Solving the above cubic gives B. If there is only one real solution for B, use it. If there is a choice of real solutions for B, choose the one with largest magnitude, to give a real factorisation of the quartic (real D). There may be more than one real factorisation. Parameter A is already known. The other coefficients are then

$$D = \sqrt{B^2 - \frac{e}{a}}$$

$$C = \frac{AB - \frac{d}{2a}}{D}$$

Hence, the differences-of-squares equation is determined. The quadratic factor equations are then

$$x^2 + (A+C)x + (B+D) = 0$$
$$x^2 + (A-C)x + (B-D) = 0$$

The three possible roots of the cubic equation give three different pairs of quadratic factor equations, that is different combinations of the four individual roots. The number of possible real factorisations depends on the roots of the quartic. If the roots are two pairs of complex conjugates, then only one factorisation has quadratics with real coefficients. If all roots are real, then in general there are three real factorisations, pairing up the roots in three different ways. In that case, the resolvent cubic will have three real roots.

The above complete solution requires a fair amount of algebra, in total requiring the solution of one cubic plus two quadratics, plus some additional work. In practice, numerical iteration methods may be as good in this case, although the analytic method above is more predictable in computation time. The analytic solution is subject to data sensitivities, that is there are combinations of coefficients, not easily predicted, which cause computational errors due to loss of accuracy in subtractive cancellation, etc. so a really effective computer program is not easily implemented. Integer-valued arguments are particularly troublesome until the program is well developed. Also, many special cases need to be considered and dealt with.

E.5 Fifth Order and Above

Although some special cases were solved by Hermite in 1858 using elliptical integrals, for general equations of higher order than fourth there is no analytic solution, and numerical methods must be employed.

For all odd-order equations, there is at least one real root, which can be bracketed and found iteratively. This is then factored out to leave an even-order equation.

For even orders above four, one method is to find a quadratic factor; this involves two-dimensional iteration which is difficult (i.e. of dubious reliability unless root estimates are available) because bracketing is not possible. Alternatively, a single root may be sought, possibly complex. Press *et al.* (1986) give a Laguerre method routine for polynomial solution. There are also several other approaches, e.g. Madsen's method.

In most engineering cases, of course, physical insight, previous experience or approximate solutions will yield good approximate root values as a starting point for iterative refinement.

Appendix F
Units

This book is, of course, in S.I. units, but Imperial ('English') units are still in everyday engineering use in the USA. The following conversion factors may therefore be useful. Note that lbm denotes pound mass, lbf denotes pound force (the weight force that acts on one pound mass due to standard gravity), kg denotes kilogram mass of course, whilst kgf denotes a 'kilogram force', the weight of one kg mass in standard gravity. The kgf is sometimes denoted the kp, the kilopond, which is in common use in Continental Europe. This must not be confused with the kip, a US unit of 1000 lb. The force units kgf and kp are not true S.I., in which forces are always in newtons. The English and international spelling of the length unit is *metre*, except for the US and Germany who spell it *meter*.

Conversion factors:

Length (SI m, metre):
1 inch	=	0.025400 m (exact by definition)
1 foot	=	0.304800 m
1 m	=	39.3701 inch

Mass (SI kg):
1 kg	=	2.20462 lb m
1 lbm	=	0.453592 kg
1 slug	=	32.17400 lb m
1 slug	=	14.5939 kg (1 lbf s^2/ft by definition)
1 snail	=	175.127 kg (1 lbf s^2/in by definition)
1 lbf s^2/ft	=	14.5939 kg
1 lbf s^2/in	=	175.127 kg

Density (SI kg/m^3):
1 lbm/ft^3	=	16.01846 kg/m^3
1 oz/ft^3	=	1.001154 kg/m^3
1 lbm/in^3	=	27.67990 kg/dm^3
1 lbf s^2/in^4	=	10.68688 kg/cm^3

Force (SI N, newton):
1 N	=	0.224809 lbf
1 N	=	7.23300 poundal
1 kgf	=	1 kp

The Shock Absorber Handbook/Second Edition   John C. Dixon
© 2007 John Wiley & Sons, Ltd

1 kgf	=	9.80665 N
1 lbf	=	4.44822 N
1 lbf	=	0.453592 kgf (kp)

Pressure (SI Pa = N/m^2, pascal):

1 Pa	=	1 N/m^2 (by definition)
1 Pa	=	1.45038×10^{-4} psi
1 Pa	=	0.0208855 lbf/ft^2
1 MPa	=	145.038 psi
1 psi	=	6.89476 kPa
1 bar	=	100 kPa (definition)
1 Ata	=	101325 Pa (standard)

Energy (SI J, joule):

1 J	=	1 N m (joule, by defintion)
1 J	=	0.737562 ft lbf
1 ft lbf	=	1.355817 J
1 Btu	=	1055.06 J

Power (SI W, watt):

1 W	=	1 J/s (1 watt, by definition)
1 ft lbf/s	=	1.35582 W
1 horsepower	=	745.70 W

Stiffness (SI N/m):

1 lbf/in	=	175.127 N/m ≈ 175 N/m
1 kN/m	=	5.71015 lbf/in

Damping (SI Ns/m):

1 lbf s/in	=	175.127 N s/m ≈ 175 Ns/m
1 kN s/m	=	5.71015 lbf s/in
1 kN s/m	=	68.5218 lbf s/ft

Valve area coefficient (SI m^2/Pa):

1 mm^2/MPa	=	10^{-12} m^4/N = 10^{-12} m^3 s^2/kg
1 mm^2/MPa	=	10.6869×10^{-6} in^2/psi
1 mm^2/MPa	=	10.6869 in^2/Mpsi

Magnetics:

1 T (tesla)	=	1 Wb/m^2 (flux density)
1 H (henry)	=	1 Wb/A (inductance)

In S.I., the units of damping coefficient are

$$\frac{N}{m/s} = N\,s/m$$

This reduces to the fundamental units kg/s. The practical-valued unit of kNs/m reduces to t/s (tonne/s) where the tonne is the metric 'ton', i.e. 1000 kg. These reduced forms are not commonly used, and kNs/m is preferred.

In Imperial units, damper velocities are usually in inches/s and forces in lbf, so the natural unit of damping coefficient is the lbf s/inch, which is a practical size of unit. The lbf s/ft is also usable, but rarer, and reduces to slug/s, although rarely expressed in the latter form. A practical passenger car damper of 2 kN s/m is 11.4 lbf s/in or 137 lbf s/ft.

The S.I. unit of kNs/m is reasonable, and is especially in its reduced (but unusual) form of t/s is compact. Perhaps for this reason there has been no move to introduce an explicit S.I. unit for damping coefficient. Such a unit, which tentatively might be named the horock, after the inventor of the telescopic damper, C.L. Horock, of the 1901 patent, would then have the S.I. symbol Ho (to be distinguished from the henry H, unit of magnetic inductance). The practical unit is then the kHo, the kilohorock. Alternatively, an S.I. unit for mass flow rate (kg/s) would be adaptable to damping coefficient, and if introduced would probably become used for the latter. The full naming of kilonewton-seconds per metre is rather laborious, but the simpler 'k.n.s.m.' is satisfactory for oral communications, and probably more convenient than a special name such as kilohorock. Even terser would be 't.p.s.' for tonnes per second.

From the practical point of view, for the foreseeable future the practical S.I. unit of damping coefficient is likely to remain the kNs/m, with t/s as a compact alternative for the more adventurous.

The introduction of magnetorheological dampers brings the magnetic units into relevance. The magnetic field (A/m) must be distinguished from the consequent flux density. The flux at all cross-sections of a simple magnetic circuit is constant, and measured in Wb (webers), or usually mWb for the practical unit. At any particular section there is a cross-sectional area giving a flux density in Wb/m^2 (weber/metre^2). An alternative name for the Wb/m^2 is the tesla, abbreviated T. The reluctance of a magnetic circuit indicates the current-turns product in A-turns required to produce a weber of flux, so is measured in A/Wb, or MA/Wb as a practical unit. The henry (H) is the unit of magnetic inductance, and is just a special name for the Wb/A (weber/ampère). All calculations could be expressed in Wb, W/m^2 and so on. The units tesla and henry are just conveniences. The permittivity of a material is usually expressed as H/m (henry/metre), but could equally be expressed as Wb/A m.

The correct use of capital letters in units is essential. Consider, for example, the possible errors in incorrect use of kg, Kg, KG and kG, which would correctly mean kilogram, kelvin gram, kelvin gauss and kilogauss, although the gauss is not SI).

Appendix G
Bingham Flow

Nomenclature:

F	N	driving or resisting force
h	m	spacing of plates
L	m	length of pipe or plate
P	Pa	driving pressure
P_{inc}	Pa	pressure to just cause yeilding
Q	m³/s	volumetric flow rate
r	m	radial position
r_Y	m	yield radius in circular pipe flow
R	m	circular pipe inner radius
τ_Y	Pa	shear yield stress
U	m/s	flow velocity
U_{max}	m/s	maximum velocity (central slug)
W	m	plate width
y	m	distance from centre plane between plates
y_Y	m	yield level position
Z		nondimensional position in shearing flow
μ	Pa s	marginal dynamic viscosity (after yielding)

G.1 Bingham Flow

A Bingham plastic or liquid is a material with a yield stress that subsequent to yield behaves as a liquid with a viscosity. This is of interest to damper analysis because electrorheological and magnetorheological fluids do this. Their yield stress depends on the electrostatic field or magnetic field, respectively.

Considering the flow of a Bingham material through a circular pipe, for a small pressure there will be no flow. The pressure must be sufficient to cause yielding at some point. Considering a circular rod element along the pipe, the driving force is proportional to the pressure and the circular end area, whereas the shear resistance is proportional to the yield stress and the circumference. Therefore, yielding will first occur at the largest radius. At higher driving pressures, there will be an unyielded central core flowing as a solid slug, with an annular shearing region around it.

G.2 Bingham Flow Between Plates

Consider two parallel rectangular plates of length L, width W, spacing h. Neglect edge effects. The Bingham liquid has a shear yield stress τ_Y and subsequent marginal viscosity μ. The driving pressure is P giving flowrate Q.

The driving force is $F = Pwh$. The resistive shear force without flow may be as large as $F = 2Lw\tau_Y$. The pressure for incipient flow is therefore

$$P_{inc} = \frac{2L\tau_Y}{h}$$

Driving pressure exceeding this gives a nonzero flow rate. For higher pressures, a central region flows as a slug, with sections above and below in flowing shear.

The central region of 'slug' flow has a total depth $2\,y_Y$ where y_Y is the yield position. Again by equating driving and resisting forces

$$2y_Y wP = 2wL\tau_Y$$

$$y_Y = \frac{L\tau_Y}{P} \quad (P > P_{inc})$$

In positive y the flow in fluid shear has a negative velocity gradient dU/dy, so here for positive marginal viscosity μ, and equating forces,

$$\tau = \tau_Y - \mu\frac{dU}{dy} = \frac{yP}{L}$$

giving

$$\frac{dU}{dy} = -\frac{P}{\mu L}(y - y_Y)$$

By standard integration, and considering the boundary conditions at $y = y_Y$ (and $h/2$), it is easy to show that

$$U_{Max} = \frac{P}{8\mu L}(h - 2y_Y)^2$$

$$U = U_{Max} - \frac{P}{2\mu L}(y - y_Y)^2$$

$$= U_{Max}(1 - z^2)$$

with

$$z = \frac{y - y_Y}{\frac{h}{2} - y_Y}$$

The volumetric flow rate is then found by integrating the velocity over the various sections, totalling

$$Q = \frac{2}{3}wU_{Max}(h + y_Y)$$

When P is known, the above equations easily find the volumetric flow rate, i.e. solve for the yield level y_Y, then U_{MAX} and finally Q.

To find the pressure at a known flow rate (the practical damper problem) is more difficult. By substitution

$$Q = \frac{Pw}{3\mu L}\left(\frac{h}{2} - \frac{L\tau_Y}{P}\right)^2 \left(\frac{L\tau_Y}{P} + h\right)$$

which gives a cubic equation for P:

$$P^3\{h^3\} + P^2\left\{-3L\tau_Y h^2 - 12\frac{\mu L Q}{w}\right\} + P^1\{0\} + P^0\{4L^3\tau_Y^3\} = 0$$

It appears that when there are several real solutions it is the maximum one that is correct. Then from P, if desired, y_Y and U_{Max} may easily be calculated.

Alternatively, the flow rate may be solved indirectly, by forming a cubic equation for the yield level:

$$y_Y^3\{4\} + y_Y^2\{0\} + y_Y\left\{-3h^3 - 12\frac{\mu Q}{w\tau_Y}\right\} + \{h^3\} = 0$$

However, it seems that this can also give a spurious real solution within the range zero to $h/2$.

G.3 Bingham Flow in a Circular Pipe

The solution for Bingham flow in a pipe is similar to that between flat plates, with the yield level y_Y being replaced by a yield radius r_Y. The method is the same, simply applying the new geometry, so considering a central cylindrical slug at uniform velocity, which is surrounded again by sections with a parabolic velocity distribution. The pipe inner radius is R, and inner diameter D. Flow is initiated when the pressure is given by

$$F = P_{inc}\pi R^2 = 2\pi R L \tau_Y$$

$$P_{inc} = \frac{4L\tau_Y}{D} = \frac{2L\tau_Y}{R}$$

For higher driving pressures, the yield radius is

$$r_Y = \frac{2L\tau_Y}{P}$$

By obtaining the differential equation for dU/dr, and a slightly more difficult integration

$$U_{Max} = \frac{P}{4\mu L}(R - r_Y)^2$$

In the shearing flow region

$$U = U_{Max}(1 - z^2)$$

with

$$z = \frac{r - r_Y}{R - r_Y}$$

Integrating for the volumetric flow rate gives

$$Q = \frac{\pi P}{24\mu L}\{3R^4 - 4r_Y R^3 + r_Y^4\}$$

The above equations solve for r_Y, U_{Max} and Q when P is known.

When the flow rate is known instead of the driving pressure, a quartic equation in P must be solved:

$$P^4\{3R^4\} + P^3\left\{-8L\tau_Y R^3 - \frac{24}{\pi}\mu L Q\right\} + P^2\{0\} + P^1\{0\} + P^0\{(2L\tau_Y)^4\} = 0$$

Use the largest real solution.

Alternatively, solving first for the yield radius r_Y, a quartic with zero cubic and quadratic terms may be formed:

$$r_Y^4\{1\} + r_Y^3\{0\} + r_Y^2\{0\} + r_Y^1\left\{-4R^3 - \frac{12\mu Q}{\pi \tau_Y}\right\} + r_Y^0\{R^4\} = 0$$

References

[1] Abd-El-Tawwab A. M., Crolla D. A. and Plummer A. R. (1998) The Characteristics of a Three-State Switchable Damper, 'Journal of Low Frequency Noise, Vibration and Active Control', **17**(2), 85–96.
[2] Ahmed A.K.W. and Rakheja S. (1992) Equivalent Linearisation Technique for the Frequency Response Analysis of Asymmetric Dampers, 'Journal of Sound and Vibration', **153**(3), 537–542.
[3] Akers A. (1984) Discharge Coefficients for Long Orifices, *Proc. Exptl. Mechanics 5th Int. Congress*, (EI Conf No. 06802), pp 642–647.
[4] Alonso M. and Comas A. (2006) Modelling a Twin Tube Cavitating Shock Absorber, *Proc. I. Mech. E.*, Part D:, 'Journal of Automobile Engineering', **220**, 1031–1040.
[5] Angermann R. (1995) Lightweight Shock Absorbers, *Technische Mitteilungen Krupp* (English Edition) December 1995, No. 2, pp 65–68.
[6] Anonymous (1992) Directive 92/7/EC, (on definition and testing of 'road friendly suspensions'), Council of the European Communities.
[7] Arndt E.M., Spies K.H. and Trauth W. (1981) Some Comments on Comfort Shock Absorber Seals, SAE 810203.
[8] Audenino A.L. and Belingardi G. (1995) Modelling The Dynamic Behaviour of a Motorcycle Damper, 'Proc. I. Mech. E., Part D', **209**(4), 249–262. (Testing and modelling).
[9] Baracat D. E. (1993) A Proposal for Mathematical Design of Shock Absorbers, SAE Paper 931691, SAE Brazil '93 Mobility Technology Conference, Sao Paulo, October 1993.
[10] Bastow D. (1980) *Car Suspension and Handling*, 1st edn., Pentech Press, pp 232–250, ISBN 0-7273-0305-8, 1980.
[11] Bastow D. (1987) *Car Suspension and Handling*, 2nd edn., Pentech Press, ISBN 0-7273-0316-3, 1987.
[12] Bastow D. and Howard G. (1993) *Car Suspension and Handling*, 3rd edn., Pentech Press and SAE, ISBN 0-7273-0318-X, or 1-56091-404-1.
[13] Batchelor G. K. (1967) *An Introduction to Fluid Dynamics*, Cambridge University Press, ISBN 052-109-817-3. Useful for some aspects of bubbles and particles in oil.
[14] Besinger F.H., Cebon D. and Cole D.J. (1991) Experimental Investigation into the Use of Semi-active Dampers on Heavy Lorries, *Vehicle System Dynamics*, **20**, Supplement, 57–71. (Also *Proc 12th IAVSD Symposium*, Conference 16282)
[15] Besinger F.H., Cebon D. and Cole D.J. (1995) Damper Models for Heavy Vehicle Ride Dynamics, 'Vehicle System Dynamics', **24**(1), 35–64.
[16] Blevins R.D. (1984) *Applied Fluid Dynamics Handbook*, Van Nostrand Reinhold, 1984, ISBN 0-442-21296-8.
[17] Bolt J.A. Derezinski S.J. and Harrington D.L. (1971) Influence of Fuel Properties on Metering in Carburetors, SAE 710207, SAE Trans V80, 1971.
[18] Bradley J. (1996) *The Racing Motorcycle*, Broadland Leisure Publications, ISBN 095-129-292-7.
[19] Brown J. (1948) Further Principles of Buffer Design, 'Product Engineering', **21**, 125–129.
[20] Brown J. (1948) Hydraulic Shock Absorber Orifice Designs and Equations, 'Product Engineering', **19**, 92–95, February and April.
[21] Brown J. (1950) Selection Factors for Mechanical Buffers, 'Product Engineering', November 1950, 156–160.
[22] Browne A.L. and Hamburg J.A. (1986) On-road Measurement of The Energy Dissipated in Automotive Shock Absorbers, *American Society of Mechanical Engineers, Applied Mechanics Division (Symposia Series)*, (Conference 09203) **80**, 167–186.

[23] Busshardt J. and Isermann R. (1992) Realisation of Adaptive Shock Absorbers by Estimating Physical Process Coefficients of a Vehicle Suspension System, *Proc. American Control Conference* (Conf. 17380), June 1992, **1**, 531–535.
[24] Cafferty S. and Tomlinson G.R. (1997) Characterization of Automotive Dampers Using Higher Order Frequency Response Functions, 'Proc. I. Mech. E., Part D', **211**, 181–203.
[25] Cafferty S., Worden K. and Tomlinson G. (1995) Characterization of Automotive Shock Absorbers using Random Excitation, 'Proc. I. Mech. E., Part D', **209**(4), 239–248. (Testing and modelling).
[26] Campbell C. (1981) *Automobile Suspensions*, Chapman and Hall, ISBN 0-412-15820-5, 1981.
[27] Carlson J. D. and Chrzan M. J. (1994a) US Patent 5,277,281, Magnetorheological Fluid Dampers, filed 18 June 1992, dated 11 Jan 1994. Includes various prototype designs and a list of related patents.
[28] Carlson J. D., Chrzan M. J. and James F. O. (1994b) Magnetorheological Fluid Devices, US Patent 5284330.
[29] Cebon D. and Newland D. E. (1984) The Artificial Generation of Road Surface Topography by the Inverse FFT Method, *8th IAVSD conference Dynamics of Vehicles on Roads and Tracks*, pp 29–42.
[30] Charalambous C., Brunning A. and Crawford I.L. (1989) The Design and Advanced Development of a Semiactive Suspension, *I. Mech. E., Conference C382/058*, pp 539–546.
[31] Choi, Seung-Bok (2003) Performance Comparison of Vehicle Suspensions Featuring Two Different Electrorheological Shock Absorbers, *Proc. I. Mech. E.*, Part D, **217**, 'Journal of Automobile Engineering', 999–1010.
[32] Cline R.C. (1958) Are Shock Absorbers Here To Stay? SAE paper S106, 5 May 1958.
[33] Cline R.C. (1974) Shock Absorbers: An Integral Part of Recreational Vehicle Developments, SAE 740678, 1974.
[34] Connor B.E. (1946) Damping in Suspensions, 'SAE Jnl (Trans)', **54**(8); 389–393.
[35] Currey N.S. (1988) Aircraft Landing Gear Design: Principles and Practice, AIAA (Education Series), ISBN 093-040-341-X.
[36] Crolla D.A., Firth G.R., Hine P.J. and Pearce P.T. (1989) The Performance of Suspensions Fitted with Controllable Dampers, *Proc 11th IAVSD Symposium*.
[37] Dalibert A. (1977) Progress in Shock Absorber Oil Technology, SAE 770850, 1977.
[38] Decker H., Schramm W. and Bethell M. R. (1990) An Optimised Approach to Suspension Control, SAE Paper 900661.
[39] Den Hartog J.P. (1985) *Mechanical Vibrations*, Dover, ISBN 0-486-64785-4.
[40] Dickerson P. and Rice W. (1969) An Investigation of Very Small Diameter Laminar Flow Orifices, 'Trans ASME, Jnl. Basic. Eng.', **91**, D, P3, 546–548.
[41] Dixon J.C. (1997) *Tires Suspension and Handling*, S.A.E. R-168, ISBN 1-56091-831-4, (1st edition 1991).
[42] Dodge L., Fluid Throttling Devices, in Yeaple (1966) below, pp 89–95, 1966.
[43] Duym S. (1997) An Alternative Force State Map for Shock Absorbers, 'Proc. I. Mech. E., Part D', **211**, 175–179.
[44] Duym S.W., Stiens R., Baron G.V. and Reybrouck K. G. (1997) Physical Modelling of the Hysteretic Behaviour of Automotive Shock Absorbers, SAE Paper 970101, SAE International Congress, Detroit, Feb 1997.
[45] Duym S. W. and Reybrouck K. (1998) Physical Characterization of Nonlinear Shock Absorber Dynamics, 'European Journal of Mechanical Engineering', **43**(4), 181–188.
[46] Duym S. W. R. (2000) Simulation Tools, Modelling and Identification, for an Automotive Shock Absorber in the Context of Vehicle Dynamics, 'Vehicle System Dynamics', **33**, 261–285.
[47] Duym S.W. and Lauwerys X.A. (2001) Methodology for Accelerating Life Tests on Shock Absorbers, SAE Paper 2001-01-1103.
[48] Eberan-Eberhorst R. and Willich J.H. (1962) Beitrag zur Theorie des Schwingungsdämpfers, ATZ., J64, H3, March 1962, pp 81–90.
[49] Elbeheiry E. M., Karnopp D. C., Elaraby M. E. and Abdelraaouf A. M. (1995) Advanced Ground Vehicle Suspension Systems - A Classified Bibliography, 'Vehicle System Dynamics', **24**, 231–258.
[50] Ellis J.T. and Karbowniczek S. (1962) Hydraulic Shock Absorbers, 'Machine Design', **34**(12), 150–157.
[51] Els P. S. and Holman T. J. (1999) Semi-Active Rotary Damper for a Heavy Off-Road Wheeled Vehicle, 'Journal of Terramechanics', **36**, 51–60.
[52] Fan Y. and Anderson R.J. (1990) Dynamic Testing and Modelling of a Bus Shock Absorber, SAE 902282, in SAE SP-843 Total Vehicle Ride, Handling and Advanced Suspensions.
[53] Fash J. W. (1994) Modeling of Shock Absorber behaviour Using Artificial Neural Networks, SAE Paper 940248, SAE International Congress, Detroit, Feb 1994.
[54] Feigel H-J. and Romano N. (1996) New Valve Technology for Active Suspension, SAE Paper 960727, SAE International Congress, Detroit, Feb 1996.
[55] Fuchs H.O. (1933) Die Kennzeichnung von Schwingungsbremsen, 'Automobiltechnische Zeitschrift', **36**, 169–172.

[56] Fukushima N., Hidaka K. and Iwata K. (1983) Optimum Characteristics of Automotive Shock Absorbers under Various Driving Conditions and Road Surfaces, 'JSAE Review', March, 62–69.

[57] Fukushima N., Iida M. and Hidaka K. (1984) Development of an Automotive Shock Absorber that Improves Riding Comfort Without Impairing Steering Stability, *Proc. 20th FISITA Conf.*, pp 218–223.

[58] Gamota D. R. and Filisko F. E. (1991) Dynamic Mechanical Studies of Electrorheological Materials: Moderate Frequencies, 'J. Rheology', **35**, 399–425.

[59] Gillespie T.D. (1992) Fundamentals of Vehicle Dynamics, SAE R-114, ISBN 1-56091-199-9, 1992.

[60] Guglielmino E. and Edge K.A. (2004) A Controlled Friction Damper for Vehicle 'Applications, Control Engineering Practice', **12**, 431–443.

[61] Guest J.J. (1925/26) The Main Free Vibrations of an Autocar, Proc. I. Auto. Engrs., V 20, No 505.

[62] Gvineriya K.I., Dzhokhadze G.D., Kiselev B.A., Kozlova I.V. and Fomchenko V.M. (1989) Calculation of the Construction Parameters of a Pneumatic Damper for Suspension Units of the Type Used in Automobiles, 'Soviet Engineering Research', **9**(3), 18–21.

[63] Hadley N.F. (1928) Shock Absorber Characteristics, 'SAE J.', **22**, paper S356, 1928, pp 356–362.

[64] Hagele K.H., Engelsdorf K., Mettner M., Panther M., Tran Q.N. and Rubel E. (1990) Continuously Adjustable Shock Absorbers for Rapid Acting Ride Control Systems, *Proc. S.A.E.*, 37–46. Also 18th FISITA Congress.

[65] Hall B.B. and Gill K.F. (1986) Performance of a telescopic dual-tube automotive damper and the implications for vehicle ride prediction, 'Proc. I. Mech. E.', **200**(D2), 115–123.

[66] Haney P. (1996) A Comparison of Modern Racing Dampers, SAE 962545, also in SAE SP-304/1, pp 221–230, 1996.

[67] Haney P. and Braun J. (1995) *Inside Racing Technology*, TV Motorsports, ISBN 0-9646414-0-2, 1995.

[68] Harris C.M. and Crede C.E. (1976) *Shock and Vibration Handbook*, 2nd edn, McGraw-Hill Book Company, ISBN 0-07-026799-5.

[69] Hayward A.T.J., Air Bubbles in Oil, in Yeaple (1966) above, pp 29–33.

[70] Hennecke D., Baier P., Jordan B. and Walek E. (1990a) The New Variable Damper System for BMW, S.A.E. paper 900662. Also S.A.E. Proceedings 1990, **99**, No. Sect 6, pp 930–942.

[71] Hennecke D., Baier P., Jordan B. and Walek E. (1990b) Further Market-oriented Development of Adaptive Damper Force Control, *Proc. S.A.E.*, 1990 pp 187–198, also 18th FISITA Congress (Conf 13957).

[72] Herr F., Mallin T., Lane J. and Roth S. (1999) A Shock Absorber Model Using CFD Analysis and Easy 5., SAE Paper 1999-01-1322, SAE International Congress, Detroit, Jan 1999.

[73] Hoffman H.J. (1958) Wirkamseit von Stossdampfern am Fahrzeug, ATZ. (Automobiltechnische Zeitschrift), J60, H10, Oct 1958, p 289.

[74] Holman T.J. (1984) Rotary Hydraulic Suspension Damper for High Mobility Off-Road Vehicles, *Proceedings of Conference on The Performance of Off-Road Vehicles and Machines* (Conference 06422), Cambridge, England, 5–11 August, pp 1065–1075.

[75] Hong S., Rakheja S. and Sankar T. S. (1989) Vibration and Shock Isolation Performance of a Pressure-Limited Hydraulic Damper, *Mechanical Systems and Signal Processing*, **3**(1), 71–86 (argues that a pressure-relief non-linear damper can perform as well as a semi-active damper).

[76] Horowitz P. and Hill W. (1989) *The Art of Electronics*, Cambridge University Press, ISBN 0-521-37095-7.

[77] Hunt J.B. (1979) *Dynamic Vibration Absorbers*, Mechanical Engineering Publications, ISBN 085-298-417-0.

[78] I.S.O. 2041:1990 (E/F), Vibration and Shock - Vocabulary, 2nd edn.

[79] ISO 2631 (1997) Mechanical Vibration and Shock—Evaluation of Human Response to whole Body Vibration.

[80] Idelchik I.E. (1986) *Handbook of Hydraulic Resistance*, 2nd ed; Hemisphere/Springer Verlag, 1986, ISBN 3-540-15962-2.

[81] Iversen H.W. (1956) Orifice Coefficients for Reynolds Numbers from 4 to 50000, 'Trans. ASME', **78**(2), 125–133.

[82] Jackson G.W. (1959) Fundamentals of the Direct Acting Shock Absorber, SAE paper 37R, National Passenger Car Body and Materials Meeting, Detroit, Mar 1959.

[83] James W.S. and Ullery F.E. (1932) An Automatic Shock Absorber, 'SAE Trans.', **30**(5), 185–191.

[84] Japanese Automobile Standards Organisation, (1983) C 602-83, Telescopic Shock Absorbers for Automobiles, revised 17 March 1983, Society of Automotive Engineers of Japan.

[85] Jennings G. (1974) A Study of Motorcycle Suspension Damping Characteristics, SAE 740628.

[86] Karadayi R. and Masada G. Y. (1986) A Non-linear Shock Absorber Model, *Proc. Symposium on Simulation and Control of Ground Vehicles and Transportation Systems*, A.S.M.E., 1986, LCCC 86-72501, pp 149–165.

[87] Karnopp D. (1983) Active Damping in Road Vehicle Suspension Systems, 'Vehicle System Dynamics', **12**(6).

[88] Karnopp D, Crosby M. and Harwood R.A. (1974) Vibration Control Using Semi-active Force Generator 'Jnl. Eng. for Industry', **96** B(2), 619–626.

[89] Karnopp D.C. and Heess G. (1991) Electronically Controllable Vehicle Suspension, Vehicle System Dynamics, **20**, 207–218.
[90] Kasteel R. Wang C-G., Qian L., Liu J-Z and Ye G-H. (2005) A New Shock Absorber Model for Use in Vehicle Dynamics Studies, 'Vehicle System Dynamics', **43**(9), 613–631.
[91] Kastner L.J. and McVeigh J.C. (1965) A Reassessment of Metering Orifices for Low Reynolds Number, 'Proc. I. Mech. E.', **180**(1, N13), 331–356.
[92] Katsuda T., Hiraiwa N., Doi S.I. and Yasuda E. (1992) Improvement of Ride Comfort by Continuously Controlled Damper, SAE SP-917, pp 73–80 (Conference 16381) (on the vehicle).
[93] Kinchin J.W. and Stock C.R. (1951/1952) Shock Absorbers, *Proc. I. Mech. E. (Auto. Div.)*, 67–86.
[94] Kindl C.H. (1933) New Features in Shock Absorbers with Inertia Control, 'SAE J.', **32**(5), 172–176.
[95] Komamura S. and Mizumukai K. (1987) History of Shock Absorbers, 'Automobile Technology', **41**(1), 1000-307, pp 126–131 (in Chinese).
[96] Kumagai K., Abe T., Bretl J.L., Ishigaki T. and Takgi R. (1991) Shock Absorber Vibration Analysis—High Frequency and Low Frequency, 1991 S.A.E. Noise and Vibration Conference (Conf 14843) pp 239–243.
[97] Kutsche T., Raulf M and Becher H-O (1997) Optimised Ride Control of Heavy Vehicles with Intelligent Suspension Control, SAE Paper 973207, SAE International Truck and Bus Meeting, Cleveland, Nov 1997.
[98] Kyoo Sung Park (1994) Performance Improvement of Semi-Active Car Suspension by Adaptive Control, *MPhil Thesis*, Cranfield University, U.K.
[99] LaJoie J.C. (1996) Damper Performance Development, SAE 962551, in SAE SP-304/1, pp 255–261.
[100] Lang H.H. (1977) A Study of the Characteristics of Automotive Hydraulic Dampers at High Stroking Frequencies, *PhD Dissertation*, Universtiy of Michigan, USA.
[101] Lee, K. (1997) Numerical Modelling for the Hydraulic Performance Prediction of Automotive Monotube Dampers, 'Vehicle System Dynamics', **28**, 25–39.
[102] Lee, C-T and Moon B-Y (2005) Study of the Simulation Model of a Displacement-Sensitive Shock Absorber of a Vehicle by Considering the Fluid Force, *Proc. I. Mech. E.*, Part D, **219**, 'Journal of Automobile Engineering', 965–975.
[103] Leih J. (1991) Control of Vibrations in Elastic Vehicles Using Saturation Non-Linear Semi-active Dampers, A.S.M.E. paper 91-WA-AES-1
[104] Lemme C.D. and Furrer F.J. (1990) Hydraulically Controlled Adjustable Dampers, SAE paper 900660, also SAE Transactions, 1990, V99, No. Sect 6, pp 920–929.
[105] Lewitske C. and Lee P. (2001) Application of Elastomeric Components for Noise and Vibration Isolation in the Automotive Industry, SAE Paper 2001-01-1447.
[106] Lichtarowicz A., Duggins R.K. and Markland E. (1965) Discharge Coefficients for Incompressible Non-Cavitating Flow Through Long Orifices, 'J. Mech. E. Sci.', **7**(2), 210–219.
[107] Lieh J. (1993) Effect of Bandwidth of Semiactive Dampers on Vehicle Ride, 'Journal of Dynamic Systems, Measurement and Control, Trans. A.S.M.E.', **115**(3), 571–575.
[108] Lion A. and Loose S. (2002) A Thermomechanically Coupled Model for Automotive Shock Absorbers: Theory, Experiments and Vehicle Simulations on Test Tracks, 'Vehicle System Dynamics', **37**(4), 2241–261.
[109] Maranville C. W. and Ginder J. M. (2003) Magnetorheological Fluid Damper Dynamics: Models and Measurements, Smart Structures and Materials, 'Proc SPIE', **5056**, 524–533.
[110] Margolis D.C. (1982) The Response of Active and Semi-active Suspensions to Realistic Feedback Signals, 'Vehicle System Dynamics', **11**, 267–282.
[111] Mayne R.W. (1973) The Effects of Fluid and Mechanical Compliance on the Performance of Hydraulic Shock Absorbers, 'Trans A.S.M.E. J. Eng. for Industry', paper 73-DET-1, also Conf E.I. 74067597.
[112] McGreehan W.F. and Schotsch M.J. (1988) Flow Characteristics of Long Orifices with Rotation and Corner Radiusing, 'Trans. ASME., Ser A, J. of Turbomachinery', **110**(2), pp 213–217.
[113] Meller T. (1978) Self-Energizing Hydropneumatic Levelling Systems, SAE 780052.
[114] Meller T. (1999) Self-Energizing Leveling Systems - Their Progress in Development and Application, SAE 1999-01-0042, SAE International Congress, Detroit, Jan 1999.
[115] Milliken W.F. and Milliken D.L. (1995) Race Car Vehicle Dynamics, SAE R-146, ISBN 1-56091-526-9.
[116] Mitschke M. and Riesenberg K.O. (1972) Stossdämpferwärmung und Dämpferwirkung, 'ATZ', **74**(4), 133–139.
[117] Morman K.N. (1984) A Modelling and Identification Procedure for the Analysis and Simulation of Hydraulic Shock Absorber Performance, ASME Winter Annual Mtg, New Orleans, Dec 1984.
[118] The Motor Manual (1919) Compiled by Staff of *The Motor*, Temple Press, London.
[119] The Motor Manual (1939) Compiled by Staff of *The Motor*, Temple Press, London.
[120] Mughal S.A. (1979) Coefficient of Discharge and Mass Flow Rate for Reed Type Valves, *Proc. Multi-phase Flow and Heat Transfer Symposium*, Florida, April 1979, **2**, 1191–1204.

[121] Nall S. and Steyn J.L. (1994) Experimental Evaluation of an Unsophisticated Two-State Semiactive Damper, 'J. Terramechanics', **31**(4), 227–238.
[122] Neugebauer O. (1969) *The Exact Sciences in Antiquity*, Dover.
[123] Newland D. E. (1984) *An Introduction to Random Vibrations and Spectral Analysis*, Longman, 2nd edn.
[124] Nickelsen J.M. (1930) Shock Absorbers, SAE Jnl., V26, paper S740, 1930, pp 740–746 and 752.
[125] Norbye J. P. (1980) *The Car and Its Wheels—A Guide to Modern Suspension Systems*, Tab Books Inc., pp 267–273, ISBN 0-8306-2058-3.
[126] Ohtake Y., Kawahara Y., Hirabayashi H., Yamamoto Y. and Iida S. (1981) A Study of Oil Seals for Shock Absorbers of Automotive Suspensions, SAE 810204.
[127] Patten W.N., Kedar P. and Abboud E. (1991) Variable Damper Suspension Design for Phase Related Road Inputs, 'Proc. A.S.M.E., D.E.', **40**, 361–374.
[128] Petek N. K. (1992a) Electronically Controlled Shock Absorber Using Electrorheological Fluid, SAE paper 920058, SAE SP-917, pp 67–72, (Conference 16485).
[129] Petek N. K. (1992b) Shock Absorber Uses Electro-rheological Fluid, 'Automotive Engineering', **100**(6), 27–30.
[130] Petek N. K. (1992c) An Electronically Controlled Shock Absorber Using Electrorheological Fluid, SAE Paper 920275, SAE International Congress and Exposition, Detroit, Feb 1992.
[131] Petek N. K., Romstadt D. J., Lizell m. B. and Weyenberg T. R. (1995) Demonstration of an Automotive Semi-Active Suspension Using Electrorheological Fluid, SAE Paper 950586, SAE International Congress and Exposition, Detroit, Feb 1995.
[132] Peterson R. R. (1953) Hydraulics Applied to the Automobile Suspension, 'Proc. National Conference on Industrial Hydraulics (USA)', **7**, 23–43.
[133] Pinkos A., Shtarkman E. and Fitzgerald T. (1993) An Actively Damped Passenger Car Suspension System with Low Voltage Electro-Rheological Magnetic Fluid, SAE Paper 930268, SAE International Congress and Exposition, Detroit, 1993.
[134] Polak P. and Burton R.T. (1971) Improving Suspension Damping, 'J. Automotive Eng.', **2**(2), 13–17.
[135] Potas M. (2004, 2005, 2006) Australian Patent Applications 2004-906197, 2004-906583, 2006-903438, US Patent Application 11/248,014, European Patent Application 05-023174.5-2425.
[136] Poyser J. (1987) Development of a Computer-controlled Suspension System, 'Int. J. Vehicle Design', **8**(1), 74–86.
[137] Press W.H., Flannery B.P., Teukolsky S.A. and Vetterling W.T. (1986) *Numerical Recipes, The Art of Scientific, Computing*, Cambridge University Press, ISBN 052-130-811-9
[138] Purdy D. J. (2000) Theoretical and Experimental Investigation into an Adjustable Automotive Damper, 'Proc I. Mech. E.', **214**(D3), 265–283.
[139] Puydak RC and Auda RS (1966) Designing to Achieve Optimum Dynamic Properties in Elastomeric Cab and Body Mounts. SAE 660439 (and SAE Transactions V75).
[140] Rabinow J. (1951) US Patent 2,575,360, Magnetic Fluid Torque and Force Transmitting Device, filed 31 Oct 1947, dated 20 Nov 1951.
[141] Rakheja S., Su H. and Sankar T.S. (1990) Analysis of a Passive Sequential Hydraulic Damper for Vehicle Suspension, 'Vehicle System Dynamics', **19**(5), 289–312.
[142] Ramos J. C., Rivas A., Biera J. Sacramento G. and Sala J. A. (2005) Development of a Thermal Model for Automotive Twin-tube Shock Absorbers, 'Applied Thermal Engineering', **25**, 1836–1853.
[143] Reybrouck K., (1994) A Non-linear Parametric Model of an Automotive Shock Absorber, SAE 940869 in SAE SP-1031, Vehicle Suspension System Advancements.
[144] Reimpell J. and Stoll H. (1996) *The Automotive Chassis*, Edward Arnold, ISBN 0-340-61443-9. (Also SAE).
[145] Roark R.J. and Young W.C. (1983) *Formulas for Stress and Strain*, 5th edn, McGraw-Hill, ISBN 0-07-085983-3.
[146] Robson J. D. (1979) Road Surface Description and Vehicle Response, 'Int. J. Vehicle Design.', **1**(1), 25–35.
[147] Rowell H.S. (1922/23) Principles of Vehicle Suspension, Proc. I. Auto. Engrs., V 17, Pt 2.
[148] Ryan J. (1996) The Case for Shaft Displacement Damping and Shaft Displaced Fluid Power, SAE 962547, in SAE SP-304/1, pp 237–240.
[149] S.A.E. (1996) Motorsports Engineering Conference Proceedings, SP-304/1 ISBN 1-56091-894-2, 1996.
[150] Savaresi S.M., Silani E., Montiglio M., Stefanini A. and Trevidi F. (2004) A Comparative Analysis of High Accuracy Black-Box Grey-Box Models of MR-Dampers for Vehicle Control, SAE Paper 2004-01-2006.
[151] Schilling R. and Fuchs H.O. (1941) Modern Passenger Car Ride Characteristics, 'ASME Trans.', **63**, A59–A66.
[152] Schubert D.W. and Racca R.H. (1974) Dynamic Characteristics of an Elastomeric-Pneumatic Isolator with Orifice-Type Relaxation Damping for Vehicular Suspension Applications, SAE 740991.
[153] Segel L. and Lang H. (1981) The mechanics of Automotive Hydraulic Dampers at High Stroking Frequencies, 'Veh. Sys. Dyn.', **10**, 79–83. (Also published in Proc. 7th IAVSD Symp 'Dynamics of Vehicles on Roads and Tracks', pp 194–214.)

[154] Shabazov S.A. and Ashikhmin V.I. (1973) The Problem of the Efflux of Liquids from Drowned External Cylindrical Nozzles, 'Int. Chem. Eng.', **13**(4), 613–615.
[155] Sharp R.S. and Crolla D.A. (1987) Road Vehicle Suspension System Design — A Review, 'Vehicle System Dynamics', **16**(3), 167–192.
[156] Sharp R.S. and Hassan S.A. (1986) Relative Performance Capabilities of Passive, Active and Semi-active Car Suspension Systems, 'Proc. I. Mech. E.', **200**, D3.
[157] Sharp R.S. and Hassan S.A. (1987) Performance and Design Considerations for Dissipative Semi-active Suspension for Automobiles, 'Proc. I. Mech. E.', **201**(D2), 149–153.
[158] Shiozaki M., Kamiya S., Kuroyanagi M., Matsui K. and Kizu R. (1991) High Speed Control of Damping Force Using Piezoelectric Elements, SAE Paper 910661, SAE International Congress and Exposition, Detroit, Feb 1991.
[159] Shulman Z. P., Korobo E. V. and Yanovskii Yu. G. (1989) The Mechanism of the Viscoelastic Behaviour of Electrorheological Suspensions, 'J. Non-Newtonian Fluid Mechanics', **33**, 181–196.
[160] Simanaitis D. J. (1976) Shock Absorbers, 'Automotive Eng.', **84**(11), 34–39.
[161] Smith C. C., Mcgehee D. Y. and Healey A. J. (1978) The Prediction of Passenger Riding Comfort from Acceleration Data, 'Journal of Dynamic Systems, Measurement and Control', **100**, 34–41. (RMQ).
[162] Smith M. C. and Wang F-C (2004) Performance Benefits in Passive Vehicle Suspensions Employing Inerters, 'Vehicle System Dynamics', **42**(4), 235–257.
[163] Soliman A.M.A., Crolla D.A. and EL Sayed F.M. (1993) Comparison of Control Strategies for the Switchable Damper Suspension System, 'Int. J. Vehicle Design', **14**(4), 308–324.
[164] Soltis M. W. (1987) 1987 Thunderbird Turbo Coupe Programmed Ride Control (PRC) Suspension, SAE Paper 870540, SAE International Congress and Exposition, Detroit, February 1987.
[165] Speckhart F.H. and Harrison E. (1968) The design of a Shock Absorber to Improve Ride Comfort by Reducing Jerk, SAE 680472.
[166] Spiegel M.R. (1968) *Mathematical Handbook*, McGraw-Hill, Schaum's Outline Series, ISBN 0-07-060224-7.
[167] Stanway R. (1989) Variable Suspension Damping using Electro-rheological Fluids, I. Mech. E., Conference C382/034, pp 547–558
[168] Steeples B., Lovett D. and Ford M. (1984) Laboratory Durability Testing of Shock Absorbers, SEE/CIT/MIRA Conference SEECO'84, Dynamics in Automotive Engineering, Cranfield, UK.
[169] Stone J.A. (1960) Discharge Coefficients and Steady State Flow Forces for Hydraulic Poppet Valves, 'Trans. ASME', **82**(**D**)(1), 144–154.
[170] Sturk M., Wu X.M. and Wong J.Y. (1995) Development and Evaluation of a High Voltage Supply Unit for Electrorheological Fluid Dampers, 'Vehicle System Dynamics', **24**(2), 101–121.
[171] Su H., Rakheja S. and Sankar T.S. (1989) Response of a nonlinear vehicle suspension with tunable shock absorber to random road excitations, A.S.M.E., Design Engineering Division, (Conference 12730), **18**(5), 185–193.
[172] Suda Y., Shiba T., Hio K., Kawamoto Y., Kondo T. and Yamagata H. (2004) Study on Electromagnetic Damper for Automobiles with Nonlinear Damping Force Characteristics, 'Vehicle System Dynamics', Supplement **41**, 637–646.
[173] Sugasawa F., Kobayashi H., Kakimoto T., Shiraishi Y. and Tateishi Y. (1985) Electronically Controlled Shock Absorber System Used as a Road Sensor which Utilises Super Sonic Waves, SAE Paper 851652, SAE Passenger Car Meeting and Exposition, Dearborn, September 1985.
[174] Surace C., Worden K. and Tomlinson G.R. (1992) On the Non-linear Characteristics of Automotive Shock Absorbers, 'Proc. I. Mech. E.', **206**(1), 1–16. (Testing and modelling)
[175] Tamura H., Komori M., Uenaba K., Kashiyama K. and Shibayama T. (1992) Trend of Bush Bearing Materials for Strut Type Shock Absorbers, SAE Special Publication SP-917, pp 57–65, (Conference 16381).
[176] Tanahashi H., Shindo K. and Nogami O. (1987) Toyota Modulated Air Suspension, SAE paper 870541.
[177] Tatarinov V. (1948) Hydraulic Shock Absorber Orifice Designs, 'Product Engineering', November, 184–188.
[178] Tatarinov V. (1949) Hydraulic Shock Absorbers, 'Product Engineering', **20**(12), 140–142.
[179] Tavner A.C.R., Turner J.D. and Hill M. (1997) A Rule-Based Technique for [Testing] Two-State Switchable Suspension Dampers, 'Proc. I. Mech. E.', **211**(D), 13–20.
[180] Thompson A. G. (1969) Optimum Damping in a Randomly Excited Non-Linear Suspension, *Proc I. Mech. E.*, **184**(2A, 8), 169–184.
[181] Thompson A. G. (1972) A Simple Formula for Optimum Suspension Damping, 'Auto. Engr.', April, p. 20.
[182] Thomson W.T. (1981) *Theory of Vibration with Applications*, George Allen and Unwin Ltd, ISBN 0-04-620012-6.
[183] Van Vliet M. and Sankar M. (1981) Computer-Aided Analysis and Experimental Verification of a Motorcycle Suspension, 'ASME Transactions, J. Mech. Design', paper 81-DET-84.
[184] Vannucci S.N. (1985), Vehicle Shock Absorber Noise Problem Diagnosis by Spectrum Analysis, *Proceedings of Intnl. Symposium on Automotive Technology and Automation*, (Conference 09882), **2**, 571–585.

[185] Verschoore R., Duquesne F. and Kermis L. (1996) Determination of the Vibration Comfort of Vehicles by Means of Simulation and Measurement, 'European Journal of Mechanical Engineering', **41**(3), 137–143.
[186] Wallaschek J. (1990) Dynamics of Non-Linear Automobile Shock-Absorbers, 'International Journal of Non-Linear Mechanics', **25**(2/3), 299–308.
[187] Warner (1996) An Analytical and Experimental Investigation of High Performance Suspension Dampers, *PhD Thesis*, Concordia University, Montreal, Canada.
[188] Warner B. and Rakheja S. (1996) An Analytical and Experimental Investigation of Friction and Gas Spring Characteristics of Racing Car Suspension Dampers, SAE 962548, in SAE SP-304/1 pp 241–254.
[189] Weaver E.W. (1929) Measure of Shock Absorber Performance, 'Auto. Ind.', 14 December 1929, pp 870–972.
[190] Weiss K. D., Duclos T. G., Carlson J. D., Chrzan M. J. and Margida A. J. (1993), High Strength Magneto- and Electro-Rheological Fluids, SAE Paper 932451.
[191] Whelan A.E. (1995) Characterizing Shock Absorbers for Ground Vehicle Simulation, 'Jnl. of Testing and Evaluation', **23**(4), 307–310. (Testing and modelling).
[192] Wilkinson P.A. (1993) Alternative Suspension Systems for Commercial Vehicles, *PhD Thesis*, University of Leeds, Department of Mechanical Engineering, UK.
[193] Winslow W. M. (1947) US Patent 2,417,850, Method and Means for Translating Electrical Impulses into Mechanical Force, filed 14 April 1942, dated 25 March 1947.
[194] Winslow W. M. (1959) US Patent 2,886,151, Field Responsive Fluid Couplings, dated 12 May 1959.
[195] Wössner F. and Dantele H. (1971) Zwei Stossdämpfersysteme, ATZ., J73, H6, June 1971, pp 223–229.
[196] Wu H.C., Yan W.Z., Mo C. and Patten W.N. (1993) Prototype Semiactive Damper, Proc. A.S.M.E., D.S.C., V52, pp 51–55 (Also Proc Conf 19927).
[197] Yamauchi H., Sugahara T., Mishima M. and Noguchi E. (2003) Theoretical Analysis and Proposition to Reduce Self-Excited Vibration of Automotive Shock Absorber, SAE Paper 2003-01-1471, SAE Noise and Vibration Conference, Traverse City, Michigan, May 2003.
[198] Yeaple F.D. (Ed.) (1966) *Hydraulic and Pneumatic Power and Control*, McGraw-Hill, LCCC 65-28832.
[199] Young D.W. (1986) Aircraft Landing Gears, 'Proc. I. Mech. E.', **200**(D2), 75–92.
[200] Yukimasa T., Motohashi H., Ohtaki M. and Chikamori H. (1985) Comments on Oil Seals for Gas Pressurised Shock Absorbers of Automotive Suspensions, S.A.E. Special Publication SP-613, pp 85–94.
[201] Yung V. Y. B. and Cole D. J. (2005) Wavelet Analysis of High-Frequency Damper Behaviour, *Proc. I. Mech. E. Part D, 'Journal of Automobile Engineering'*, **219**, 977.
[202] Yung V. Y. B. and Cole D. J. (2006) Modelling High Frequency Force Behaviour of Hydraulic Automotive Dampers, 'Vehicle System Dynamics', **44**(1), 1–31.
[203] Zucrow M.J. (1928) Properties of Submerged Carburettor Jets, 'Autom. Ind.', 6 October 1928.

INDEX

After a page number, the letter 'e' means a significant equation, 'f' means a figure, 't' means a table. The index does not list every mention of a term. The Nomenclature and References are not indexed.

A

A-arm	14f
absorber, resonant	86
absorbing of shocks	1
acceleration sensitive valves	240, 244f
adaptive dampers	42, 296–301
adjustable dampers	20, 33f, 289
adjustable, electric	33f
adjustables, rotary	250–253f
aerodynamic ride stiffness	125
air suspension	31f, 32f
properties	375–378
aircraft oleo damper	241f
Andre Telecontrol	11f
Andre–Hartford scissor	5f
Andrex oil bath	7f, 8f
annular area of piston	269, 315e
anti-roll bar stiffness factor	94–95
Arcton damper	180f
area coefficient	204, 291, 293f, 295
Armstrong double telescopic	15f, 20f
lever arm	18f
asymmetry of force	129e
automatic valves/dampers	295–299
axle vibrations	122
inclined dampers	167f

B

Belleville washer stack	222
bellows valve	252
bend losses in pipe	200
Bernoulli's equation	188–189
Bilstein	33f
bimetallic strip compensator	238f–239f
Bingham flow	212, 318f, 397–400
Bond number	187e
BSA Telefork	40f
bubble rising speed	188e
shape	187
distinguishing diameter	187e
size, N molecules	179
surface tension pressure	179
velocity coefficient	188
bubbles, appearance in liquid	181, 187
bump camber coefficient, struts	153
wishbones	153
bushes, see also mountings	
amplitude effect	71
asymmetrical	44–45f

C

Campbell	32f
cavitation	274–276
characteristic equation, damped	103
undamped	97
Chrysler Oriflow damper	262f
Citroen air suspension	31f
coil inductance	329
Colebrook diagram	194f
Colebrook's equation	194e
colloidal ferrofluids	216
compliance in series, see also series stiffness	43

compressibility effects	276–278	passenger	111, 112f, 113
related to acceleration	277–278	tyre	111–116
compression velocity	260	displays of $F(V)$ curve	261f
configurations of damper	17–33	double piston	288f
continuity, principle of	188	telescopic	15f, 16f
continuous velocity pick off	356	wishbones	150–153
cost effectiveness of active valve	296	double-acting shim valve	249, 251f
of damper specified	335	double-tube damper	23f, 24f, 27f, 28f, 29f, 30f
Coulomb damping	74–77	drop test	47, 48t
on ride height	76	drowned efflux	206e
damping	74–77	dual path mounting	46f, 47f
friction	15	durability testing	338, 357
creeping motion	268	dynamic index, pitch	91
cubic equation solution	386–389	pressure	192
CVP	356		
cyclical characteristics	278–282f	**E**	
		Einstein's equation	175e
D		electromagnetic damper	31, 35f
		electromechanical testers	342
damper characteristics, main		electrorheological, *see also* ER	
parameters	259	electrorheological fluids	214t–216
matching, of pairs	357	elliptic springs	1, 5f
model, polynomial	141	emulsification	179–184
models in ride	88	emulsified oil damper	26f
listed	88	emulsion	179–186
optimisation	129–132	compressibility	185f, 186e
specification, listed	333	density	181–184, 182e, 183f
damping asymmetry	126, 131–134	viscosity	186e
coefficient C	66e, 67	end fittings specified	334
coefficient required	67, 126	frequencies	88–89
factor α	66e, 67	energy dissipated per cycle	349
ratio ζ	67, 68e	entry losses	200f
from motion trace	68e	equivalent linear damping	
optimum	67	coefficient	263
DampMatic	249	mass (compressibility)	278
Darcy–Weisbach equation	193e	ER controlled valve	319, 320f
data processing in testing	347–348	ER damper	314–319
de Carbon telescopic	25f	capacitance	318e
de Guzmann-Carrancio	174e	design	314f
Delco ride levelling	35f	fast switching	319f
Delco-Lovejoy	22	fibrous microstructure	313
diameter specified	334	fluid properties	311
diffuser	202	material model	314f
loss factor ε	202	materials	309–314
recovery coefficient ζ_R	203	mechanism of viscosity	
disc valve	219f, 220f	change	312
discharge coefficient C_d	203–207	stick slip	313f
as K	205	evolution of damper, stages	1–2
with cavitation	206e	exit losses	203
discomfort loop	128f		

expansion number (bubble)	187e	general damper configuration	271
extension limiter, hydraulic	40f	Girling Arcton damper	180
velocity of damper	260f	Greeves dry friction	6f
		Guzmann-Carrancio	174e

F

fade	172, 239, 276
failure speeds	52
fast adaptive dampers	289
FBD of piston	270f
floating secondary piston	248f
fluid momentum	189–191
foam	180
force cancellation by resonance	351
forced vortex	210
force-momentum analysis of valve	190
force-velocity curve specified	334
Foster snubber	4
free body diagram of piston	270f
free vortex	210, 211f, 212e
frequency domain ride analysis	117
frequency ratio	71e
of highest T	73
sensitive valves	245
selective damping	249
friction force	268e
types, listed	15
mechanical	265
piston, rod, seal	265, 266
strut	266, 267
front girder fork	159f
leading link	158f
trailing link	157f
froth	180
FSD	249
Fuchs	17, 21
Fukushima	52, 129f–131f, 246, 247f

G

Gabriel snubber	7f
gas absorbability in oil	177e
gas absorption coefficient C_{GA}	178
volume coefficient C_{GLV}	182
bulk modulus	177
compressibility	177
density	176e
force	268e, 269e
springing	19
stiffness	269e
viscosity	176e

H

Hadamard	188
Hagen-Poiseuille equation	193e, 254, 255
handling qualities of vehicle	120
straight line	120–121
Q_H	124–128
Hartford	3, 4, 5f, 6f
Telecontrol	4, 6f
head valve	257
heave and pitch undamped 2-dof	95
damped, 1-dof	92
undamped, 1-dof	90
model parameters	112f
with link compliance	139f
heave-and-pitch damped	2-dof
full analysis	102–104
damped 2-dof simplified	100–102
damped 2-dof, characteristic equation	103
Heaviside D-operator	65, 96, 102
Henry's equation	177e
Houdaille rotary vane	9f, 10f
human tolerance to vibration	115ef
Hydragas	20, 32f
hydraulic dampers, success factors	9, 11, 15
diameter	195
testers	344
Hydrolastic	20, 32f
Hydromat (Boge)	37f

I

induced velocity of vortex	210f, 211e
inlet edge radius, effect on losses	200e, 206
instantaneous stroke	246f
instrumentation for testing	345
in-use thickening (MR)	323
iron pentacarbonyl	321
ISO 2631	115
IUT	323

J

jacking force	286–287

K

Kendall's equation	175e
Kinchin and Stock	17f–22f, 43f–45f, 55
Kindl	55, 244f
kinematic viscosity	173e, 377e
kinetic energy correction factor	198t, 315e
knee of $F(V)$ curve	12, 140e, 228f, 232
Komamura and Mizumukai	19f, 242f, 243f, 248f, 253f–254f
Koni	27f

L

Lancia Lambda pillar	12f
LDE	323
leaf spring friction	3f
springs	3f, 5f, 11
length range specified	334
lever arm, parallel piston	19f, 169
single acting	21f
life of damper specified	335
lifetime dissipated energy	323
liquid expansion compensator	239f
solid suspensions	212–214
literature review	54–60
loading function for tolerance	115e
Lovejoy hydraulic	12, 13f, 22f
Lucas parallel piston	13f, 19f
Luvax rotary vane	18f

M

magnetorheological, see also MR	
magnetorheological fluids	59, 214–216, 321–324
magnification factor	70
Mannesman Sachs levelling	38f, 39f
manufacture	53
mass (load) variation	63
maximum area of valve	291, 292f, 294
mean damping coefficient	259, 263e
power dissipation	349, 350f
mechanical friction	265–268
minor losses	199
modal damping coefficient	101e
factor	101e
ratio	101e
frequencies	97e
inertia	100e
positions	99e, 100e
stiffness	101e
mode shapes	98e
momentum correction factor	99f, 198t
of fluid	189–191
Moody diagram	194f
Moody's equation	194e
motion ratio adjustment of damping	301
displacement method	137–138
effect	139–142, 141f
installation	301
scaling effect	142t
motorcycle front suspensions	156–160
rear suspensions	160–165
mountings, see also bushes	
mountings	42–47
MR damper coefficient	329
design	324f, 330f, 331f
analysis	327t
complete model	332f
magnetic circuit	325f
fluid properties v. solid content	322t
properties	322t
materials	214–216, 321–324
multi stage valve	257

N

Nikuradse	197
Nivomat (Boge)	37f
noise	287

O

oil additives	171
compressibility	172e
density	171e
properties	170t
listed	171
specified	335
surface tension	179
thermal capacity	175
conductivity	176
expansion	172e
vapour pressure	176e
viscosity	173–175
causes of	173
temperature effect	174e
on-road testing	357
operating range	130f
operating temperature range	170
optimum suspension damping	82, 124–128

Index 413

stiffness	125
orifice flow	203–207, 204f
orifices, combined	207–209
parallel	207, 208f
series	207, 208f
triple	208, 209f
Oriflow damper	262f

P

parallel hole	290, 291f, 294
piston damper	13f, 19f
passenger on seat	118–119
effect on vehicle	119
passenger-tyre discomfort loop	128f
peak velocity pick off	356
Penske racing damper sections	358f, 359f
Peterson	23, 55
piezoelectric valves	249, 250f
pipe flow	191
regime laminar/turbulent	192f
Reynolds number	191e, 192e
noncircular sections	195–196
friction factor f	194f
laminar pressure loss	193e
laminar/turbulent transition	194
turbulent pressure loss	192e
velocity profiles	196–199
piston and rod forces, listed	269
mass	270
seals	54
pitch vibration	121
Poisson's ratio	283–284
position sensitive valves	240, 242f
position-dependent dampers	35, 240–243
Prandtl	197, 377e
pressure loss coeff K as C_d	205
pressure-rate valves	243–245f
pressures and forces	272–273
production quantity	1
progressivity factor	259, 264e
pullrods	155
purging	337
pushrods	155
PVP	356

Q

quartic characteristic equation	103
quadratic damping	77–79
equation solution	65, 385–386

quality factor (ER & MR)	215
quarter car model (repudiated)	111
see heave model	
quartic equation solution	103e, 389–391

R

Rabinow patent abstract	305
range of a damper	53
ratio of damping ratios, heave and pitch	93
frequencies, heave and pitch	90, 92
relative roughness	194
remote valve systems	42f
resistance reducing with length	205
resonant absorber	86–87
ride motions	49–50
quality Q_R	122
ride-handling compromise	121, 124–128
parameter f_{SRH}	126, 127e
quality loop	125f, 126f
ride-levelling	33–35
rigid arm suspension	148–150
listed types	148
rim impact event	133
rising rate factor	144e, 147e
RMQ of Gaussian distribution	116
RMS of Gaussian distribution	116
road modelling	105–110
roughness, correlation of tracks	109–110
ISO model	106, 107f
reference spectral density	108t
spectral	106–110
sinusoidal	106
testing categories list	338
rocker design	142–148, 143f–145f
deviation angle	145e
rod valve	219f, 221, 222f
roll centre height, struts	155
height, wishbones	153
roll vibration	121
damped	94–95
undamped	93–94
rotary adjustables	250–252, 253f

S

safety certification	338
scalogram	288f
scissor action discs	3, 4f, 5f, 6f

damper	129f	top mounting	46f, 47f
Scotch yoke	343f	sudden contraction	201f
seat, vibration isolating	118f	expansion	201f
selection tables, high/low damping	300t	supersaturated solution gas-in-oil	179
semi-active dampers	289	surface roughness	194t
series hole	290, 291f, 294	suspension workspace	114
stiffness	79–85	liquid–solid	212–214
settlement velocity	214e	Sutherland's equation	177e
shim valve	223–225, 223f, 224f	T	
bending modes	224f	Telecontrol	4, 6f, 11f
double acting	249–250, 251f	Telefork	35, 40f
SI units	393–395	telescopic damper, basic types	34f
sink rate on dampers	105e	conventional	23f, 24f
sinusoidal test theory	348–351	general form	37–42, 41f
sliding forks	156f	standard	24f
pillar	12f	temperature compensation for fade	237–239
slow adaptive dampers	289, 299	effect	276
snubber	4, 7f	rise in testing	349
solenoid actuator	251f	test procedure	352t, 352–354
solution of equations	385–391	sheets	381 383
spatial frequency, road	106e	testing categorisation list	337
resonant	106e	thermal expansion of materials	172, 237
specific damping coefficient	68	Thompson	128
stiffness	67	time domain ride analysis	113–117
Speckhart and Harrison	56, 240, 244f	timetable, early development	7
speed, operating	47–53	transfer factor	131e, 259, 264e
spool valve	219f, 222f–223	transient response traces	339–341
F–M analysis	190	testing	338–342
Stabilus anti-roll system	22f	transients, acceleration	50
steering dampers	21, 34f	roll	51
static damper forces	268	transmissibility factor	69e, 70f, 71e, 72ef
steering damper	5f	greatest	73
problems	124	peak	73, 73t, 74f
vibrations	124	road to passenger	117f
Stokes' equation	187e, 214	triangular test	354–356
strain analysis	283–286	Truffault	3, 4f, 5f
effect on cylinder volume	286	tube valve	252, 255f
negative axial	284		
stress analysis	283–286	U	
stroke, free and installed	135	ultrasonic ride height measurement	57, 297f
sensitive valves	245–249	units (SI) and conversions	393–395
utilisation	135		
operating	47–53, 52t, 53f	V	
strut side forces	266, 267f		
suspension	153–155	valve area coefficient	291–295
low motion ratio	154	characteristics listed	225–227
front	29f, 30f	flow rates	271–272

force–momentum analysis	190f	profile reciprocal index	197
fully closed pressure	291, 293f, 294	*vena contracta*	201f, 204f
fully open pressure	291	vibration, forced damped, 1-dof	71–74
power dissipation	227e	undamped, 1-dof	68–71
switching speed	298	free damped, 1-dof	63–68
types listed	219–220, 219f	2-dof	85–86
adjustable	290–294	undamped, 1-dof	61–63
area fraction open	233	2-dof	85
variation with P	228	viscosity of a mixture	175e
basic models	227–230	of a suspension	175e
bellows	252	kinematic	173e
general model	231–233	von Karman	197
in head	257	vortex	209–212
knee 232		radial pressure gradient	211
linear analysis	273–274	strength	211
multi-stage	257	valve	246, 247f, 248f, 249
piezoelectric	249, 250f		
simple area analysis	218et		
solution of flow rate	235–237	W	
stages 232, 233f			
transition pressure ratio	234e	wall forces	285f
region 233		wallowing	67
volumetric flow ratio	234	Walther viscosity equation	175e
variable stiffness	295f	water, properties	379–380
vortex 209–212, 246–249		wear	67, 168
valves in pistons	20f	Weisbach	193
vane damper, double acting	9f	wheel hop	119–120
vapour pressure	176, 274, 275	white noise test	356
variable hole	291	Winslow patent abstract	304f, 307
stiffness valve	295f	wishbone suspension	14f, 150–153
vehicle ride damping coefficients, C_{D0} etc.	102e	Woodhead emulsified	26f
		Woodhead–Monroe telescopic	16f
stiffness coefficients C_{K0} etc.	96e, 102e		
velocities and forces	260, 261f	X,Y,Z	
velocity coefficient	188e, 204	zeitgeist, historical	2

Printed in the USA/Agawam, MA
July 23, 2013